GUIDE TO WORLD SCREW THREADS

GUIDE TO WORLD SCREW THREADS

Edited by

P. A. Sidders

Industial Press Inc.

INDUSTRIAL PRESS INC.
200 Madison Avenue
New York, New York 10016

Library of Congress Catalog Card Number: 71-185990

ISBN 0-89381-1092-9

Printed and Bound by Edwards Brothers, Inc.,
Ann Arbor, Michigan.

20 19 18 17 16 15 14 13 12

CONTENTS

PREFACE TO THE FIRST AMERICAN EDITION

This volume has appeared in England under the title Machinery's Screw Thread Book, and is now published in an American edition with the title, GUIDE TO WORLD SCREW THREADS which is more indicative of the contents. Some additional data have been included in this edition. Information is now given for USA standard thread forming and thread cutting self-tapping screws, with details of thread forms, thread dimensions, and hole sizes for different types of these screws. Some details have also been added concerning screw thread inserts. To complement the section on threaded fasteners, information on washers has been added, covering American standard plain, spring lock, and internal and external tooth washers, also British standard plain and spring washers with metric dimensions.

In view of the worldwide move to metrication, and the increasing interest in the subject in this country, it is hoped that the section on metric screw threads will prove useful. Details of ISO metric threads are included, which are coming into use in countries that have adopted the metric system.

At this time it is important to note that in May 1970, the Governing Board of the Industrial Fasteners Institute, 1505 East Ohio Building, 1717 East 9th Street, Cleveland, Ohio 44114, met to establish a policy on metric threads in view of possible metrication in this country. Part of the policy states that the:

> "Industrial Fasteners Institute recognizes that important segments of American industry may elect to convert to metric. In such event, the IFI position is that a new system of standard fasteners designed for optimum performance capability should be developed and stated in metric units. The new system would be a complete 'fresh start' approach not necessarily related to any existing system. The purpose would be to achieve optimum performance capability in the use of all materials used in the fastening function and to attain maximum simplification in the number and styles of standard fasteners."

The Institute is engaged in a 4-phase study to develop an optimum metric fastener system. The separate phases are concerned with determining a series of basic diameters, thread pitches, and a thread form; dimensional design of mechanical fasteners; mechanical properties; and related parts, such as washers. At the time of publication of this book, the study was still in progress and no standard data were available for inclusion in the volume.

PREFACE TO THE 20th BRITISH EDITION

Since the publication of the 19th Edition important developments i
screw thread standardization have taken place both internationally an
nationally.

In the international field, ISO Committee TC1 has completed the IS
Metric system of screw threads, with the exception of miniature thread
for watchmaking and similar purposes, and the resulting Recommendations are being actively implemented in all countries that have adopte
the metric system. In addition to comprehensive series of diameter-pitc
combinations, the system includes versatile tolerancing arrangements
similar to the ISO system of limits and fits for holes and shafts, as well a
a complete gauging system.

The ISO Metric series of miniature threads is approaching completion
and the relevant Recommendations are now in draft form. These thread
will use the ISO Basic Profile modified to have a larger truncation at th
thread root, in order to provide greater core strength and to suit th
method of manufacture. They will be designated by the letter "S" followe
by the nominal diameter.

International standardization of fasteners with ISO Metric thread ha
also proceeded apace, and implementation of the agreements reached i
well advanced.

In this country the national change to the metric system was initiate
in May 1965 by the announcement of the President of the Board of Trad
that "it would be desirable for this country to change to the metric system"
In November of the same year, the B.S.I. called a conference on futur
screw thread policy, which approved the following statement:

> "IN VIEW OF THE WORLD TREND TOWARDS THE METRI
> SYSTEM, AND HAVING PARTICULAR REGARD TO THE DE-
> CLARED U.K. NATIONAL POLICY FOR ITS ADOPTION, I
> IS STRONGLY RECOMMENDED THAT BRITISH INDUSTR
> SHOULD ADOPT THE ISO METRIC SCREW THREAD SYSTEM."

The International Standardization Organization (I.S.O.) – formed i
1946 – took up the task of international screw thread standardization
and its first success was an agreement that for all triangular screw thread
(with the exception of pipe threads) the Unified thread profile would b
used. However, in spite of continued efforts it proved impossible to recon-
cile the inch and metric systems of measurement and to agree upon
single series of diameter-pitch combinations. For metric systém users th
ISO Committee TC1 therefore completed the development of the IS
Metric screw thread system, while the Unified thread was given IS
status as the standard inch-system thread.

At the 1961 meeting in Helsinki, TC1 approved a versatile new tolerancing system for ISO Metric threads. Although this system produces tolerances very similar to the Unified thread tolerance system, the U.S.A. declared at the 1964 New Delhi meeting that they were unable to accept it for inch threads.

Work to develop an internationally standardized thread gauging system had already started, but the completion of the ISO Metric thread and its tolerancing system was the signal for metric countries to commence changing their national standards and practices in favour of the new and genuinely international system. Within the last few years, the U.K. has also initiated a similar process.

Another ISO Committee (TC5) is responsible for the standardization of pipe threads, and for this application the British Standard pipe thread – B.S. 21 – has been accepted as the international standard. The original inch dimensions and agreed metric equivalents are published in ISO Recommendations R7 and R228.

Further development of small screw threads for watches and instruments has also taken place and with the co-operation of the horological and fine mechanical industries the scope of the ISO Metric screw thread system has been extended to cover the needs of those industries.

As a result of the national metrication programme started in the U.K. in 1965, a national policy was adopted in that year whereby first priority is given to the ISO Metric thread system, and B.S.W., B.S.F. and B.A. threads are rendered obsolescent.

Work on screw thread metrication is continuing, and in this printing of the 20th edition of Machinery's Screw Thread Book, reference is made to some recent developments on pages 145 and 165.

FOREWORD

The screw thread obviously arose as an ingenious solution of an oft recurring practical problem. Its origin is obscure, but it is logical to assume that it was discovered independently by a number of craftsmen. The screw thread has been used for temporary fastenings of personal armour, and for connecting parts of ancient cannon. It also formed an important part of the early printing press.

Screws cut by die plates and holes screwed by longitudinally grooved screws have been known for centuries. Such primitive taps were cut by hand, and no standardisation of pitch was possible. However, an attempt to standardise diameters and pitches, by originating hand screw-cutting tools from a carefully preserved set of master tools, was made by Holtzappfel, an expert mechanic who was in business at the beginning of the nineteenth century in the Charing Cross area of London. But standardisation in the strict sense of independent reproducibility may be said to have commenced with the development of the engineers' lathe at the hands of Maudslay. A natural outcome of this was the series of threads with a definite axial form proposed by Sir Joseph Whitworth in a paper before the Institution of Civil Engineers on June 15th, 1841. Whitworth based his pitches and flank angle on the values most usually used at the time. Nevertheless when William Sellers in the U.S.A. in 1864 proposed the 60-deg. flank angle, he considered the Whitworth angle to be difficult to verify. For the general run of threads, 60-deg. bids to become the universally adopted thread angle.

A considerable step forward in the unification of screw threads was initiated by the International Conference in Zurich in 1898 which resulted in the formulation of the S.I. (Système International) Metric Threads. This was generally accepted by the Continental countries as a basis for their own standards. For horological and instrument work the Thury thread, developed at a conference in Geneva during 1877, was used and was the basis for the British Association (B.A.) thread.

In more recent years, the most notable advance in the standardization of inch-system threads was the joint development by America, Britain and Canada of the Unified thread. This achieved fairly wide usage in the U.K. but was not adopted by important sections of industry.

While it was accepted that some sections of British industry might need to continue to use Unified (inch) threads, in view of their close association with the U.S.A., it was agreed that nationally the Unified thread should be regarded as a second choice. Furthermore it was decided that, from that date, Whitworth, B.S.F. and B.A. threads should be regarded as obsolescent. In view of these developments in U.K. policy, the old British Standard Metric (S.I.) thread has been declared obsolete, and B.S. 1095 has been withdrawn.

It will be noted that the ISO Metric screw thread system includes only one complete graded pitch series – the Coarse Series, although ISO R262 gives a directive for the selection of fasteners with fine pitch threads (Table G18). More recently the Society of Motor Manufacturers and Traders Ltd, have made recommendations for a limited fine pitch series, and these recommendations are now being considered for inclusion in a British Standard. In this connection, however, it should be borne in mind that the coarse pitch series threads (Tables G15–17) will be suitable for the great majority of applications.

agents for standards throughout the world †

National standardizing bodies, especially the Member Bodies of the International Organization for Standardization (ISO), serve as national representatives for one another. Each acts within its own country as sole sales agent and information center for the standards of the other national standardizing bodies.

Hence, standards issued by the American National Standards Institute, Inc., 1430 Broadway, New York, New York, 10018, the United States Member Body of ISO, may be purchased abroad from the respective national standardizing body in each of the following countries

Albania: STASH, STASH, Bureau de Standardisation aupres de la Commission du Plan d'Etat de la Republique Populaire D'Albania, Tirana

Argentina: IRAM, IRAM, Instituto Argentino de Racionalizacion de Materiales, Chile 1192, Buenos Aires

Australia: SAA, AS, Standards Association of Australia, P. O. Box 458, North Sydney NSW 2060

Austria: ONA, ONORM, Oesterreichischer Normenausschuss, Bauernmarkt 18, A-1010 Wien

Barbados W.I.: Economic Planning Unit, Government Headquarters, Bay Street

Belgium: IBN, NBN,* Institut Belge de Normalisation, 29 Avenue de la Brabanconne, Bruxelles 4

Brazil: ABNT; NB, EB,* Associacao Brasileira de Normas Tecnicas, Caixa Postal 1680, Rio de Janeiro

Bulgaria: ISMIU, BDS, Institut de Normalisation, Mesures et Appareils de Mesure, 8 rue Sveta, Sofia

Burma: UBARI, UBS, Union of Burma Applied Research Institute, Junction of Kaba Aye Pagoda-Kanbe Roads, Rangoon

Canada: CSA, CSA, Canadian Standards Association, 178 Rexdale Blvd. Rexdale 603, Ontario, Canada

Ceylon: Bureau of Ceylon Standards, 141/2 Vajira Road, Colombo 5

Chile: INDITECNOR, INDITECNOR, Instituto Nacional de Investigaciones Tecnologicas y Normalizacion, Plaza Bulnes 1302, Depto. 62, Casilla de correo 995, Santiago

China: CNS, CNS, National Bureau of Standards, No. 1 First Street, Chen Kung Road 4, Tainan, Taiwan

Colombia: ICONTEC, Instituto Colombiano de Normas Tecnicas, Apartado Aereo, 14237 Bogota D.E.

Costa Rica: Instituto Centroamericano de Investigaciones y Tecnologia Industrial, 4a Calle y Avenida la Reforma, Zona 10, Guatemala, Guatemala

Cyprus: Ministry of Commerce and Industry of the Republic of Cyprus, Nicosia

Czechoslovakia: CSN, CSN, Urad pro normalisaci a mereni, Vaclavske namesti c. 19, Praha 1, Nove Mesto

Denmark: DS, DS, Dansk Standardiseringsraad, Aurehojvej 12, DK-2900, Hellerup

Ethiopia: Ethiopian Standards Institution, P. O. Box 1769, Addis Ababa

El Salvador: Instituto Centroamericano de Investigaciones y Technologia Industrial, 4a Calle y Avenida la Reforma, Zona 10, Guatemala, Guatemala

Finland: SFS, SFS, Suomen Standardisoimisliitto r.y. Bulevardi 5A 7, Helsinki

France: AFNOR, NF, Association Francaise de Normalisation, Tour Europe, Cedex 7, 92 Paris – La Defense

Germany: DNA, DIN, Deutscher Normenausschuss, 4-7 Burggrafenstrasse, 1 Berlin 30

Greece: ENO, ENO, Ministry of Industry, Director of Standardization P, Kanigos Street, Athens

Ghana: ISIG, Institute of Standards and Industrial Research, Ghana Academy of Sciences, P O Box M-32, Accra

Guatemala: ICAITI, Instituto Centroamericano de Investigaciones y Tecnologia Industrial, 4a Calle y Avenida la Reforma, Zona 10, Apartado Postal 1552

Honduras: Instituto Centroamericano de Investigaciones y Tecnologia Industrial, 4a Calle y Avenida la Reforma, Zona 10, Guatemala, Guatemala

Hong Kong: Federation of Hong Kong Industries, 31-37 des Voeux Road C

Hungary: MSZH, MSZ, Office Hongrois de Normalisation, Ulloi-ut 25, Budapest IX

India: ISI, IS, Indian Standards Institution, "Manak Bhavan 9 Bahadur Shah Zafer Merg, New Delhi 1

Indonesia: DNI, NI, Jajasan "Dana Normalisasi Indonesia," Djalan Braga 38, Bandung

Iran: ISIRI, ISIRI, Institute of Standards and Industrial Research of Iran, Ministry of Economy, P.O. Box 2937, Teheran

Iraq: IOS, Iraqi Organization for Standardization, Ministry of Industry, P.O. Box 11185, Baghdad

Ireland: IIRS, I.S., Institute for Industrial Research and Standards, Glasnevin House, Ballymun Road, Dublin

Israel: SII, SI, Standards Institution of Israel, University Street Ramat-Aviv

† Courtesy of the American National Standards Institute

* Standards bearing other designations have also been approved by this standards body.

Italy: UNI, UNI, Ente Nazionale Italiano de Unificazione, Piazza Armando Diaz 2, 120123 Milano

Japan: JISC, JIS, Japanese Industrial Standards Committee, Agency of Ind. Science and Technology, Ministry of International Trade and Industry, 3-1 Kasumigaseki Chiyoda-Ku, Tokyo.

Korea: KBS, KS, Korean Bureau of Standards, Chong, Ro 1 - 19, Seoul

Kuwait: Ministry of Commerce and Industry, P.O. Box 2944

Lebanon: LIBNOR, Lebanese Standards Institution, P.O. Box 2806, Beirut

Madagascar: Ministere d'Etat Charge de l'Agriculture de l'Expansion Rurale et du Ravitailleement, Direction de la Production, Tananarive

Malaysia: Standards Institute of Malaysia, Kuala Lumpur

Malta: Department of Industry, Standards Lab, Industrial Estate, Marsa

Mexico: DGN, DGN, Direccion General de Normas, Av. Cuauhtemac No. 80, Mexico 7, D.F.

Morocco: SNIMA, Service de Normalisation Industrielle Marocaine. Ministere du Commerce de l'Industrie des Mines de l'Artisanat et de la Marine Marchande, Rabat

Netherlands: NNI, NEN, Stiching Nederlands Normalisatie-instituut, Polakweg, 5 Rijswijk (ZH)

New Zealand: SANZ, Standards Association of New Zealand, Private Bag, Wellington

Nicaragua: Instituto Centroamericano de Investigaciones y Tecnologia Industrial, 4a Calle y Avenida la Reforma, Zona 10, Guatemala, Guatemala

Norway: NSF, NS, Norges Standardiseringsforbund, Haakon VII's gt. 2, Oslo 1

Pakistan: PSI, PS, Pakistan Standards Institution, 39 Garden Road, Saddar, Karachi 3

Panama: Instituto Centroamericano de Investigaciones y Tecnologia Industrial, 4a Calle Y Avenida la Reforma, Zona 10, Guatemala, Guatemala

Paraguay: INTECNOR, Instituto Nacional de Tecnologia y Normalizacion, Brasil y Jose Berges, Annuncion

Peru: ITINTEC, Instituto de Investigacion Tecnologica Industrial y de Normas Tecnicas, Apartado No. 145, Av. Republica de Chile 698, Lima

Philippine Islands: Bureau of Standards of the Philippines, P. O. Box 3719 Manila

Poland: PKN, PN, Polski Komitet Normalizacyjny, Centralny Osrodek Informacji Normalizacyjneji, Warszawa ul Wiejskazo

Portugal: IGPAI, NP, Reparticao de Normalizacac, Avenida de Berna 1, Lisbon

Republic of South Africa: SABS, SABS, South African Bureau of Standards, Private Bag 191, Pretoria

Rhodesia: Standards Association of Central Africa, Sara House, Hatfield Road, Salisbury, Southern Rhodesia

Romania: IRS, Institutul Roman de Standardizare, Str. Edgar Quinet, 6, Bucarest 1

Singapore: SIRU, Singapore Institute of Standards and Industrial Research, P.O. Box 2611, Singapore

Spain: IRATE, UNE, Instituto Nacional de Racionalizacion Del Trabajo, 150 Serrano St., Madrid 6

Sweden: SIS, SIS,* Sveriges Standardiseringskommission, Box 3295, Stockholm 3

Switzerland: SNV, SNV, * Association Suisse de Normalisation, 4, Kirchenweg, 8032 Zurich

Syria: Industrial Testing and Research Centre, P.O. Box 845, Damascus

Thailand: CTNSS, Centre for Thai National Standard Specifications, Applied Scientific Research Corporation of Thailand, Bangkok

Tunisia: Direction de l'Industrie, Secretariat d'Etat a l'Economie Nationale, 195 rue de la Kasbah, Tunis

Turkey: TSE, TS, Turk Standardlari Enstitusu, Necatibey Caddesi, Ankara

United Arab Republic: ECS, EOS, Egyptian Organization for Standardization, 2 Latin America Street, Garden City, Cairo

United Kingdom: BSI, BS, British Standards Institution, 2 Park Street, London W.1 A 2BS

Uruguay: UNIT, UNIT, Instituto Uruguayo de Normas Tecnicas, Agraciada 1464, Piso 9, Montevideo

USSR: GOST, GOST, Komitet Standartov, Mer i Izmeritel 'nyh Priborov pri, 38 Kvartal Jugo-Zapada, Korpus 189-a, Moskva V-421

Venezuela: COVENIN, COVENIN, Comision Venezolana de Normas Industriales, Direccion de Industrias, Ministerio de Fomento, Caracas

Yugoslavia: JZS, JUS, Jugoslavenskizavod za Standardizaciju, Cara Urosa ul 54, Post pregr 933, 11001 Beograd

*Standards bearing other designations have also been approved by this standards body.

INTRODUCTION

A parallel screw thread is a continuous helical ridge of uniform section and uniform axial spacing formed on the exterior of a circular cylinder or on the interior of a circular hole such that the axial displacement of a point which travels along a helix connecting corresponding points of successive ridge profiles is always proportional to its angular displacement about the axis. When formed on the exterior, the thread is called an external thread, a screw, or a male thread; and when formed on the interior it is called an internal thread, nut, or female thread. A single helical ridge is called a single-start thread; *n* helical ridges running side by side around a cylinder give rise to an *n*-start thread or multi-start thread.

A taper thread may be defined in a similar way, but is formed on the exterior of a right circular cone or on the interior of a right circular conical hole.

The form or sectional profile of a screw thread is usually defined in axial section and comprises the shape of one complete contour of the thread between adjacent ridges. The bottom of the groove between the ridges is called the root, and the most prominent part of the ridge is called the crest. A crest or root may be either curved or flat in axial section. The straight part of the contour of the ridge that connects a crest to a root is called the flank. A complete thread form has two thread flanks which may be equally inclined to the axis, as in the Unified form, or inclined at unequal angles as in the buttress form. The included angle between the flanks is called the thread angle, and the angle between a flank and a plane perpendicular to the axis, the flank angle. The axial distance between corresponding points on adjacent thread forms is the axial pitch, or more commonly "the pitch."

The relative axial displacement of two threaded and engaged members when one member is rotated one complete revolution relative to the other is known as the lead. For a single-start thread, lead = pitch; for an *n*-start thread, lead = $n \times$ pitch.

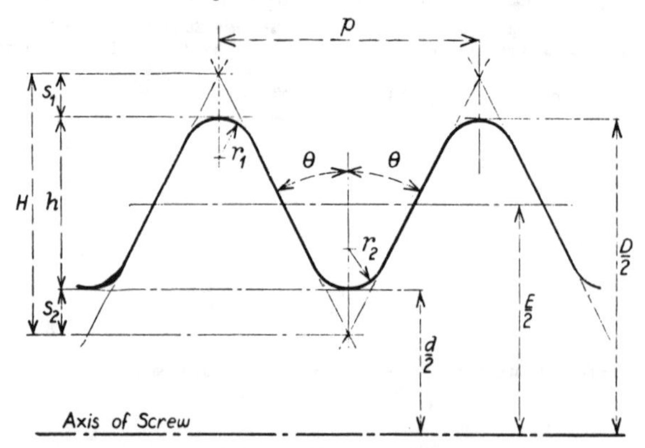

Fig. 1. Elements of Basic Form of a Symmetrical Thread

A parallel screw thread is completely defined when the following elements (Fig. 1) are known:

1. Pitch (p)
2. Flank angle (θ) or angles measured in an axial plane
3. Depth of Truncation of triangular height at root (S_2)
4. ,, ,, ,, ,, ,, ,, ,, crest (S_1)
5. Radius at root (r_2)
6. Radius at crest (r_1)
7. Effective Diameter (E)
8. Number of Starts (n)
9. Whether screw is right or left hand. (A right-handed screw will travel forward in an axial direction in its nut, when rotated in a clockwise direction.)

In practice certain other elements are specified, such as minor diameter (d), major diameter (D) and depth of thread (h). The relations existing between the elements for a screw having equal flank angles may be shown by reference to Fig. 1.

$$D = E + \tfrac{1}{2}p \cot \theta - 2S_1;$$
$$d = E - \tfrac{1}{2}p \cot \theta + 2S_2;$$
$$h = \tfrac{1}{2}p \cot \theta - (S_1+S_2).$$

The distance H is known as the height of the fundamental triangle or triangular height and is given by

$$H=\tfrac{1}{2}\, p \cot \theta.$$

Where S_1 and S_2 are specified as fractions of the triangular height, with $S_1=k_1\, H$, and $S_2=k_2H$

$$D = E + (1 - 2k_1)\, H;$$
$$d = E - (1 - 2k_2)\, H;$$
$$h = (1 - k_1 - k_2)\, H.$$

If the screw is taper, i.e. when the minor, effective and major diameters progressively increase along the axis, there must be known the additional particulars:

10. The cone angle of taper.
11. The position at which it is agreed to measure the effective diameter.
12. Whether pitch is measured along the axis or along the pitch-cone generator.

Some useful definitions are:

Major Diameter (*of parallel thread*). The diameter of an imaginary cylinder, coaxial with the screw, which envelops the thread with helical contact on the crests of the screw or at the roots of the nut.

Major Diameter (*of a taper thread*). The diameter in a selected position of an imaginary circular cone, coaxial with the screw, which envelops the thread with conico-helical contact at the crests of the screw or at the roots of the nut or coupling.

**Effective Diameter* (*of parallel thread*). The diameter of an imaginary cylinder, coaxial with the screw, which intersects the flanks of the thread to give equal intercepts of thread thickness and space thickness in axial section.

**Effective Diameter* (*of taper thread*). The diameter in a selected position of an imaginary circular cone, coaxial with the screw and having a vertex angle equal to the taper angle of the thread, which intersects the flanks of the thread to give equal intercepts of thread thickness and space thickness in axial section.

Minor Diameter (*of parallel thread*). The diameter of an imaginary cylinder, coaxial with the screw or nut, which makes a continuous helical contact at the roots of the thread on the screw or at the crests of thread on the nut.

Minor Diameter (*of taper thread*). The diameter in a selected position of an imaginary circular cone, coaxial with the screw, which has a conico-helical contact at the roots of the screw or at the crests of the nut or coupling.

Helix Angle (λ). The helix angle at any point on the flank at distance q from the axis is given by

$$\tan \lambda = \frac{l}{2 \pi q}$$

This angle is sometimes called the rake angle.

Mean Helix Angle (λ_m) is the helix angle at the effective diameter E and is therefore given by:

$$\tan \lambda_m = \frac{l}{\pi E}$$

(An approximate formula for finding the mean helix angle in minutes is:

$$\lambda_m = \frac{l}{E} \times 1094{\cdot}0$$

This for the usual run of thread is correct to within 2 mins, at most. For λ between 0° and 3° it is correct to the nearest minute.)

As a basis of standard systems of screw threads and series of screws it has been found convenient to specify basic dimensions for each member of the series. These represent the maximum theoretical metal conditions to which, in practice, a tolerance is applied. In some cases the basic dimensions based on the nominal diameter are modified before the application of tolerances by an "allowance" to provide a special class of fit or to allow for plating subsequent to manufacture. Examples of this are found in the Unified Series and the B.A. screws. In general, the basic dimensions, or basic dimensions modified by an allowance, become the maximum for the screw and minimum for the nut.

---

*These are the "simple effective diameters" based on the assumption that no pitch or flank angle differences exist between the mating threads. For "virtual effective diameter" due to such errors see pages 211 and 212.

Section A

Section A

BRITISH THREADS OF WHITWORTH FORM

Note: Since the metric system has now been officially adopted as th primary system of weights and measures in the U.K., it has been agree that future screw thread usage should be based on the ISO Metric thread In consequence, B.S.W. and B.S.F. threads have been declared obsolescent These decisions do not affect the position of the British Standard pip thread, which has been accepted as the international standard pipe threac

Whitworth Screw-Thread Form

After its introduction by Sir Joseph Whitworth in 1841, the Whitwort form of thread rapidly became adopted throughout the country, and fo certain purposes, on the Continent. The basic form is shown in Fig. and has the following proportions:

Thread angle $(2\theta)=55$ deg; Flank angle $(\theta)=27$ deg 30 mins Triangular height $(H)=0{\cdot}9604911\ p$; Shortening at crest and root $(S$ $=\frac{1}{6}\ H=0{\cdot}1600819\ p$; Depth of thread $(h)=0{\cdot}6403274\ p$; Radius a crest and root $(r)=0{\cdot}1373292\ p$; Depth of rounding $(a)=0{\cdot}0739176\ p$ Values of basic dimensions for a wide range of pitches are given i Tables A1 and A2, in inches and mm. respectively.

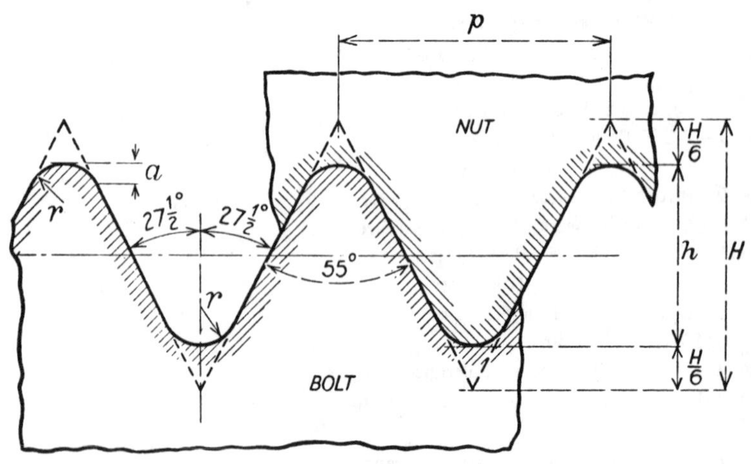

Fig. 1. Whitworth Thread Form

Whitworth Series

There are two series of Whitworth screw threads in general use. Th first of these to be formulated is that designated as "British Standar Whitworth" (B.S.W.), and represents a useful selection of the range o diameters and corresponding pitches originally proposed by Sir Josep Whitworth. At the beginning of the century a need was felt, owing t the extension in engineering activities, for screws of finer pitch, and a second series, the British Standard Fine (B.S.F.) was introduced by th British Engineering Standards Association in 1908. Both these series ar now specified in B.S. 84: 1956.

British Standard Whitworth (B.S.W.)

The basic sizes of screws and nuts for B.S.W. are given in Table A3. These dimensions form the datum to which the various allowances and tolerances are applied. As specified in B.S. 84: 1956, there are three classes of screw, viz. close, medium, and free; and three classes of nut, namely, close, medium and normal.

Allowances. No allowance is applied to the close class of screw, but for sizes up to $\frac{3}{4}$-in. nominal diameter in the medium and free classes, an allowance is applied to screws in the unplated or 'before plating' condition. Hence such an allowance must be subtracted from the basic dimensions of the screw for major, effective and minor diameters before the tolerances are subtracted to obtain the corresponding minimum dimensions.

Where the screw is subsequently plated, the maximum limits of the final dimensions correspond to the basic dimensions.

The allowances for the medium and free classes of B.S.W. screw are as follows:

B.S.W. Screws (Medium and Free Classes), Unplated and Before Plating. Allowance on Major, Effective and Minor Diameters (*to be subtracted from the Basic Sizes given in Table A3 before Tolerances from Table A5 are applied*).

Nom. Diam. (ins.)	Allowance (ins.)	Nom. Diam. (ins.)	Allowance (ins.)	Nom. Diam. (ins.)	Allowance (ins.)
$\frac{1}{8}$	0·0012	$\frac{3}{8}$	0·0014	$\frac{5}{8}$	0·0017
$\frac{3}{16}$	0·0012	$\frac{7}{16}$	0·0015	$\frac{11}{16}$	0·0017
$\frac{1}{4}$	0·0012	$\frac{1}{2}$	0·0015	$\frac{3}{4}$	0·0018
$\frac{5}{16}$	0·0013	$\frac{9}{16}$	0·0016	over $\frac{3}{4}$	no allowance

Tolerances. The tolerances are derived from a 3-term formula based on a combination of major diameter D, pitch p and length of engagement L. This formula is

$$e=0{\cdot}002\sqrt[3]{D}+0{\cdot}003\sqrt{L}+0{\cdot}005\sqrt{p}.$$

The value e constitutes the effective diameter tolerance for the "medium" class, the others being derived as follows:

Screws

Tolerance on	Close Class	Medium Class	Free Class
Minor Diameter ..	$\frac{2}{3}e+0{\cdot}013\sqrt{p}$	$e+0{\cdot}02\sqrt{p}$	$1\frac{1}{2}e+0{\cdot}02\sqrt{p}$
Effective ,, ..	$\frac{2}{3}e$	e	$1\frac{1}{2}e$
Major ,, ..	$\frac{2}{3}e+0{\cdot}01\sqrt{p}$	$e+0{\cdot}01\sqrt{p}$	$1\frac{1}{2}e+0{\cdot}01\sqrt{p}$

Nuts

Tolerance on	Close Class	Medium Class	Free Class
Effective Diameter ..	$\frac{2}{3}e$	e	$1\frac{1}{2}e$

Minor Diameter (all classes):
- $0{\cdot}2p+0{\cdot}004$ in.; for 26 t.p.i. and finer.
- $0{\cdot}2p+0{\cdot}005$ in.; for 24 and 22 t.p.i.
- $0{\cdot}2p+0{\cdot}007$ in.; for 20 t.p.i. and coarser.

Tolerances calculated on the foregoing basis for the usual sizes ar given in Table A5. Here the value of L is taken to equal D, a length whic is approximately equal to the depth of a standard nut.

TAPPING SIZES FOR BRITISH STANDARD WHITWORTH

A comprehensive table of tapping sizes for the complete range o British Standard Whitworth are given in Section L. The drills liste are the standard sizes, and the corresponding depths of thread are give

BRITISH STANDARD FINE SERIES (B.S.F.)

The basic form of thread is Whitworth standard as detailed on page 1 and in Table A1. The pitches were derived from the following formulæ

$p=\frac{1}{10}$ (nominal diameter)$^{\frac{2}{3}}$.
for screws of up to 1 in. diameter

$p=\frac{1}{10}$ (nominal diameter)$^{\frac{5}{8}}$.
for screws above 1 in. and up to 3 in. diameter.

The basic sizes of the standard series as specified in B.S. 84: 1956 ar given in Table A7, and the corresponding tolerances for close, medium an free classes are given in Table A8. These tolerances are derived from th B.S. 3-term formula given on page 17.

Allowances. The remarks concerning allowances for B.S.W. as given o page 17 also apply to the B.S.F. series, and the corresponding allowance are as follows:

B.S.F. SCREWS (MEDIUM AND FREE CLASSES), UNPLATED AND BEFOR PLATING. ALLOWANCE ON MAJOR, EFFECTIVE AND MINOR DIAMETER (*to be subtracted from the Basic Sizes given in Table A7 before Tolerance from Table A8 are applied*).

Nom. Diam. (ins.)	Allowance (ins.)	Nom. Diam. (ins.)	Allowance (ins.)	Nom. Diam. (ins.)	Allowanc (ins.)
$\frac{3}{16}$	0·0011	$\frac{5}{16}$	0·0012	$\frac{9}{16}$	0·0015
$\frac{7}{32}$	0·0011	$\frac{3}{8}$	0·0013	$\frac{5}{8}$	0·0016
$\frac{1}{4}$	0·0011	$\frac{7}{16}$	0·0014	$\frac{11}{16}$	0·0017
$\frac{9}{32}$	0·0012	$\frac{1}{2}$	0·0015	$\frac{3}{4}$	0·0018

No allowance for sizes above $\frac{3}{4}$-inch nominal diameter.

TRUNCATED WHITWORTH FORM

The truncated form of Whitworth thread is obtained from the corres ponding normal Whitworth form in the following way.

The crests of the screw are truncated by an amount equal to the dept of rounding (0·073917p) thus reducing the maximum major diameter b $2\times0{\cdot}073917p=0{\cdot}147835p$. Similarly the crests of the thread at th minor diameter of the nut are truncated by the same amount, this increas ing the minimum minor diameter of the nut by 0·147835p. The majo diameter of nut, the minor diameter of screw and the effective diamete of both remain unaffected. The dimensions arrived at in this way co stitute the basic dimensions to which the allowances and tolerances ar applied.

The effective diameters of screw and nut, and the minor diameter of the screw are subject to the usual B.S.W. and B.S.F. allowances and tolerances for close, medium and free classes (see Table A5). The tolerances for major diameter of screw and the minor diameter of nut are the same for all classes of fit and are calculated from:

Bolts $0{\cdot}052165p+0{\cdot}003$ (for all pitches).
Nuts $0{\cdot}052165p+0{\cdot}004$ (for 26 t.p.i. and finer).
$0{\cdot}052165p+0{\cdot}005$ (for 24 and 22 t.p.i.).
$0{\cdot}052165p+0{\cdot}007$ (for 20 t.p.i. and coarser).

BRITISH BRASS THREAD

The origin of the British "Standard" Brass thread is obscure; for although there exists no British Standard for it, this thread has been in use for many years for gas fittings, brass tubing and general brass work. It may have been the result of an attempt at standardisation by the manufacturers of early gas-burner fittings. Whatever the diameter, it consists of 26 threads to the inch of Whitworth form. (A Whitworth form of thread of the same pitch is adopted for brass in Germany in diameters of $\frac{1}{4}$, $\frac{5}{16}$, $\frac{3}{8}$, $\frac{7}{16}$, $\frac{1}{2}$, $\frac{5}{8}$, $\frac{3}{4}$, $\frac{7}{8}$ and 1 inch).

British "Standard" Brass Thread is still extensively used for fittings on some British gas appliances, but below $\frac{1}{4}$-in., is being replaced by B.A. threads. Where the Brass Thread is used, it should be treated as a "special" Whitworth thread, with dimensions of form as given in Table A1 and tolerances for the required fit calculated by the usual formulæ. The dimensions for a fairly full range of the diameters in use are given in Table A9, and where possible drills from the standard range have been indicated which give approximately 80 and 90 per cent. depth of thread.

MODEL ENGINEERS' THREAD (M.E.)

A series of threads suitable for model makers was established in 1912 by a meeting of the Society of Model and Experimental Engineers under the presidency of the late Mr. Percival Marshall, editor of *The Model Engineer*. This series is also found convenient in certain applications by instrument makers, and the corresponding taps and dies are regularly obtainable. The regular M.E. series are given in Table A10. In addition the following combinations of diameter and pitch are in regular use.

Diameter (ins.)	*t.p.i.*
$\frac{3}{32}$	60
$\frac{1}{8}$	60
$\frac{5}{32}$	60
$\frac{3}{16}$	60
$\frac{7}{32}$	60
$\frac{1}{4}$	60
$\frac{9}{32}$	— 40, 26
$\frac{5}{16}$	60, 40, 26
$\frac{3}{8}$	60, 40, 26
$\frac{7}{16}$	— 40, 32
$\frac{1}{2}$	— 40, 32
$\frac{9}{16}$	— — 32

British Standard Electrical Conduit Thread

This thread which is of Whitworth standard form is specified in B.S. 31: 1940 for use on electrical conduit, couplers and all associated fittings. The basic sizes are given in Table A11, and the B.S. tolerances in Table A12.

British Standard Pipe Thread

The British Standard Pipe Thread has been derived from the original Whitworth gas and water pipe thread and is in general use for gas, water and steam pipes and pipe fittings. It is also used in connection with other piped services such as air and oil, although for the latter, the A.P.I. thread is now gradually finding general acceptance, especially within the petroleum industry. Originally it was for parallel threads only, this type being now specified in B.S. 2779: 1956, and mainly employed for small diameter pipes or for comparatively low pressures, and for mechanical purposes where the problem of fluid-tightness does not occur. The parallel form is also used for most sizes of hydraulic pipe.

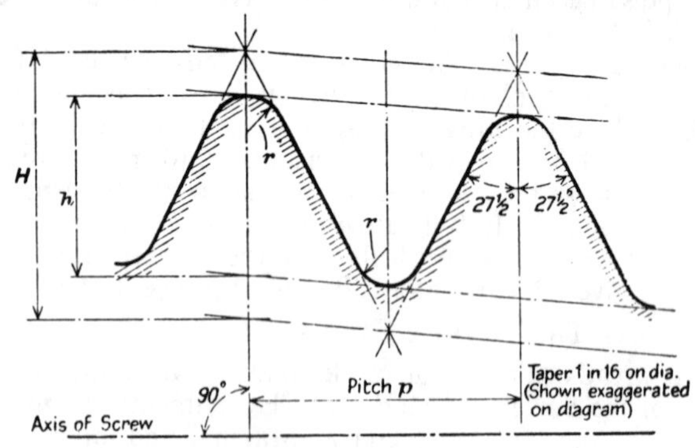

Fig. 2. British Standard Taper Pipe Thread Form

For pipework generally, except for hydraulic purposes, the British Standard Taper Pipe Thread is to be preferred. The basic form of this is shown in Fig. 2 and 3, specified by B.S. 21: 1957, and for heavy copper pipes by B.S. 61: Part 2: 1946. Both the parallel and taper varieties have the Whitworth profile in axial section with the flanks equally inclined* to a plane to which the axis of the pipe or threaded component is normal. Basic major, effective and minor diameters are the same in both, but for taper threads, these are measured in a transverse plane (gauge plane) whose distance (gauge length) from the end of the pipe is specified. The terms used in connection with the taper thread are shown in Fig. 3. The washout thread is that part of the thread which is not

*When first employed, this thread had flanks equally inclined to the pitch cone, similar to the present threads for gas cylinders.

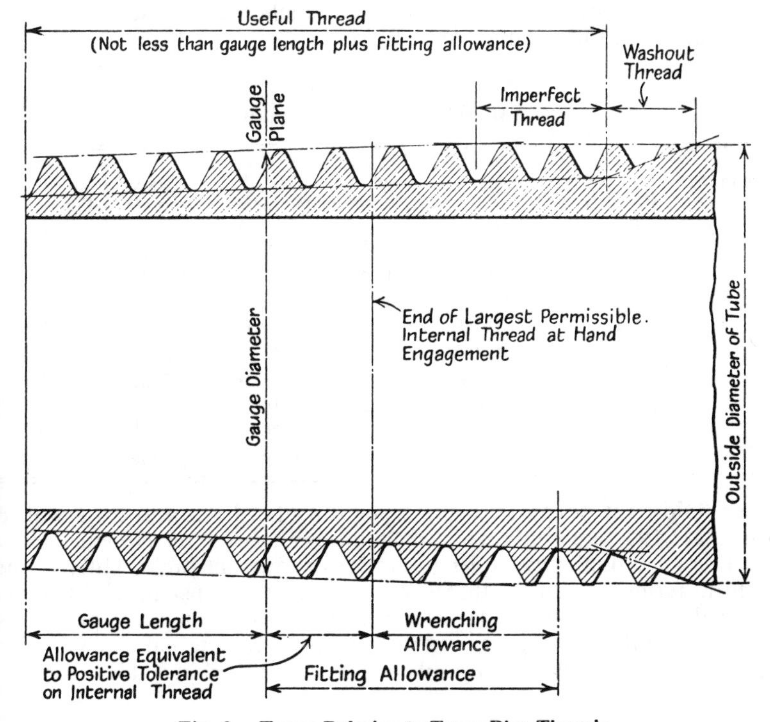

Fig. 3. Terms Relating to Taper Pipe Threads

perfectly formed at the root. The length of useful thread includes any imperfect thread but excludes the washout thread. Basic dimensions are given in Table A13.

For the parallel British Standard Pipe thread which is intended for general engineering purposes, the tolerances are specified in B.S. 2779: 1956, yielding three classes of fit, viz. close, medium and free. Tolerances are given in Table A15.

Important Note: For certain intermediate sizes of pipe thread, it wa
the practice of the engineering trade to adhere to the dimensions of th
Original Whitworth Pipe Threads, for which there exists no officia
standard. These dimensions are as follows:

ORIGINAL WHITWORTH PIPE THREADS

Nominal Size	Threads per Inch	Pitch	Depth of Thread	Major Diameter	Effective Diameter	Minor Diameter
$\frac{5}{8}$	14	0·07143	0·0457	0·9022	0·8565	0·8107
$\frac{7}{8}$	14	0·07143	0·0457	1·1890	1·1433	1·0975
$1\frac{5}{8}$	11	0·09091	0·0582	2·0210	1·9628	1·9046
$1\frac{7}{8}$	11	0·09091	0·0582	2·2450	2·1868	2·1286
$2\frac{1}{8}$	11	0·09091	0·0582	2·4670	2·4088	2·3506
$2\frac{3}{8}$	11	0·09091	0·0582	2·7940	2·7358	2·6776
$2\frac{5}{8}$	11	0·09091	0·0582	3·1240	3·0658	3·0076
$2\frac{7}{8}$	11	0·09091	0·0582	3·3670	3·3088	3·2506
$3\frac{1}{4}$	11	0·09091	0·0582	3·6980	3·6398	3·5816

For fluid-tight purposes, using taper threads, the tolerances specifie
in B.S. 21: 1957 are applicable. In the case of taper threads these consis
of a tolerance on the gauge length and, as shown in Table A14, may b
stated in terms of turns of thread or as a linear dimension. For paralle
threaded pipe couplings the tolerance on basic diametral dimension
is one-sixteenth of the gauge length tolerance for internal screws. Pip
threads recommended for hydraulic work are detailed in Table A16.

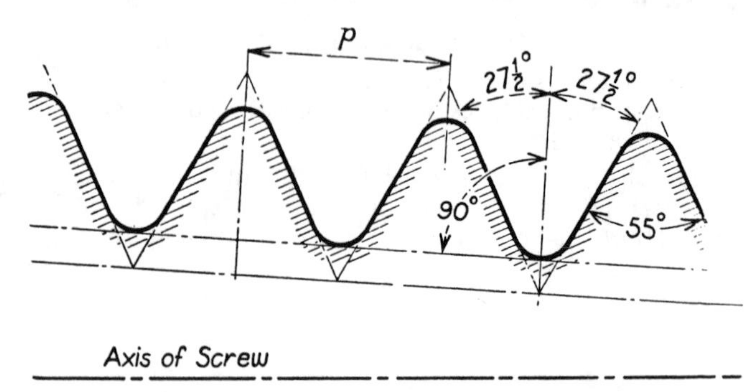

Fig. 4. Taper Thread Form for Gas Cylinders

TAPER THREADS FOR GAS CYLINDERS

The form of thread for the internal screw in the outlet neck of ga
cylinders and for the corresponding external thread of the valves use
with the cylinders is shown in Fig. 4, as specified by B.S. 341: 1962
It will be noticed that the flanks of the thread in axial section are equall

inclined to the pitch-cone generator, the taper of the cone being 1 in 8 measured on diameter for sizes 0·715, 1, and 1·25; and 1 in 5·625 for size 0·6. The included thread flank angle is 55 deg. and the amount of truncation and radius at root and tip correspond to standard Whitworth proportions. The thread is right-handed. Major, effective, and minor diameters are measured at the small end for the external thread, and at the large end for the internal thread. Corresponding dimensions and limits will be found in Table A17 and Table A19a.

Copper-Tube Screw Thread

Work in connection with the standardisation of threads for copper pipes was first undertaken by a joint committee of the Institution of Heating and Ventilating Engineers and the National Association of Master Heating and Domestic Engineers. This was subsequently incorporated, with modifications, in Engineering Standards Committee Report No. 61 of 1913, and is now specified by B.S. 61, part 2: 1946.

The form of thread is standard Whitworth, i.e. with flanks making equal angles with the axis and with an included angle of 55 deg. between them. The copper thread is a taper thread with a taper of $\frac{1}{16}$ on diameter. It thus closely resembles the British Taper Pipe Thread, its gauging dimentions being specified in the same way, so that Figs. 2 and 3 are applicable so both B.S.P. and Copper Thread. Its basic sizes are set out in Table A18.

Miscellaneous Whitworth Thread Applications

Condenser Tube Glands. Screwed with 14 t.p.i.

Drum Plugs for mild steel heavy duty drums. Large bung is screwed $2\frac{1}{4}$ in. B.S.F.—6 t.p.i. with 0·500 in. minimum length of thread; the small bung, $\frac{3}{4}$-in. B.S.P.—14 t.p.i. with 0·378 in. minimum length of thread.

Endoscope Lamp Housings: for type C (external thread): for No. 1 D = 5·40 mm. – 0·03 mm. + 0·00 mm. with 48 t.p.i.; for No. 2 lamp, D = 4·55 mm. – 0·03 mm. + 0·00 mm. with 60 t.p.i.; for No. 3 lamp, D = 3·75 mm. – 0·02 mm. + 0·00 mm. with 72 t.p.i. Length of thread =2·05/2·00 mm. in each case. *Lamps Types A and B,* Type A, D=0·137 in., 70 t.p.i.×0·158 in. long, with corresponding thread in holder 0·139 in. + 0·0005 – 0·002 in. (Type B is screwed *No.* 8 B.A.×0·171 in. long.)

Microscope Objectives and Nose Pieces. Objective, D=0·7982/0·7952 ins. or 20·274/20·198 mm., E=0·7804 in. or 19·822 mm. maximum; d= 0·7626 in. or 19·370 mm. maximum. Pitch = 0·02778 in. (approx.)= 0·7056 mm. Length of thread=0·125 in. Length of plain pilot=0·100 in. by 0·7626/0·7623 in. diam. *Nosepiece,* D=0·8000 in. or 20·320 mm. minimum, E=0·7822 in. or 19·868 mm. minimum, d=0·7644/0·7674 in. or 19·416/19·492 mm.

Oil Switchgear (Round copper connections to the main contacts). For $\frac{1}{2}$-inch diameter copper studs, 12 t.p.i.; For $\frac{5}{8}$, $\frac{3}{4}$, $\frac{7}{8}$, 1, $1\frac{1}{8}$, or $1\frac{1}{2}$ inch diameters, 11 t.p.i.

Optical Instruments (*Fine Motion and Adjustment*). Whitworth form, but of tooth depth 10 per cent. greater than standard to provide clearance, the radius at the root being correspondingly smaller. The combinations of major diameters and pitches recommended by "British Optical Standards" are as follows:

D (*ins.*)	=	0·10;	0·15;	0·15;	0·20;	0·20;
t.p.i.	=	60;	60;	48;	60;	48;
D (*ins.*)	=	0·25;	0·30;	0·35;	0·40;	0·45
t.p.i.	=	48;	48;	36;	36;	36

Optical Instruments (Metal tubing, Cells and other Fittings).

Bore of Tube (ins.)	: 0·50 to 0·70;	0·75 to 0·95;	1·00 to 1·20;
t.p.i.	: 60 ;	60 and 48;	60, 48 and 36;
Bore of Tube (ins.)	: 1·25 to 1·50;	1·55 to 1·75;	1·80 to 2·50;
t.p.i.	: 60, 48, 36 and 30;	48, 36 and 30;	36 and 30;
Bore of Tube (ins.)	: 2·60 to 3·00		
t.p.i.	: 36		

Nominal major diameter of screw=bore of tube+Q, where Q= 0·025 in. for 60 t.p.i.; 0·030 in. for 48 t.p.i.; 0·040 in. for 36 t.p.i.; 0·050 in. for 30 t.p.i.

Pitches of 18 and 30 t.p.i. to be used when others are not practicable. When the product $D \times$t.p.i.$>$about 72, screws of Whitworth form when screwed home on to square shoulders tend to jam.

Camera Lens Flange Fittings (*R.P.S. Standard*). The corresponding diameters and pitches are:

24 t.p.i. for D=1, 1·25, 1·375, 1·5, 1·625, 1·75, 1·875, 2·0, 2·25, 2·5 and 3.·0 ins.
12 t.p.i. for D=3·5, 4, 5 ins. and advancing by 1·0 inch increments.

Water Well Casing. Taper of thread=$\frac{3}{16}$ in. per ft. on diameter. Flanks equally inclined to axis.

Nom. Bore ins.	t.p.i.	Gauge Length	Effect. Diam. at Gauge Length	Nom. Bore ins.	t.p.i.	Gauge Length	Effect. Diam. at Gauge Length
4	10	$1\frac{3}{8}$	4·3916	13	8	$3\frac{1}{8}$	13·8749
6	10	$1\frac{5}{8}$	6·5165	15	8	$3\frac{1}{8}$	15·8749
8	8	$2\frac{1}{8}$	8·5003	18	8	$3\frac{1}{8}$	18·8749
10	8	$2\frac{5}{8}$	10·6251	21	8	$3\frac{5}{8}$	21·8437
12	8	$2\frac{5}{8}$	12·6251	24	8	$3\frac{5}{8}$	24·8437

Table A1. WHITWORTH FORM

Basic Dimensions (Inches)

Threads per inch	Pitch	Triangular Height	Shortening	Depth of Thread	Depth of Rounding	Radius
	(p)	(H)	(s)	(h)	(a)	(r)
72	0·013889	0·013340	0·002223	0·008894	0·001027	0·001907
60	0·016667	0·016009	0·002668	0·010672	0·001232	0·002289
56	0·017857	0·017151	0·002859	0·011434	0·001320	0·002452
48	0·020833	0·020010	0·003335	0·013340	0·001540	0·002861
40	0·025000	0·024012	0·004002	0·016008	0·001848	0·003433
36	0·027778	0·026681	0·004447	0·017787	0·002053	0·003815
32	0·031250	0·030015	0·005003	0·020010	0·002310	0·004292
28	0·035714	0·034303	0·005717	0·022869	0·002640	0·004905
26	0·038462	0·036942	0·006157	0·024628	0·002843	0·005282
24	0·041667	0·040021	0·006670	0·026681	0·003080	0·005722
22	0·045455	0·043659	0·007277	0·029106	0·003360	0·006242
20	0·050000	0·048025	0·008004	0·032016	0·003696	0·006866
19	0·052632	0·050553	0·008425	0·033702	0·003890	0·007228
18	0·055556	0·053361	0·008894	0·035574	0·004107	0·007629
16	0·062500	0·060031	0·010005	0·040020	0·004620	0·008583
14	0·071429	0·068607	0·011434	0·045738	0·005280	0·009809
12	0·083333	0·080041	0·013340	0·053360	0·006160	0·011444
11	0·090909	0·087317	0·014553	0·058212	0·006720	0·012484
10	0·100000	0·096049	0·016008	0·064033	0·007392	0·013733
9	0·111111	0·106721	0·017787	0·071147	0·008213	0·015259
8	0·125000	0·120061	0·020010	0·080041	0·009240	0·017166
7	0·142857	0·137213	0·022869	0·091475	0·010560	0·019618
6	0·166667	0·160082	0·026680	0·106721	0·012320	0·022888
5	0·200000	0·192098	0·032016	0·128065	0·014784	0·027466
4·5	0·222222	0·213442	0·035574	0·142295	0·016426	0·030518
4	0·250000	0·240123	0·040020	0·160082	0·018479	0·034332
3·5	0·285714	0·274426	0·045738	0·182950	0·021119	0·039237
3·25	0·307692	0·295535	0·049256	0·197024	0·022744	0·042255
3	0·333333	0·320163	0·053361	0·213442	0·024639	0·045776
2·875	0·347826	0·334084	0·055681	0·222723	0·025710	0·047767
2·75	0·363636	0·349269	0·058212	0·232846	0·026879	0·049938
2·625	0·380952	0·365901	0·060984	0·243934	0·028159	0·052316
2·5	0·400000	0·384196	0·064033	0·256131	0·029567	0·054932

Table A2. WHITWORTH FORM

Basic Dimensions (mm.)

Threads per inch	Pitch	Tri-angular Height	Shorten-ing	Depth of Thread	Depth of Rounding	Radius
	(p)	(H)	(s)	(h)	(a)	(r)
72	0·3528	0·3388	0·0565	0·2259	0·0261	0·0484
60	0·4233	0·4066	0·0678	0·2711	0·0313	0·0581
56	0·4536	0·4356	0·0726	0·2904	0·0335	0·0623
48	0·5292	0·5083	0·0847	0·3388	0·0391	0·0727
40	0·6350	0·6099	0·1017	0·4066	0·0469	0·0872
36	0·7056	0·6777	0·1130	0·4518	0·0521	0·0969
32	0·7938	0·7624	0·1271	0·5083	0·0587	0·1090
28	0·9071	0·8713	0·1452	0·5809	0·0671	0·1246
26	0·9769	0·9383	0·1564	0·6256	0·0722	0·1342
24	1·0583	1·0165	0·1694	0·6777	0·0782	0·1453
22	1·1546	1·1089	0·1848	0·7393	0·0853	0·1585
20	1·2700	1·2198	0·2033	0·8132	0·0939	0·1744
19	1·3369	1·2840	0·2140	0·8560	0·0988	0·1836
18	1·4111	1·3554	0·2259	0·9036	0·1043	0·1938
16	1·5875	1·5248	0·2541	1·0165	0·1173	0·2180
14	1·8143	1·7426	0·2904	1·1617	0·1341	0·2491
12	2·1167	2·0330	0·3388	1·3553	0·1565	0·2907
11	2·3091	2·2179	0·3696	1·4786	0·1707	0·3171
10	2·5400	2·4396	0·4066	1·6264	0·1878	0·3488
9	2·8222	2·7107	0·4518	1·8071	0·2086	0·3876
8	3·1750	3·0495	0·5083	2·0330	0·2347	0·4360
7	3·6286	3·4852	0·5809	2·3235	0·2682	0·4983
6	4·2333	4·0661	0·6777	2·7107	0·3129	0·5814
5	5·0800	4·8793	0·8132	3·2529	0·3755	0·6976
4·5	5·6444	5·4214	0·9036	3·6143	0·4172	0·7752
4	6·3500	6·0991	1·0165	4·0661	0·4694	0·8720
3·5	7·2571	6·9704	1·1617	4·6469	0·5364	0·9966
3·25	7·8154	7·5066	1·2511	5·0044	0·5777	1·0733
3·0	8·4667	8·1321	1·3554	5·4214	0·6258	1·1627
2·875	8·8348	8·4857	1·4143	5·6572	0·6530	1·2133
2·75	9·2364	8·8714	1·4786	5·9143	0·6827	1·2684
2·625	9·6762	9·2939	1·5490	6·1959	0·7152	1·3288
2·5	10·1600	9·7586	1·6264	6·5057	0·7510	1·3953

Table A3.

BRITISH STANDARD WHITWORTH SERIES (B.S.W.)

Basic Sizes in Inches

Nom. Diam.	Threads per inch	Major Diam.	Effect. Diam.	Minor Diam.	Cross-Sectional Area at Root	Mean Helix Angle
$\frac{1}{8}$	40	0·1250	0·1090	0·0930	0·0068	4° 11′
$\frac{3}{16}$	24	0·1875	0·1608	0·1341	0·0141	4° 43′
$\frac{1}{4}$	20	0·2500	0·2180	0·1860	0·0272	4° 10′
$\frac{5}{16}$	18	0·3125	0·2769	0·2413	0·0457	3° 39′
$\frac{3}{8}$	16	0·3750	0·3350	0·2950	0·0683	3° 24′
$\frac{7}{16}$	14	0·4375	0·3918	0·3461	0·0941	3° 19′
$\frac{1}{2}$	12	0·5000	0·4466	0·3932	0·1214	3° 24′
$\frac{9}{16}$	12	0·5625	0·5091	0·4557	0·1631	2° 59′
$\frac{5}{8}$	11	0·6250	0·5668	0·5086	0·2032	2° 55′
$\frac{11}{16}$	11	0·6875	0·6293	0·5711	0·2562	2° 38′
$\frac{3}{4}$	10	0·7500	0·6860	0·6220	0·3039	2° 39′
$\frac{7}{8}$	9	0·8750	0·8039	0·7328	0·4218	2° 31′
1	8	1·0000	0·9200	0·8400	0·5542	2° 28′
$1\frac{1}{8}$	7	1·1250	1·0335	0·9420	0·6969	2° 31′
$1\frac{1}{4}$	7	1·2500	1·1585	1·0670	0·8942	2° 15′
$1\frac{1}{2}$	6	1·5000	1·3933	1·2866	1·3000	2° 11′
$1\frac{3}{4}$	5	1·7500	1·6219	1·4938	1·7530	2° 15′
2	4·5	2·0000	1·8577	1·7154	2·3110	2° 11′
$2\frac{1}{4}$	4	2·2500	2·0899	1·9298	2·9250	2° 11′
$2\frac{1}{2}$	4	2·5000	2·3399	2·1798	3·7320	1° 57′
$2\frac{3}{4}$	3·5	2·7500	2·5670	2·3840	4·4640	2° 2′
3	3·5	3·0000	2·8170	2·6340	5·4490	1° 51′
$3\frac{1}{8}$	3·5	3·1250	2·9421	2·7591	5·9789	1° 46′
$3\frac{1}{4}$	3·25	3·2500	3·0530	2·8560	6·4060	1° 50′
$3\frac{3}{8}$	3·25	3·3750	3·1780	2·9810	6·9794	1° 46′
$3\frac{1}{2}$	3·25	3·5000	3·3030	3·1060	7·5770	1° 42′
$3\frac{5}{8}$	3·25	3·6250	3·4280	3·2310	8·1991	1° 38′
$3\frac{3}{4}$	3·0	3·7500	3·5366	3·3232	8·6740	1° 43′
$3\frac{7}{8}$	3·0	3·8750	3·6616	3·4481	9·3379	1° 40′
4	3·0	4·0000	3·7866	3·5732	10·0300	1° 36′
$4\frac{1}{8}$	3·0	4·1250	3·9116	3·6981	10·7411	1° 33′
$4\frac{1}{4}$	2·875	4·2500	4·0273	3·8046	11·3687	1° 34′
$4\frac{3}{8}$	2·875	4·3750	4·1523	3·9296	12·1280	1° 32′
$4\frac{1}{2}$	2·875	4·5000	4·2773	4·0546	12·9100	1° 29′
$4\frac{3}{4}$	2·75	4·7500	4·5172	4·2843	14·4162	1° 28′
5	2·75	5·0000	4·7672	4·5344	16·1500	1° 23′
$5\frac{1}{4}$	2·625	5·2500	5·0061	4·7621	17·8110	1° 23′
$5\frac{1}{2}$	2·625	5·5000	5·2561	5·0122	19·7301	1° 19′
6	2·5	6·0000	5·7439	5·4878	23·6500	1° 16′

Table A4.

BRITISH STANDARD WHITWORTH SERIES (B.S.W.)

Basic Dimensions in mm.

Nom. Diam. (ins.)	Threads per inch	Major Diam.	Effect. Diam.	Minor Diam.
$\frac{1}{8}$	40	3·1750	2·7686	2·3622
$\frac{3}{16}$	24	4·7625	4·0843	3·4061
$\frac{1}{4}$	20	6·3500	5·5372	4·7244
$\frac{5}{16}$	18	7·9375	7·0333	6·1290
$\frac{3}{8}$	16	9·5250	8·5090	7·4930
$\frac{7}{16}$	14	11·1125	9·9517	8·7909
$\frac{1}{2}$	12	12·7000	11·3436	9·9873
$\frac{9}{16}$	12	14·2875	12·9311	11·5748
$\frac{5}{8}$	11	15·8750	14·3967	12·9184
$\frac{11}{16}$	11	17·4625	15·9842	14·5059
$\frac{3}{4}$	10	19·0500	17·4244	15·7988
$\frac{7}{8}$	9	22·2250	20·4191	18·6131
1	8	25·4000	23·3680	21·3360
$1\frac{1}{8}$	7	28·5750	26·2509	23·9268
$1\frac{1}{4}$	7	31·7500	29·4259	27·1018
$1\frac{1}{2}$	6	38·1000	35·3898	32·6796
$1\frac{3}{4}$	5	44·4500	41·1963	37·9425
2	4·5	50·8000	47·1856	43·5712
$2\frac{1}{4}$	4	57·1500	53·0835	49·0169
$2\frac{1}{2}$	4	63·5000	59·4335	55·3669
$2\frac{3}{4}$	3·5	69·8500	65·2018	60·5536
3	3·5	76·2000	71·5518	66·9036
$3\frac{1}{8}$	3·5	79·3750	74·7293	70·0811
$3\frac{1}{4}$	3·25	82·5500	77·5462	72·5424
$3\frac{3}{8}$	3·25	87·7250	80·7212	75·7174
$3\frac{1}{2}$	3·25	88·9000	83·8962	78·8924
$3\frac{5}{8}$	3·25	92·0750	87·0712	82·0674
$3\frac{3}{4}$	3·0	95·2500	89·8296	84·4093
$3\frac{7}{8}$	3·0	98·4250	93·0046	87·5817
4	3·0	101·6000	96·1796	90·7593
$4\frac{1}{8}$	3·0	104·7750	99·3546	93·9317
$4\frac{1}{4}$	2·875	107·9500	102·2934	96·6368
$4\frac{3}{8}$	2·875	111·1250	105·4684	99·8118
$4\frac{1}{2}$	2·875	114·3000	108·6434	102·9868
$4\frac{3}{4}$	2·75	120·6500	114·7369	108·8212
5	2·75	127·0000	121·0869	115·1738
$5\frac{1}{4}$	2·625	133·3500	127·1549	120·9573
$5\frac{1}{2}$	2·625	139·7000	133·5049	127·3099
6	2·5	152·4000	145·8951	139·3901

Table A9. BRITISH BRASS THREAD

Dimensions in Inches

Nom. Size	Major Diam.	Effective Diam.	Minor Diam.	Tapping Drills 90%		Tapping Drills 80%	
1/8	0·1250	0·1004	0·0757	No. 46	(0·0810)	No. 44	(0·0860)
5/32	0·15625	0·1316	0·1070	2·85 mm.	(0·1122)	2·95 mm.	(0·1161)
3/16	0·1875	0·1629	0·1382	3·65 mm.	(0·1437)	3·75 mm.	(0·1476)
1/4	0·2500	0·2254	0·2007	No. 5	(0·2055)	No. 4	(0·2090)
9/32	0·28125	0·2566	0·2320	6·0 mm.	(0·2362)	No. C	(0·2420)
5/16	0·3125	0·2879	0·2632	6·8 mm.	(0·2677)	No. I	(0·2720)
0·320	0·3200	0·2954	0·2707	7·0 mm.	(0·2756)	No. K	(0·2810)
3/8	0·3750	0·3504	0·3257	8·4 mm.	(0·3307)	8·5 mm.	(0·3346)
7/16	0·4375	0·4129	0·3882	10·0 mm.	(0·3937)	10·1 mm.	(0·3976)
0·4724	0·4724	0·4478	0·4231	10·9 mm.	(0·4291)	11·0 mm.	(0·4331)
1/2	0·5000	0·4754	0·4507	11·6 mm.	(0·4567)	11·7 mm.	(0·4606)
0·5512	0·5512	0·5266	0·5019	12·9 mm.	(0·5079)	13·0 mm.	(0·5118)
9/16	0·5625	0·5379	0·5132	13·2 mm.	(0·5197)	13·3 mm.	(0·5236)
5/8	0·6250	0·6004	0·5757				
11/16	0·6875	0·6629	0·6382				
3/4	0·7500	0·7254	0·7007				
13/16	0·8125	0·7879	0·7632				
7/8	0·8750	0·8504	0·8257				
15/16	0·9375	0·9129	0·8882				
1	1·0000	0·9754	0·9507				
1 1/8	1·1250	1·1004	1·0757				
1 1/4	1·2500	1·2254	1·2007				
1 3/8	1·3750	1·3504	1·3257				
1 7/16	1·4375	1·4129	1·3882				
1 1/2	1·5000	1·4754	1·4507				
1 5/8	1·6250	1·6004	1·5757				
1 3/4	1·7500	1·7254	1·7007				
1 7/8	1·8750	1·8504	1·8257				
2	2·0000	1·9754	1·9507				

Table A10. MODEL ENGINEER'S THREAD (M.E.)

Dimensions in Inches

Nom. Diam.	Threads per inch	Pitch	Major Diam.	Effective Diam.	Minor Diam.	Tapping Drill
$\frac{1}{8}$	40	0·02500	0·1250	0·1090	0·0930	No. 40
$\frac{5}{32}$	40	,,	0·1563	0·1403	0·1243	30
$\frac{3}{16}$	40	,,	0·1875	0·1715	0·1555	$\frac{5}{32}$
$\frac{7}{32}$	40	,,	0·2188	0·2028	0·1868	$\frac{3}{16}$
$\frac{1}{4}$	40	,,	0·2500	0·2340	0·2180	$\frac{7}{32}$
$\frac{9}{32}$	32	0·03125	0·2813	0·2613	0·2413	C
$\frac{5}{16}$	32	,,	0·3125	0·2925	0·2725	J
$\frac{3}{8}$	32	,,	0·3750	0·3550	0·3350	R
$\frac{7}{16}$	26	0·03846	0·4375	0·4129	0·3882	10 mm
$\frac{1}{2}$	26	,,	0·5000	0·4754	0·4507	$\frac{29}{64}$

Table A11.

BRITISH STANDARD ELECTRICAL CONDUIT THREAD

Basic Sizes in Inches

Nom. Diam.	Threads per inch	Pitch	Depth of Thread	Major Diam.	Effect Diam.	Minor Diam.	Length of Thread on End: Limits
$\frac{1}{2}$	18	0·05556	0·03555	0·5000	0·4644	0·4289	$\frac{3}{8}$–$\frac{7}{16}$
$\frac{5}{8}$	18	,,	,,	0·6250	0·5894	0·5539	$\frac{7}{16}$–$\frac{1}{2}$
$\frac{3}{4}$	16	0·06250	0·04000	0·7500	0·7100	0·6700	$\frac{1}{2}$–$\frac{9}{16}$
1	16	,,	,,	1·0000	0·9600	0·9200	$\frac{5}{8}$–$\frac{11}{16}$
$1\frac{1}{4}$	16	,,	,,	1·2500	1·2100	1·1700	$\frac{11}{16}$–$\frac{3}{4}$
$1\frac{1}{2}$	14	0·07143	0·04575	1·5000	1·4543	1·4085	$\frac{3}{4}$–$\frac{13}{16}$
2	14	,,	,,	2·0000	1·9543	1·9085	$\frac{7}{8}$–$\frac{15}{16}$
$2\frac{1}{2}$	14	,,	,,	2·5000	2·4543	2·4085	1–$1\frac{1}{16}$

'able A12.

BRITISH STANDARD ELECTRICAL CONDUIT THREAD

Tolerances in Units of 0·0001-inch

Nom. Size	Threads per inch	Major Diameter		Effective Diameter	Minor Diameter	
		Conduit (—)	Coupling (+)	Conduit (—) Coupling (+)	Conduit (—)	Coupling (+)
½	18	106	141	71	141	106
⅝	18	106	141	71	141	106
¾	16	113	150	75	150	113
1	16	113	150	75	150	113
1¼	16	113	150	75	150	113
1½	14	120	160	80	160	120
2	14	120	160	80	160	120
2½	14	120	160	80	160	120

able A13. **BRITISH PIPE THREAD (B.S.P.)**

Basic Sizes in Inches

Nom. Size	Threads per inch	Pitch	Depth of Thread	Major Diam.	Effect. Diam.	Minor Diam.	Gauge Length
⅛	28	0·03571	0·0229	0·3830	0·3601	0·3372	0·1563
¼	19	0·05263	0·0337	0·5180	0·4843	0·4506	0·2367
⅜	19	0·05263	,,	0·6560	0·6223	0·5886	0·2500
½	14	0·07143	0·0457	0·8250	0·7793	0·7336	0·3214
*⅝	14	,,	,,	0·9020	0·8563	0·8106	0·3214
¾	14	,,	,,	1·0410	0·9953	0·9496	0·3750
*⅞	14	,,	,,	1·1890	1·1433	1·0976	0·3750
1	11	0·09091	0·0582	1·3090	1·2508	1·1926	0·4091
1¼	11	,,	,,	1·6500	1·5918	1·5336	0·5000
1½	11	,,	,,	1·8820	1·8238	1·7656	0·5000
*1¾	11	,,	,,	2·1160	2·0578	1·9996	0·6250

[*continued on page* 36

*These sizes are not used for British Standard Taper Pipe Threads but are recognized in B.S. 2779, Fastening Threads of B.S.P. Sizes.

Table A13 *continued.*

BRITISH PIPE THREAD (B.S.P.)

Basic Sizes in Inches

Nom. Size	Threads per inch	Pitch	Depth of Thread	Major Diam.	Effect. Diam.	Minor Diam.	Gauge Length
2	11	0·09091	0·0582	2·3470	2·2888	2·2306	0·6250
*2¼	11	,,	,,	2·5870	2·5288	2·4706	0·6875
2½	11	,,	,,	2·9600	2·9018	2·8436	0·6875
*2¾	11	,,	,,	3·2100	3·1518	3·0936	0·8125
3	11	0·09091	0·0582	3·4600	3·4018	3·3436	0·8125
3½	11	,,	,,	3·9500	3·8918	3·8336	0·8750
4	11	,,	,,	4·4500	4·3918	4·3336	1·0000
5	11	,,	,,	5·4500	5·3918	5·3336	1·1250
6	11	,,	,,	6·4500	6·3918	6·3336	1·1250

*These sizes are not used for British Standard Taper Pipe Threads, but are recognized i B.S. 2779, Fastening Threads of B.S.P. Sizes.

Table A13M.

BRITISH PIPE THREAD (B.S.P.)

Basic Sizes in mm.

Nom. Size	Threads per inch	Pitch	Depth of Thread	Major Diam.	Effect. Diam.	Minor Diam.	Gauge Length
in.							
⅛	28	0·907	0·581	9·728	9·147	8·566	4·0
¼	19	1·337	0·856	13·157	12·301	11·445	6·0
⅜	19	1·337	0·856	16·662	15·806	14·950	6·4
½	14	1·814	1·162	20·955	19·793	18·631	8·2
¾	14	1·814	1·162	26·441	25·279	24·117	9·5
1	11	2·309	1·479	33·249	31·770	30·291	10·4
1¼	11	2·309	1·479	41·910	40·431	38·952	12·7
1½	11	2·309	1·479	47·803	46·324	44·845	12·7
2	11	2·309	1·479	59·614	58·135	56·656	15·9
2½	11	2·309	1·479	75·184	73·705	72·226	17·5
3	11	2·309	1·479	87·884	86·405	84·926	20·6
3½	11	2·309	1·479	100·330	98·851	97·372	22·2
4	11	2·309	1·479	113·030	111·551	110·072	25·4
5	11	2·309	1·479	138·430	136·951	135·472	28·6
6	11	2·309	1·479	163·830	162·351	160·872	28·6

The metric dimensions given above are the standardized conversions of the inch sizes show in Table A13.

Table A14.

BRITISH STANDARD TAPER PIPE (B.S.P.Tr.)
Tolerances and Allowances
TURNS OF THREAD

B.S.P. Size (Nominal Bore of Tube)	Gauge Length (Distance of Gauge Plane from Pipe End)		Position of Gauge Plane on Internal Screws	Length of useful Thread on Pipe End not less than: (for Basic Gauge Length)	Fitting Allowance
	Basic	Tol. Plus and Minus	Tol. Plus and Minus		
in.	in.	in.	in.	in.	in.
$\frac{1}{8}$	$4\frac{3}{8}$	1	$1\frac{1}{4}$	$7\frac{1}{8}$	$2\frac{3}{4}$
$\frac{1}{4}$	$4\frac{1}{2}$	1	$1\frac{1}{4}$	$7\frac{1}{4}$	$2\frac{3}{4}$
$\frac{3}{8}$	$4\frac{3}{4}$	1	$1\frac{1}{4}$	$7\frac{1}{2}$	$2\frac{3}{4}$
$\frac{1}{2}$	$4\frac{1}{2}$	1	$1\frac{1}{4}$	$7\frac{1}{4}$	$2\frac{3}{4}$
$\frac{3}{4}$	$5\frac{1}{4}$	1	$1\frac{1}{4}$	8	$2\frac{3}{4}$
1	$4\frac{1}{2}$	1	$1\frac{1}{4}$	$7\frac{1}{4}$	$2\frac{3}{4}$
$1\frac{1}{4}$	$5\frac{1}{2}$	1	$1\frac{1}{4}$	$8\frac{1}{4}$	$2\frac{3}{4}$
$1\frac{1}{2}$	$5\frac{1}{2}$	1	$1\frac{1}{4}$	$8\frac{1}{4}$	$2\frac{3}{4}$
2	$6\frac{7}{8}$	1	$1\frac{1}{4}$	$10\frac{1}{8}$	$3\frac{1}{4}$
$2\frac{1}{2}$	$7\frac{9}{16}$	$1\frac{1}{2}$	$1\frac{1}{2}$	$11\frac{9}{16}$	4
3	$8\frac{15}{16}$	$1\frac{1}{2}$	$1\frac{1}{2}$	$12\frac{15}{16}$	4
$3\frac{1}{2}$	$9\frac{5}{8}$	$1\frac{1}{2}$	$1\frac{1}{2}$	$13\frac{5}{8}$	4
4	11	$1\frac{1}{2}$	$1\frac{1}{2}$	$15\frac{1}{2}$	$4\frac{1}{2}$
5	$12\frac{3}{8}$	$1\frac{1}{2}$	$1\frac{1}{2}$	$17\frac{3}{8}$	5
6	$12\frac{3}{8}$	$1\frac{1}{2}$	$1\frac{1}{2}$	$17\frac{3}{8}$	5

[*continued on page* 38

Table A14
continued.

BRITISH STANDARD TAPER PIPE (B.S.P.Tr.)

Tolerances and Allowances

LINEAR MEASURE

B.S.P. Size (Nominal Bore of Tube)	Gauge Length (Distance of Gauge Plane from Pipe End)		Position of Gauge Plane on Internal Screws	Length of useful Thread on Pipe End not less than (for Basic Gauge Length)	Fitting Allow-ance
	Basic	Tol. Plus and Minus	Tol. Plus and Minus		
in.	in.	in.	in.	in.	in.
$\frac{1}{8}$	0·1563	0·0357	0·0446	0·2545	0·0982
$\frac{1}{4}$	0·2367	0·0526	0·0658	0·3814	0·1447
$\frac{3}{8}$	0·2500	0·0526	0·0658	0·3947	0·1447
$\frac{1}{2}$	0·3214	0·0714	0·0893	0·5178	0·1964
$\frac{3}{4}$	0·3750	0·0714	0·0893	0·5714	0·1964
1	0·4091	0·0909	0·1136	0·6591	0·2500
$1\frac{1}{4}$	0·5000	0·0909	0·1136	0·7500	0·2500
$1\frac{1}{2}$	0·5000	0·0909	0·1136	0·7500	0·2500
2	0·6250	0·0909	0·1136	0·9204	0·2954
$2\frac{1}{2}$	0·6875	0·1364	0·1364	1·0511	0·3636
3	0·8125	0·1364	0·1364	1·1761	0·3636
$3\frac{1}{2}$	0·8750	0·1364	0·1364	1·2386	0·3636
4	1·0000	0·1364	0·1364	1·4091	0·4091
5	1·1250	0·1364	0·1364	1·5795	0·4545
6	1·1250	0·1364	0·1364	1·5795	0·4545

able A7.

BRITISH STANDARD FINE THREAD (B.S. 84—1956)

Basic Sizes of Screws and Nuts (Inches)

Nom. Diam.	Threads per inch	Major Diam.	Effective Diam.	Minor Diam.	Cross-Sectional Area at Bottom of Thread	Mean Helix Angle
$\frac{3}{16}$	32	0·1875	0·1675	0·1475	0·0171	3° 24′
$\frac{7}{32}$	28	0·2188	0·1959	0·1730	0·0235	3° 19′
$\frac{1}{4}$	26	0·2500	0·2254	0·2008	0·0317	3° 7′
$\frac{9}{32}$	26	0·2812	0·2566	0·2320	0·0423	2° 44′
$\frac{5}{16}$	22	0·3125	0·2834	0·2543	0·0508	2° 55′
$\frac{3}{8}$	20	0·3750	0·3430	0·3110	0·0760	2° 39′
$\frac{7}{16}$	18	0·4375	0·4019	0·3663	0·1054	2° 31′
$\frac{1}{2}$	16	0·5000	0·4600	0·4200	0·1385	2° 28′
$\frac{9}{16}$	16	0·5625	0·5225	0·4825	0·1828	2° 10′
$\frac{5}{8}$	14	0·6250	0·5793	0·5336	0·2236	2° 15′
$\frac{11}{16}$	14	0·6875	0·6418	0·5961	0·2791	2° 2′
$\frac{3}{4}$	12	0·7500	0·6966	0·6432	0·3249	2° 11′
$\frac{13}{16}$	12	0·8125	0·7591	0·7057	0·3911	2° 0′
$\frac{7}{8}$	11	0·8750	0·8168	0·7586	0·4520	2° 2′
1	10	1·0000	0·9360	0·8720	0·5972	1° 57′
$1\frac{1}{8}$	9	1·1250	1·0539	0·9828	0·7586	1° 55′
$1\frac{1}{4}$	9	1·2500	1·1789	1·1078	0·9639	1° 43′
$1\frac{3}{8}$	8	1·3750	1·2950	1·2150	1·1590	1° 46′
$1\frac{1}{2}$	8	1·5000	1·4200	1·3400	1·4100	1° 36′
$1\frac{5}{8}$	8	1·6250	1·5450	1·4650	1·6860	1° 29′
$1\frac{3}{4}$	7	1·7500	1·6585	1·5670	1·9280	1° 34′
2	7	2·0000	1·9085	1·8170	2·5930	1° 22′
$2\frac{1}{4}$	6	2·2500	2·1433	2·0366	3·2580	1° 25′
$2\frac{1}{2}$	6	2·5000	2·3933	2·2866	4·1060	1° 16′
$2\frac{3}{4}$	6	2·7500	2·6433	2·5366	5·0540	1° 9′
3	5	3·0000	2·8719	2·7438	5·9130	1° 16′
$3\frac{1}{4}$	5	3·2500	3·1219	2·9938	7·0390	1° 10′
$3\frac{1}{2}$	4·5	3·5000	3·3577	3·2154	8·1200	1° 12′
$3\frac{3}{4}$	4·5	3·7500	3·6077	3·4654	9·4320	1° 7′
4	4·5	4·0000	3·8577	3·7154	10·8400	1° 3′
$4\frac{1}{4}$	4	4·2500	4·0899	3·9298	12·1300	1° 7′

Table A8.

BRITISH STANDARD FINE SERIES (B.S. 84—1956)

Tolerances* in Units of 0·0001 inch

(C=Close class; M=Medium class; F=Free class; N=Normal class

Nom. Diam. (ins.)	Threads per inch	Major Diam. Screws only (—)			Effective Diam. Screws (—) & Nuts (+)			Minor Diam. Screws only (—)			Minor Diam. Nuts (+)
		C	M	F	C	M	F&N	C	M	F	C,M,N
$\frac{3}{16}$	32	40	51	68	22	33	50	45	68	85	102
$\frac{7}{32}$	28	43	55	72	24	36	53	49	74	91	111
$\frac{1}{4}$	26	45	57	76	25	37	56	50	76	95	117
$\frac{9}{32}$	26	46	59	78	26	39	58	51	78	97	117
$\frac{5}{16}$	22	48	62	83	27	41	62	55	84	105	141
$\frac{3}{8}$	20	51	66	88	29	44	66	58	89	111	170
$\frac{7}{16}$	18	55	71	94	31	47	70	62	94	117	181
$\frac{1}{2}$	16	58	75	99	33	50	74	65	100	124	195
$\frac{9}{16}$	16	59	77	102	34	52	77	66	102	127	195
$\frac{5}{8}$	14	63	81	108	36	54	81	71	107	134	213
$\frac{11}{16}$	14	64	83	111	37	56	84	72	109	137	213
$\frac{3}{4}$	12	68	88	117	39	59	88	77	117	146	237
$\frac{13}{16}$	12	69	89	119	40	60	90	78	118	148	237
$\frac{7}{8}$	11	72	92	123	42	62	93	81	122	153	252
1	10	76	98	131	44	66	99	85	129	162	270
$1\frac{1}{8}$	9	79	102	137	46	69	104	89	136	171	292
$1\frac{1}{4}$	9	81	105	141	48	72	108	91	139	175	292
$1\frac{3}{8}$	8	85	110	148	50	75	113	96	146	184	320
$1\frac{1}{2}$	8	87	112	151	52	77	116	98	148	187	320

NOTE.—Nuts of the previously denoted "Free Class" are now termed "Normal."

*For Allowances see page 17.

Table A6. OTHER B.S.W. SIZES IN USE

Dimensions in Inches

Nom. Diam.	Threads per inch	Pitch	Major Diam.	Effect. Diam.	Minor Diam.
$\frac{1}{16}$	60	0·01667	0·0625	0·0518	0·0411
$\frac{5}{64}$	56	0·01786	0·0781	0·0667	0·0553
$\frac{3}{32}$	48	0·02083	0·0938	0·0805	0·0672
$\frac{7}{64}$	48	0·02083	0·1094	0·0961	0·0828
$\frac{5}{32}$	32	0·03125	0·1563	0·1363	0·1163
$\frac{7}{32}$	24	0·04167	0·2188	0·1921	0·1654
$\frac{9}{32}$	20	0·05000	0·2813	0·2493	0·2173
$\frac{11}{32}$	18	0·05556	0·3438	0·3082	0·2726
$\frac{13}{16}$	10	0·10000	0·8125	0·7485	0·6845
$\frac{15}{16}$	9	0·11111	0·9375	0·8664	0·7953
$1\frac{3}{8}$	6	0·16667	1·3750	1·2683	1·1616
$1\frac{5}{8}$	5	0·20000	1·6250	1·4969	1·3688
$1\frac{7}{8}$	4·5	0·22222	1·8750	1·7327	1·5904

Table A5.

BRITISH STANDARD WHITWORTH SERIES (B.S.W.)
Tolerances* in Units of 0·0001 inch (B.S. 84—1956)

(C=Close class; M=Medium class : F=Free class : N=Normal class)

Nom. Diam. (ins.)	Threads per inch	Major Diam. Screws only (—)			Effective Diam. Screws (—); Nuts (+)			Minor Diam. Screws only (—)			Minor Diam. Nuts (+)
		C	M	F	C	M	F&N	C	M	F	C,M,N
1/8	40	35	45	59	19	29	43	40	61	75	90
3/16	24	43	55	72	23	35	52	50	76	93	133
1/4	20	48	61	80	26	39	58	55	84	103	170
5/16	18	52	66	87	28	42	63	59	89	110	181
3/8	16	55	70	93	30	45	68	62	95	118	195
7/16	14	59	75	100	32	48	73	67	101	126	213
1/2	12	63	81	106	34	52	77	72	110	135	237
9/16	12	65	82	109	36	53	80	74	111	138	237
5/8	11	67	86	114	37	56	84	76	116	144	252
11/16	11	68	88	116	38	58	86	77	118	146	252
3/4	10	72	92	122	40	60	90	81	123	153	270
7/8	9	76	97	129	43	64	96	86	131	163	292
1	8	80	103	137	45	68	102	91	139	173	320
1 1/8	7	86	110	145	48	72	107	97	148	183	356
1 1/4	7	87	112	149	49	74	111	98	150	187	356
1 1/2	6	94	121	161	53	80	120	106	162	202	403
1 3/4	5	102	131	174	57	86	129	115	175	218	470
2	4·5	108	138	184	61	91	137	122	185	231	514
2 1/4	4	114	146	194	64	96	144	129	196	244	570
2 1/2	4	116	150	199	66	100	149	131	200	249	570
2 3/4	3·5	123	157	210	70	104	157	139	211	264	641
3	3·5	125	161	214	72	108	161	141	215	268	641

NOTE.—Nuts of the previously denoted "Free Class" are now termed "Normal."

*For Allowances see page 17.

Table A9. BRITISH BRASS THREAD

Dimensions in Inches

Nom. Size	Major Diam.	Effective Diam.	Minor Diam.	Tapping Drills 90%	Tapping Drills 80%
1/8	0·1250	0·1004	0·0757	No. 46 (0·0810)	No. 44 (0·0860)
5/32	0·15625	0·1316	0·1070	2·85 mm. (0·1122)	2·95 mm. (0·1161)
3/16	0·1875	0·1629	0·1382	3·65 mm. (0·1437)	3·75 mm. (0·1476)
1/4	0·2500	0·2254	0·2007	No. 5 (0·2055)	No. 4 (0·2090)
9/32	0·28125	0·2566	0·2320	6·0 mm. (0·2362)	No. C (0·2420)
5/16	0·3125	0·2879	0·2632	6·8 mm. (0·2677)	No. I (0·2720)
0·320	0·3200	0·2954	0·2707	7·0 mm. (0·2756)	No. K (0·2810)
3/8	0·3750	0·3504	0·3257	8·4 mm. (0·3307)	8·5 mm. (0·3346)
7/16	0·4375	0·4129	0·3882	10·0 mm. (0·3937)	10·1 mm. (0·3976)
0·4724	0·4724	0·4478	0·4231	10·9 mm. (0·4291)	11·0 mm. (0·4331)
1/2	0·5000	0·4754	0·4507	11·6 mm. (0·4567)	11·7 mm. (0·4606)
0·5512	0·5512	0·5266	0·5019	12·9 mm. (0·5079)	13·0 mm. (0·5118)
9/16	0·5625	0·5379	0·5132	13·2 mm. (0·5197)	13·3 mm. (0·5236)
5/8	0·6250	0·6004	0·5757		
11/16	0·6875	0·6629	0·6382		
3/4	0·7500	0·7254	0·7007		
13/16	0·8125	0·7879	0·7632		
7/8	0·8750	0·8504	0·8257		
15/16	0·9375	0·9129	0·8882		
1	1·0000	0·9754	0·9507		
1 1/8	1·1250	1·1004	1·0757		
1 1/4	1·2500	1·2254	1·2007		
1 3/8	1·3750	1·3504	1·3257		
1 7/16	1·4375	1·4129	1·3882		
1 1/2	1·5000	1·4754	1·4507		
1 5/8	1·6250	1·6004	1·5757		
1 3/4	1·7500	1·7254	1·7007		
1 7/8	1·8750	1·8504	1·8257		
2	2·0000	1·9754	1·9507		

Table A10. MODEL ENGINEER'S THREAD (M.E.)

Dimensions in Inches

Nom. Diam.	Threads per inch	Pitch	Major Diam.	Effective Diam.	Minor Diam.	Tapping Drill
$\frac{1}{8}$	40	0·02500	0·1250	0·1090	0·0930	No. 40
$\frac{5}{32}$	40	,,	0·1563	0·1403	0·1243	30
$\frac{3}{16}$	40	,,	0·1875	0·1715	0·1555	$\frac{5}{32}$
$\frac{7}{32}$	40	,,	0·2188	0·2028	0·1868	$\frac{3}{16}$
$\frac{1}{4}$	40	,,	0·2500	0·2340	0·2180	$\frac{7}{32}$
$\frac{9}{32}$	32	0·03125	0·2813	0·2613	0·2413	C
$\frac{5}{16}$	32	,,	0·3125	0·2925	0·2725	J
$\frac{3}{8}$	32	,,	0·3750	0·3550	0·3350	R
$\frac{7}{16}$	26	0·03846	0·4375	0·4129	0·3882	10 mm
$\frac{1}{2}$	26	,,	0·5000	0·4754	0·4507	$\frac{29}{64}$

Table A11.

BRITISH STANDARD ELECTRICAL CONDUIT THREAD

Basic Sizes in Inches

Nom. Diam.	Threads per inch	Pitch	Depth of Thread	Major Diam.	Effect Diam.	Minor Diam.	Length of Thread on End: Limits
$\frac{1}{2}$	18	0·05556	0·03555	0·5000	0·4644	0·4289	$\frac{3}{8}$–$\frac{7}{16}$
$\frac{5}{8}$	18	,,	,,	0·6250	0·5894	0·5539	$\frac{7}{16}$–$\frac{1}{2}$
$\frac{3}{4}$	16	0·06250	0·04000	0·7500	0·7100	0·6700	$\frac{1}{2}$–$\frac{9}{16}$
1	16	,,	,,	1·0000	0·9600	0·9200	$\frac{5}{8}$–$\frac{11}{16}$
$1\frac{1}{4}$	16	,,	,,	1·2500	1·2100	1·1700	$\frac{11}{16}$–$\frac{3}{4}$
$1\frac{1}{2}$	14	0·07143	0·04575	1·5000	1·4543	1·4085	$\frac{3}{4}$–$\frac{13}{16}$
2	14	,,	,,	2·0000	1·9543	1·9085	$\frac{7}{8}$–$\frac{15}{16}$
$2\frac{1}{2}$	14	,,	,,	2·5000	2·4543	2·4085	1–$1\frac{1}{16}$

'able A14M.

BRITISH STANDARD TAPER PIPE (B.S.P.Tr.)
Tolerances and Allowances
Linear Measure in mm.

B.S.P. Size (Nominal Bore of Tube)	Gauge Length (Distance of Gauge Plane from Pipe End)		Position of Gauge Plane on Internal Screws	Length of useful Thread on Pipe End not less than: (for Basic Gauge Length)	Fitting Allowance
	Basic	Tol. Plus and Minus	Tol. Plus and Minus		
in.					
1/8	4·0	0·9	1·1	6·5	2·5
1/4	6·0	1·3	1·7	9·7	3·7
3/8	6·4	1·3	1·7	10·1	3·7
1/2	8·2	1·8	2·3	13·2	5·0
3/4	9·5	1·8	2·3	14·5	5·0
1	10·4	2·3	2·9	16·8	6·4
1 1/4	12·7	2·3	2·9	19·1	6·4
1 1/2	12·7	2·3	2·9	19·1	6·4
2	15·9	2·3	2·9	23·4	7·5
2 1/2	17·5	3·5	3·5	26·7	9·2
3	20·6	3·5	3·5	29·8	9·2
3 1/2	22·2	3·5	3·5	31·4	9·2
4	25·4	3·5	3·5	35·8	10·4
5	28·6	3·5	3·5	40·1	11·5
6	28·6	3·5	3·5	40·1	11·5

he metric dimensions given above are the standardized conversions of the inch sizes shown
Table A14.

Table A15. BRITISH STANDARD PARALLEL PIPE

Tolerances in Units of 0·0001-inch

(*C* = *Close class; M* = *Medium class; F* = *Free class; N* = *Normal class*)

Nom. Size (in.)	Threads per inch	Major Diam.			Effect. Diam.			Minor Diam.			
		Pipe (–)			Pipe (–); Coupling (+)			Pipe (–)			Couplings (+)
		C	M	F	C	M	F&N	C	M	F	C, M, N
1/8	28	47	61	83	28	42	64	53	80	102	111
1/4	19	55	72	96	32	49	73	62	95	119	175
3/8	19	56	73	98	33	50	75	63	96	121	175
1/2	14	64	83	111	37	56	84	72	109	137	213
5/8	14	65	83	112	38	56	85	73	109	138	213
3/4	14	67	87	116	40	60	89	75	113	142	213
7/8	14	67	88	118	40	61	91	75	114	144	213
1	11	73	95	128	43	65	98	82	125	158	252
1 1/4	11	76	99	133	46	69	103	85	129	163	252
1 1/2	11	78	102	137	48	72	107	87	132	167	252
1 3/4	11	79	103	139	48	73	109	88	133	169	252
2	11	79	104	140	49	73	110	88	134	171	252
2 1/4	11	80	105	142	50	74	112	89	135	172	252
2 1/2	11	81	106	144	50	76	113	90	136	174	252
2 3/4	11	81	107	145	51	76	115	90	137	175	252
3	11	82	107	146	51	77	116	91	137	176	252
3 1/2	11	82	109	148	52	79	118	92	139	178	252
4	11	83	110	150	53	80	120	92	140	180	252
5	11	85	112	153	55	82	123	94	142	183	252
6	11	86	114	156	56	84	126	95	144	186	252

Table A16. HYDRAULIC PIPE THREADS

Outside Diam. (approx.) (in.)	Thread size = B.S.P. (Parallel) Nom. (in.)	Threads per inch	Tolerances in Units of 0·0001 inch				
			Major Diam.	Effective Diam.		Minor Diam.	
			Pipe (—)	Pipe (—)	Flange (+)	Pipe (—)	Flange (+)
17/32	1/4	19	81	58	58	104	175
11/16	3/8	19	85	62	62	108	175
27/32	1/2	14	94	67	67	120	213
1 1/16	3/4	14	97	70	70	123	213
1 11/32	1	11	104	74	74	134	252
1 11/16	1 1/4	11	107	77	77	137	252
1 29/32	1 1/2	11	110	80	80	140	252
2 3/8	2	11	114	83	83	142	252
3	2 1/2	11	118	88	88	148	252
3 1/2	3	11	122	92	92	152	252
4	3 1/2	11	123	93	93	153	252
4 1/2	4	11	126	96	96	156	252
5	*	11	128	98	98	158	252
5 1/2	5	11	131	101	101	161	252
6 5/8	†	8	143	108	108	179	320

*Screwed B.S. Whit. to Major Diam. = 4·950 in.; Effective Diam. = 4·8918 in.; and Minor Diam. = 4·8336 in.

†Screwed B.S. Whit. to Major Diam. = 6·575 in.; Effective Diam. = 6·4950 in.; and Minor Diam. = 6·4150 in.

Table A17. TAPER THREADS FOR GAS CYLINDERS
(1 in 8 Taper) Dimensions in Inches

Nom. Size	Threads per inch (along cone)	External Threads (Valve Stems) Dimensions at Small End			Internal Threads (Cylinder Necks) Dimensions at Large End		
		Major Diam.	Effect. Diam.	Minor Diam.	Major Diam.	Effect. Diam.	Minor Diam.
(in.)		max./ min.	max./ min.	max./ min.	max./ min.	max./ min.	max./ min.
0·715	14	0·7150	0·6692	0·6234	0·8037	0·7525	0·7094
		0·7070	0·6639	0·6127	0·7930	0·7472	0·7014
1·000	14	1·0000	0·9542	0·9084	1·1047	1·0535	1·0104
		0·9920	0·9489	0·8977	1·0940	1·0482	1·0024
1·250	11	1·2500	1·1917	1·1334	1·3890	1·3237	1·2689
		1·2395	1·1847	1·1194	1·3750	1·3167	1·2584

Table A18. COPPER TUBE THREAD
Dimensions in Inches

Nom. Size	Threads per inch	Pitch	Major Diam.	Effect. Diam.	Minor Diam.	Gauge Length	Gauge Length Tolerance ±
1/8	28	0·03571	0·248	0·2251	0·2022	0·2344	0·0357
1/4	20	0·0500	0·389	0·3570	0·3250	0·2812	0·0500
3/8	,,	,,	0·514	0·4820	0·4500	0·3750	,,
1/2	,,	,,	0·639	0·6070	0·5750	0·3750	,,
5/8	,,	,,	0·764	0·7320	0·7000	0·4688	,,
3/4	20	0·0500	0·889	0·8570	0·8250	0·4688	0·0500
7/8	,,	,,	1·014	0·9820	0·9500	0·5625	,,
1	,,	,,	1·155	1·1230	1·0910	0·5625	,,
1 1/4	,,	,,	1·405	1·3730	1·3410	0·6562	,,
1 1/2	20	0·0500	1·655	1·6230	1·5910	0·6562	0·0500
1 3/4	16	0·0625	1·929	1·8890	1·8490	0·7500	0·0625
2	,,	,,	2·179	2·1390	2·0990	0·7500	0·0625
2 1/4	,,	,,	2·429	2·3890	2·3490	0·7500	0·0938
2 1/2	,,	,,	2·679	2·6390	2·5990	0·7500	0·0938
2 3/4	,,	,,	2·929	2·8890	2·8490	0·7500	0·0938
3	16	0·0625	3·203	3·1630	3·1230	0·7812	0·0938
3 1/4	,,	,,	3·453	3·4130	3·3730	0·7812	0·0938
3 1/2	,,	,,	3·727	3·6870	3·6470	0·7812	0·0938
3 3/4	,,	,,	3·977	3·9370	3·8970	0·7812	0·0938
4	16	0·0625	4·251	4·2110	4·1710	0·7812	0·0938

Table A19. WHITWORTH FORM

Functions of Diameter and Length of Engagement for Tolerance Calculation

D or *L* (in.)	$0{\cdot}002\sqrt[3]{D}$	$0{\cdot}003\sqrt{L}$	*D* or *L* (in.)	$0{\cdot}002\sqrt[3]{D}$	$0{\cdot}003\sqrt{L}$
$\frac{1}{8}$ (0·1250)	0·00100	0·00106	$3\frac{1}{8}$ (3·1250)	0·00292	0·00530
$\frac{3}{16}$ (0·1875)	0·00114	0·00130	$3\frac{1}{4}$ (3·2500)	0·00296	0·00541
$\frac{1}{4}$ (0·2500)	0·00126	0·00150	$3\frac{3}{8}$ (3·3750)	0·00300	0·00551
$\frac{5}{16}$ (0·3125)	0·00136	0·00168	$3\frac{1}{2}$ (3·5000)	0·00304	0·00561
$\frac{3}{8}$ (0·3750)	0·00144	0·00184	$3\frac{5}{8}$ (3·6250)	0·00307	0·00571
$\frac{7}{16}$ (0·4375)	0·00152	0·00198	$3\frac{3}{4}$ (3·7500)	0·00311	0·00581
$\frac{1}{2}$ (0·5000)	0·00159	0·00212	$3\frac{7}{8}$ (3·8750)	0·00314	0·00591
$\frac{9}{16}$ (0·5625)	0·00165	0·00225	4 (4·0000)	0·00317	0·00600
$\frac{5}{8}$ (0·6250)	0·00171	0·00237	$4\frac{1}{8}$ (4·1250)	0·00321	0·00609
$\frac{11}{16}$ (0·6875)	0·00177	0·00249	$4\frac{1}{4}$ (4·2500)	0·00324	0·00618
$\frac{3}{4}$ (0·7500)	0·00182	0·00260	$4\frac{3}{8}$ (4·3750)	0·00327	0·00628
$\frac{7}{8}$ (0·8750)	0·00191	0·00281	$4\frac{1}{2}$ (4·5000)	0·00330	0·00636
1 (1·0000)	0·00200	0·00300	$4\frac{3}{4}$ (4·7500)	0·00336	0·00654
$1\frac{1}{8}$ (1·1250)	0·00208	0·00318	5 (5·0000)	0·00342	0·00671
$1\frac{1}{4}$ (1·2500)	0·00215	0·00335	$5\frac{1}{4}$ (5·2500)	0·00348	0·00687
$1\frac{1}{2}$ (1·5000)	0·00229	0·00367	$5\frac{1}{2}$ (5·5000)	0·00353	0·00704
$1\frac{3}{4}$ (1·7500)	0·00241	0·00397	6 (6·0000)	0·00363	0·00735
2 (2·0000)	0·00252	0·00424	7 (7·0000)	0·00383	0·00794
$2\frac{1}{4}$ (2·2500)	0·00262	0·00450	8 (8·0000)	0·00400	0·00849
$2\frac{1}{2}$ (2·5000)	0·00271	0·00474	10 (10·0000)	0·00431	0·00949
$2\frac{3}{4}$ (2·7500)	0·00280	0·00497	12 (12·0000)	0·00458	0·01039
3 (3·0000)	0·00288	0·00520	18 (18·0000)	0·00524	0·01273

Table A19a. TAPER THREADS FOR GAS CYLINDERS

(1 in 5·625 Taper) Dimensions in Inches

Nom. Size	Threads per inch (along cone)	External Threads (Valve Stems) Dimensions at Small End			Internal Threads (Cylinder Necks) Dimensions at Large End		
		Major Diam.	Effect. Diam.	Minor Diam.	Major Diam.	Effect. Diam.	Minor Diam.
(in.)		max./ min.	max./ min.	max./ min.	max./ min.	max./ min.	max./ min.
0·6	14	0·6000 0·5920	0·5543 0·5493	0·5086 0·4986	0·7666 0·7556	0·7149 0·7099	0·6722 0·6642

Table A20. WHITWORTH FORM

Functions of Pitch for Tolerance Calculation

Threads per inch	$\sqrt{Pitch}$	$0{\cdot}02\sqrt{p}$	$0{\cdot}005\sqrt{p}$	$0{\cdot}013\sqrt{p}$
72	0·11785	0·0024	0·00059	0·00153
60	0·12910	0·0026	0·00065	0·00168
56	0·13363	0·0027	0·00067	0·00174
48	0·14434	0·0029	0·00072	0·00188
40	0·15811	0·0032	0·00079	0·00206
36	0·16667	0·0033	0·00083	0·00217
32	0·17678	0·0035	0·00088	0·00230
28	0·18898	0·0038	0·00094	0·00246
26	0·19612	0·0039	0·00098	0·00255
24	0·20412	0·0041	0·00102	0·00265
22	0·21320	0·0043	0·00107	0·00277
20	0·22361	0·0045	0·00112	0·00291
19	0·22942	0·0046	0·00115	0·00298
18	0·23570	0·0047	0·00118	0·00306
16	0·25000	0·0050	0·00125	0·00325
14	0·26726	0·0053	0·00134	0·00347
12	0·28868	0·0058	0·00144	0·00375
11	0·30151	0·0060	0·00151	0·00392
10	0·31623	0·0063	0·00158	0·00411
9	0·33333	0·0067	0·00167	0·00433
8	0·35355	0·0071	0·00177	0·00460
7	0·37796	0·0076	0·00189	0·00491
6	0·40825	0·0082	0·00204	0·00531
4·5	0·47140	0·0094	0·00236	0·00613
4·0	0·50000	0·0100	0·00250	0·00650
3·5	0·53452	0·0107	0·00267	0·00695
3·25	0·55470	0·0111	0·00277	0·00721
3·0	0·57735	0·0115	0·00289	0·00751
2·875	0·58978	0·0118	0·00295	0·00767
2·75	0·60302	0·0121	0·00302	0·00784
2·625	0·61721	0·0123	0·00309	0·00802
2·5	0·63246	0·0126	0·00316	0·00822

Section B

Section B
BRITISH THREADS OF NON-WHITWORTH FORM

British Association (B.A.) Thread

The British Association thread is that generally used in this country for small screws in instrument and electrical work, and to some extent for clock parts. The thread angle and depth of thread are the same as the corresponding dimensions of the Thury thread, and its formulation was first proposed by the British Association in 1884, to be adopted by that body in 1903. The pitches were calculated from $p=(0{\cdot}9)^N$ where N is the designating number of the screw, and the basic major diameters from $D=6p^{1{\cdot}2}$. The present British specification is B.S. 93: 1951 which covers the range 0 to 16 B.A.

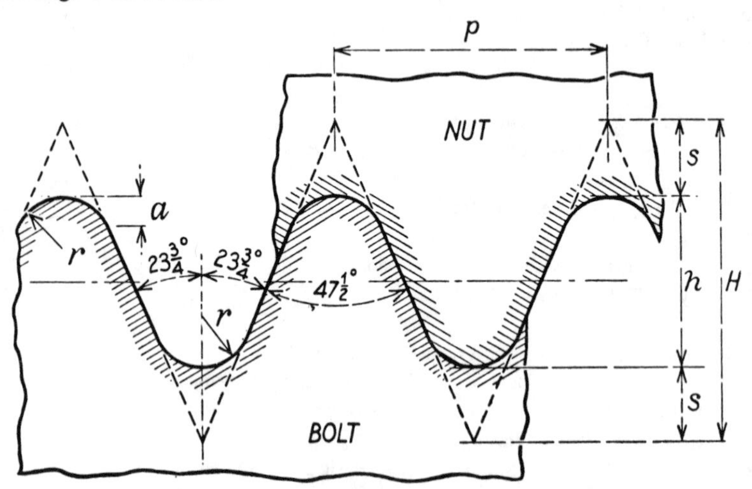

Fig. 1. British Association (B.A.) Thread Form

The B.A. thread has the form shown in Fig. 1, with angles and proportions as follows:

Thread angle $(2\theta) = 47$ deg. 30 mins.; Flank angle $(\theta) = 23$ deg. 45 mins.; Triangular height $(H) = 1{\cdot}1363365\ p$; Shortening at crest and root $(S) = 0{\cdot}2681683\ p$; Radius at crest and root $(r) = 0{\cdot}1808346\ p$; Depth of thread $(h) = 0{\cdot}6\ p$; Depth of rounding $(a) = 0{\cdot}1080041\ p$.

Basic thread form dimensions in mm. are given in Table B1, and the basic dimensions for the series are given in Tables B2 and B3, the former in inches and the latter in mm.

Note: Since the metric system has now been officially adopted as the primary system of weights and measures in the U.K., it has been agreed that future screw thread usage should be based on the ISO Metric thread. In consequence, the B.A. thread has been declared obsolescent.

Tolerances for B.A. Threads. The tolerances specified by B.S.93: 1951 provide for two classes of bolt fit—"close" and "normal"—the "close" fit applying to the range 0 to 10 B.A., and the "normal" over the range 0 to 16 B.A. Only one class of fit is specified for the nut. The tolerances are derived from the following formulæ:

Close-class bolt tolerances

Effective diameter: $0{\cdot}08p + 0{\cdot}02$ mm.
Major ,, : $0{\cdot}15p$
Minor ,, : $0{\cdot}16p + 0{\cdot}04$ mm.

These are applied to the basic dimensions for 0 to 10 B.A. as given in Tables B2 and B3, which therefore become the maximum values.

Normal-class bolt tolerances

Effective diameter			: $0{\cdot}10p + 0{\cdot}025$ mm.
Major	,,	(0 to 10 B.A.)	: $0{\cdot}20p$
,,	,,	(11 to 16 B.A.)	: $0{\cdot}25p$
Minor	,,	(all)	: $0{\cdot}20p + 0{\cdot}050$ mm.

In the case of the range 0 to 10 B.A., these tolerances are applied to dimensions that are 0·025 mm. less than the basic dimensions given in Table B3, thus allowing 0·025 mm. for subsequent plating. No such allowance is made for the range 11 to 16 B.A., for which tolerances are applied to the unmodified basic values of Table B3.

Nuts (all)

Effective diameter tolerances: $0{\cdot}12p + 0{\cdot}03$ mm.
Minor ,, ,, : $0{\cdot}375p$.

These are applied to the basic dimensions of Table B3, unmodified. Values of tolerances in mm. and inches will be found in Tables B4 and B5.

British Standard Cycle Thread (B.S.C.)

At the beginning of the century a need was experienced for a thread form suitable for the various threaded parts of cycles and motor cycles.

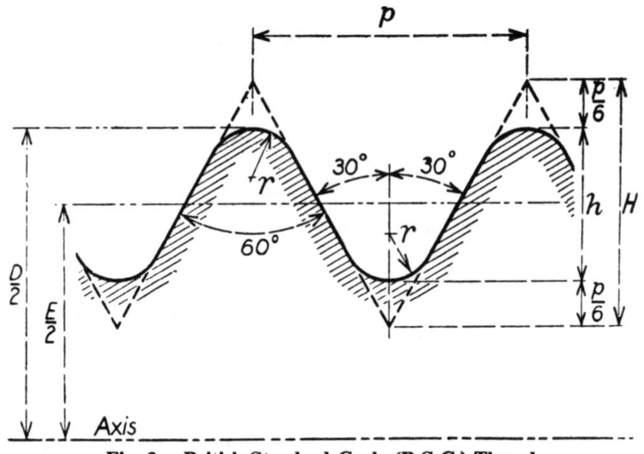

Fig. 2. British Standard Cycle (B.S.C.) Thread

A form was therefore specified by the Cycle Engineers Institute and was known as C.E.I. thread form. It is now specified by the British Standards Institution in B.S. 811: 1950 and known as the B.S.C.

The profile is shown in Fig. 2, which has the following proportions:

Angle of thread (2θ) = 60 deg.; Flank angle (θ) = 30 deg.; Triangular height (H) = 0·8660254 p; Shortening at crest and root (S) = 0·1666667 p; Radius at crest and root (r) = 0·1666667 p; Depth of thread (h) = 0·5326921 p. Basic form dimensions related to pitch are given in Table B6, and basic sizes in Table B7.

Three classes of fit are specified by B.S. 811 viz., close, medium and free, all of these classes being applied to bolts and nuts, but only the medium class to spokes and nipples and to threads for special applications. The appropriate tolerances for B.S.C. nuts and bolts will be found in Table B8.

The tolerances for spokes and nipples are calculated from the following formulæ:

Spokes Major Diameter: 0·2 p + 0·0004 in.
Effective ,, : 0·1 p + 0·0010 in.
Minor ,, : 0·2 p + 0·0020 in.

Nipples Effective Diameter: 0·12 p + 0·0012 in.
Minor ,, : 0·375 p.

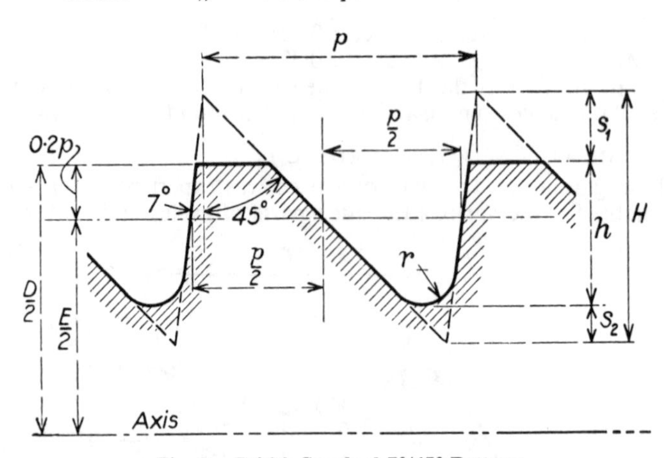

Fig. 3. British Standard 7°/45° Buttress

7/45° British Standard Buttress Thread

This thread is specified by B.S. 1657: 1950 and has the basic form shown in Fig. 3. Its proportions are as follows:

Thread angle ($\theta_p + \theta_t$) = 52 deg.; Pressure flank angle (θ_p) = 7 deg.; Trailing flank angle (θ_t) = 45 deg.; Triangular height (H) = 0·89064 p; Shortening at crest (S_1) = 0·24532 p; Shortening at root (S_2) = 0·13946 p; Radius at root (r) = 0·12055 p; Depth of thread (basic) = h = 0·50586 p.

The pitch cylinder is located 0·2 p below the major diameter of the screw, hence the basic effective diameter is given by

$$E = D - 0{\cdot}4\ p.$$

Basic form dimensions are given in Table B9.

For clearance purposes it is recommended that an allowance be applied to the nominal major, effective and minor diameters of the screw. The dimensions thus obtained constitute the maximum dimensions of the screw and to them are applied the tolerances for the required class of fit.

The minimum dimensions of the nut correspond to the basic nominal dimensions and to these are applied the nut tolerances for the required class of fit.

The same amount of tolerance is applied, for a given class of fit, to the major diameter of the screw, the effective diameter of both screw and nut, and to the minor diameter of nut. The recommended value of tolerance for the major diameter of the nut and for the minor diameter of screw is twice the corresponding effective diameter tolerance.

Should however, for certain purposes, the tolerances tabulated be considered too great for the major diameter of the screw and the minor diameter of the nut, it is recommended that the tolerances be as follows:

Nom. Diam. (in.)	1 to $1\frac{1}{2}$;	$> 1\frac{1}{2}$ to $2\frac{1}{2}$;	$> 2\frac{1}{2}$ to 4.
Tolerance (in.)	0·002;	0·0025;	0·0030.

The standard tolerances are derived from the formula:

$$0{\cdot}002\ \sqrt[3]{D} + 0{\cdot}0173\ \sqrt{p}.$$

The combinations of pitch and nominal diameter specified by B.S. 1657: 1950 range from 1 to 24 ins. diameter, and from 20 to 1 t.p.i. Table B10 gives a useful list from 1 to 4 ins. inclusive. The allowances given there are those recommended for normally stressed interchangeable threads. In special cases it may be found desirable to reduce these allowances.

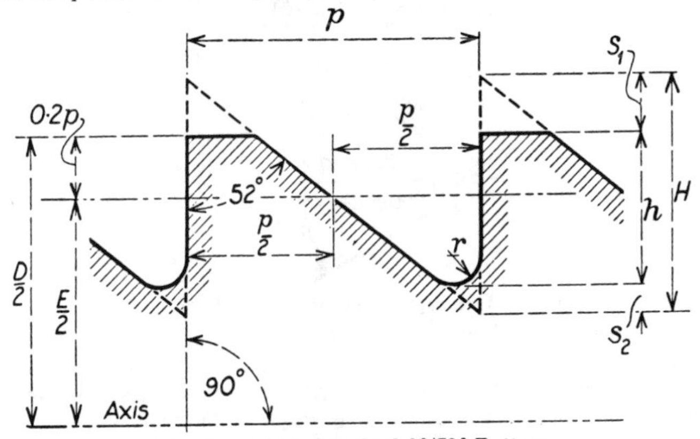

Fig. 4. British Standard 0°/52° Buttress

0°/52° British Standard Buttress Thread

For certain applications of buttress threads it is desirable either for ease of production or for the prevention of undue stress to use a pressure flank angle of zero. A suitable form is specified in B.S. 1657: 1950 with the basic profile shown in Fig. 4. Here the theoretical triangular height is 0·78129 p, the shortening at the tip is 0·19064 p, the shortening at the root is 0·09767 p and the depth of thread is 0·49298 p. The pitch cylinder is located 0·2 p below the major diameter so that Effective Diameter = Major Diameter minus 0·4 p. The radius at the root is equal to 0·09298 p. The tolerances and allowances are as already given for the 7°/45° form in Table B10, for the same combination of nominal diameter and pitch. The dimensions of the basic form are given in Table B11.

Cordeaux Thread for Telegraph Insulators

The Cordeaux thread derives its name from John Henry Cordeaux who in 1877 obtained a patent for using cup insulators with an internal thread. Before this invention, insulators were provided with grouted-in iron stems and difficulty was encountered in removing the rusted-on nuts from the lower end of the stem. Cordeaux's patent overcame this. It is believed that the first Cordeaux thread was cut in wood by hand and then copied, a process which apparently unintentionally gave the thread unequal flank angles, viz. 35 and 29 deg. When the B.S. 16 was prepared in 1937, so many insulators and spindles were in service that for the purpose of interchangeability, the old thread form was retained.

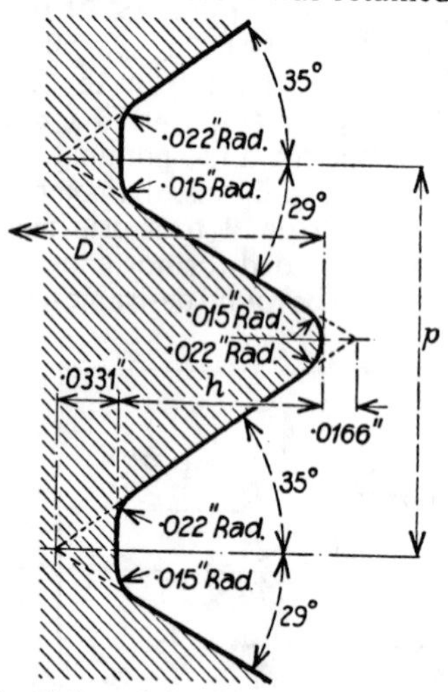

Fig. 5. Telegraph Insulator Thread (Cordeaux)

The Cordeaux thread is made in two nominal sizes—½ and ⅝ inch. The 29 deg. angle flanks of the male screw are uppermost when the insulator is erected. The basic form of the thread is shown in Fig. 5, and the dimensions and tolerances are given in Table B12.

Water Well Casing Threads (Square Form)

Two forms of thread are specified for water-well casing threads by B.S. 879: 1939. One is the square form shown in Fig. 6 and the other is a taper form with the thread flanks equally inclined to the pipe axis with a taper of $\frac{3}{16}$ inch per foot on diameter. This latter is dealt with in Section A.

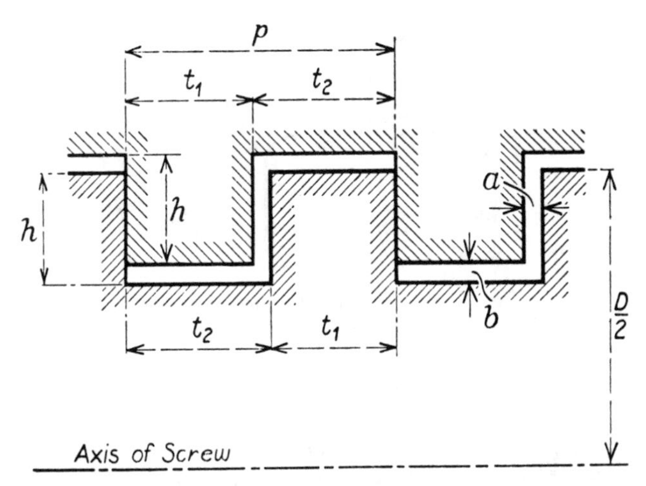

Fig. 6. Water-well Casing Thread—Square Form

Basic dimensions of the square form are given in Table B13, and the major diameters of pipe for two kinds of connection, viz. screwed and socketed and flush butt joints are given in Table B14.

ROUND OR KNUCKLE THREADS

Suction Hose Coupling Thread (Fire)

Pitch = $\frac{1}{3}$-inch; Radius at root = 0·094-inch; Radius at crest = 0·07267-inch; Depth of thread = 0·209-inch; Thickness of thread = 0·1453 + 0·000 – 0·010-inch.

The couplings are specified for a range of 7 sizes, viz. 3, $3\frac{1}{2}$, 4, $4\frac{1}{2}$, 5, $5\frac{1}{2}$ and 6 inches nominal diameter (△), the basic dimensions of each size being obtained as follows:

Male Thread

Maximum Major Diameter = △ + 0·625-inch
Minimum ,, ,, = △ + 0·605-inch
Maximum Minor ,, = △ + 0·207-inch

Female Thread

Minimum Major Diameter = △ + 0·668-inch
Minimum Minor ,, = △ + 0·250-inch
Maximum ,, ,, = △ + 0·270-inch.

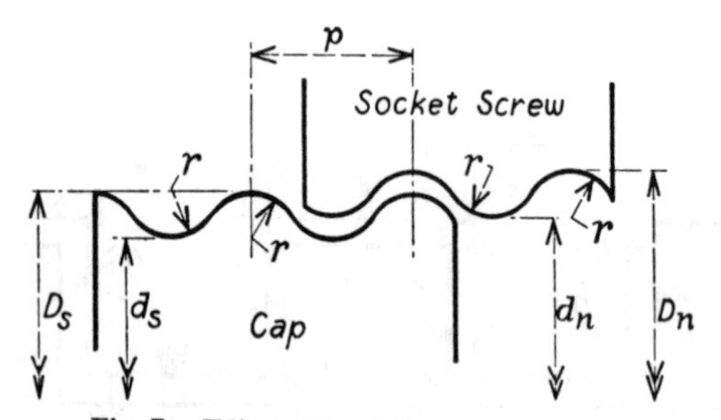

Fig. 7. Edison Thread for Electric Lamps

Edison Screw Threads for Lamps

The use of a screw for effecting an electrical and mechanical connection between an electric lamp and its socket was amongst the earliest methods adopted when the employment of electricity for lighting became general, especially in the United States. In this country its use is mainly confined to miniature lamps and for lamps of 150 watts and upwards. Five sizes of Edison screw are specified in this country, by B.S. 98: 1962. The corresponding dimensions of these are given in Tables B15 and B16, the form being depicted in Fig. 7.

Bottle Closure Threads (Glass Container Finishes)

A form of thread for glass bottle necks and for metal and plastic caps is shown in Fig. 8. The dimensions for the thread form as specified in B.S. 1918: 1953 are given in Tables B17 and B18. There are four standard pitches corresponding to 5, 6, 8 and 12 threads per inch. The thread forms for these pitches are not geometrically similar.

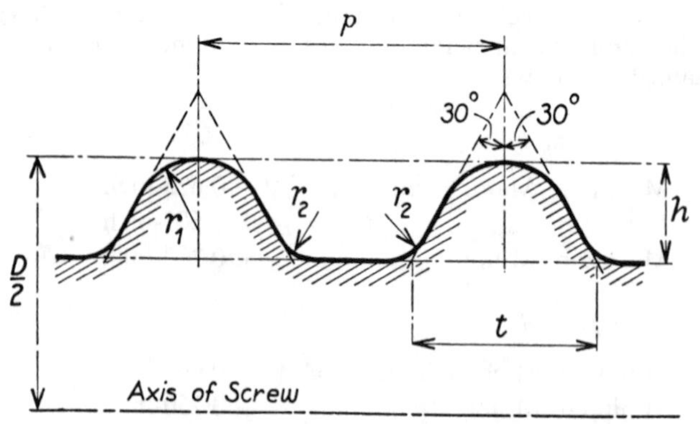

Fig. 8. Bottle Closure Thread (Glass Container Finishes)

THREAD FOR SPARKING PLUGS (B.S. 45: 1952)

The basic form of thread is Metric (S.I.) i.e. the thread angle is 60 deg. and the shortening at crest and root is equal to one-eighth of the triangular height. The thread is of 14 mm. nominal diameter and 1·25 mm. pitch with the recommended dimensions of plug and hole as follows:

Plug
Major Diameter: 13·975 mm. (maximum); 13·790 mm. (minimum)
Effective Diameter: 13·163 mm. (maximum); 13·038 mm. (minimum)

Hole
Effective Diameter: 13·338 mm. (maximum); 13·213 mm. (minimum)
Minor Diameter: 12·586 mm. (maximum); 12·461 mm. (minimum)

BRITISH STANDARD METRIC THREAD

In 1943, the British Standards Institution issued B.S. 1095: 1943, a War Emergency Standard Specification with the title "Metric Screw Threads—Système Internationale." To avoid confusion this standard has now been withdrawn since it no longer represents the standard metric thread used in the U.K. However, information on the form and theoretical proportions of the thread is given below for reference purposes.

The form of this thread for maximum metal conditions may be represented as in Fig. 9, and has theoretical proportions as follows:

Thread Angle (2θ) = 60 deg.; Flank Angle (θ) = 30 deg.; Triangular height (H) = 0·8660254 p; Shortening at crest (S_1) = 0·1082532 p; Shortening at root (S_2) = 0·1894431 p; Depth of thread (h) = 0·5683292 p; Major diameter (D) = Nominal diameter; Effective diameter = D – 0·6495191 p; Minor diameter (d) = D – 1·1366584 p. The crests are flat, and the roots may be flat at the dimension of S_2, or, due to tool wear, the roots may have a maximum radius of 0·126 p, producing a concavity below the basic flat at S_2 by an amount of 0·063 p.

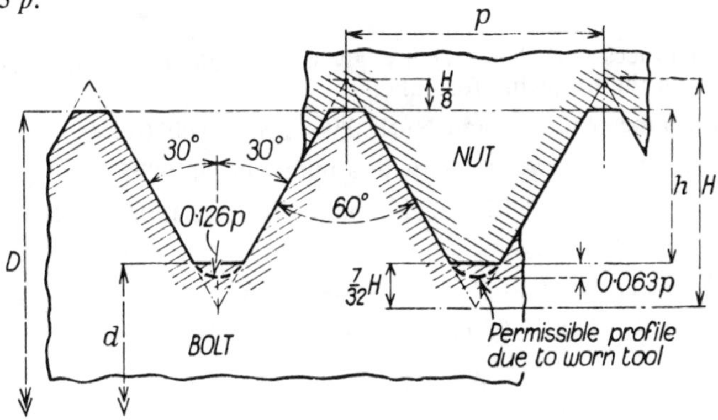

Fig. 9. British Standard Metric Thread (S.I.)

Thread profile dimensions are given in Tables B19 and B20, and the maximum-metal dimensions in B21 and B22. The B.S. tolerances are to be found in Tables B23 and B24.

ACME THREADS (B.S. 1104: 1957)

This standard specifies four classes of Acme thread, namely, 2G, 3G, 4G and 5G. The classes 2G, 3G and 4G are equivalent to the similarly designated classes of the American Acme, (see p. 108), but with the differences in diameter and pitch combinations for the $\frac{7}{16}$, 1, $1\frac{1}{8}$ and $1\frac{1}{4}$-in. diameters as indicated in Table B25, for which the tolerances are given in Table B26.

The class 5G may be regarded as additional to the American series, and has maximum-metal dimensions derived from the following formulæ, where N = the nominal diameter.

Screw:

Maximum Major Diameter, $D_s = N$;

Maximum Effective Diameter, $E_s = N - p/2$;

Maximum Minor Diameter, $d_s = N - p - 0{\cdot}20$ in. (10 t.p.i. and coarser),

$d_s = N - p - 0{\cdot}010$ in. (finer than 10 t.p.i.);

Nut:

Minimum Major Diameter, $D_n = N + 0{\cdot}020$ in. (10 t.p.i. and coarser),

$D_n = N + 0{\cdot}010$ (finer than 10 t.p.i.);

Minimum Effective Diameter, $E_n = N - p/2$;

Minimum Minor Diameter, $d_n = N - p$.

The maximum-metal dimensions for class 5G are given in Table B27.

Tolerances

The tolerances for class 5G are given in Table B28, and these are derived from the following formulæ:

Effective Diameter (Screw and Nut): $0{\cdot}008\sqrt{p} + 0{\cdot}0016\sqrt{N}$.

Major Diameter, Screw: $0{\cdot}05\,p$, with a minimum of 0·005 in.

" " Nut: 0·020 in. (10 t.p.i. and coarser); 0·01 in. (finer than 10 t.p.i.).

Minor Diameter, Screw: $0{\cdot}012\sqrt{p} + 0.0024\sqrt{N}$.

" " Nut: $0{\cdot}05\,p$, with a minimum of 0·005 in.

Table B1. **BRITISH ASSOCIATION (B.A.)**

Basic Form Dimensions in mm.

B.A. No.	Pitch	Triangular Height	Shortening	Depth of Thread	Depth of Rounding	Radius
0	1·0000	1·13634	0·26817	0·60000	0·10800	0·18083
1	0·9000	1·02271	0·24135	0·54000	0·09720	0·16275
2	0·8100	0·92044	0·21722	0·48600	0·08748	0·14647
3	0·7300	0·82953	0·19576	0·43800	0·07884	0·13201
4	0·6600	0·74998	0·17699	0·39600	0·07128	0·11935
5	0·5900	0·67044	0·15822	0·35400	0·06372	0·10669
6	0·5300	0·60226	0·14213	0·31800	0·05724	0·09584
7	0·4800	0·54544	0·12872	0·28800	0·05184	0·08680
8	0·4300	0·48863	0·11531	0·25800	0·04644	0·07776
9	0·3900	0·44317	0·10459	0·23400	0·04212	0·07052
10	0·3500	0·39772	0·09386	0·21000	0·03780	0·06329
11	0·3100	0·35227	0·08313	0·18600	0·03348	0·05606
12	0·2800	0·31818	0·07509	0·16800	0·03024	0·05063
13	0·2500	0·28409	0·06704	0·15000	0·02700	0·04521
14	0·2300	0·26136	0·06168	0·13800	0·02484	0·04159
15	0·2100	0·23863	0·05632	0·12600	0·02268	0·03797
16	0·1900	0·21590	0·05095	0·11400	0·02052	0·03436
17	0·1700	0·19318	0·04559	0·10200	0·01836	0·03074
18	0·1500	0·17045	0·04023	0·09000	0·01620	0·02712
19	0·1400	0·15909	0·03754	0·08400	0·01512	0·02532
20	0·1200	0·13636	0·03218	0·07200	0·01296	0·01270
21	0·1100	0·12500	0·02950	0·06600	0·01188	0·01989
22	0·1000	0·11363	0·02682	0·06000	0·01080	0·01808

Table B2. BRITISH ASSOCIATION (B.A.) SERIES

Basic Dimensions in Inches

B.A. No.	Pitch	Threads per inch	Depth of Thread	Major Diam.	Effective Diam.	Minor Diam.
0	0·039370	25·4000	0·023622	0·2362	0·2126	0·1890
1	0·035433	28·2222	0·021260	0·2087	0·1874	0·1661
2	0·031890	31·3580	0·019134	0·1850	0·1659	0·1468
3	0·028740	34·7945	0·017244	0·1614	0·1441	0·1268
4	0·025984	38·4849	0·015591	0·1417	0·1262	0·1106
5	0·023228	43·0508	0·013937	0·1260	0·1120	0·0980
6	0·020866	47·9245	0·012520	0·1102	0·0976	0·0850
7	0·018898	52·9167	0·011339	0·0984	0·0870	0·0756
8	0·016929	59·0698	0·010157	0·0866	0·0764	0·0661
9	0·015354	65·1282	0·009213	0·0748	0·0656	0·0563
10	0·013780	72·5714	0·008268	0·0669	0·0587	0·0504
11	0·012205	81·9355	0·007323	0·0591	0·0518	0·0445
12	0·011024	90·7143	0·006614	0·0512	0·0445	0·0378
13	0·009843	101·6000	0·005906	0·0472	0·0413	0·0354
14	0·009055	110·4348	0·005433	0·0394	0·0339	0·0283
15	0·008268	120·9524	0·004960	0·0354	0·0305	0·0256
16	0·007480	133·6842	0·004488	0·0311	0·0266	0·0220
17	0·006693	149·4118	0·004016	0·0276	0·0236	0·0197
18	0·005906	169·3333	0·003543	0·0244	0·0209	0·0173
19	0·005512	181·4286	0·003307	0·0213	0·0179	0·0146
20	0·004724	211·6667	0·002835	0·0189	0·0161	0·0134
21	0·004331	230·9091	0·002598	0·0165	0·0140	0·0114
22	0·003937	254·0000	0·002362	0·0146	0·0122	0·0098

Table B3. **BRITISH ASSOCIATION (B.A.) SERIES**

Basic Dimensions in mm.

B.A. No.	Major Diam.	Effective Diam.	Minor Diam.	Cross Sectional area at Core (sq. mm.)	Mean Helix Angle	Threads per mm.
0	6·00	5·400	4·80	18·10	3° 22′	1·000000
1	5·30	4·760	4·22	13·99	3° 27′	1·111111
2	4·70	4·215	3·73	10·93	3° 30′	1·234568
3	4·10	3·660	3·22	8·14	3° 38′	1·369863
4	3·60	3·205	2·81	6·20	3° 45′	1·515152
5	3·20	2·845	2·49	4·87	3° 47′	1·694915
6	2·80	2·480	2·16	3·66	3° 54′	1·886793
7	2·50	2·210	1·92	2·89	3° 58′	2·083333
8	2·20	1·940	1·68	2·22	4° 2′	2·325581
9	1·90	1·665	1·43	1·61	4° 16′	2·564103
10	1·70	1·490	1·28	1·29	4° 17′	2·857143
11	1·50	1·315	1·13	1·00	4° 17′	3·225807
12	1·30	1·130	0·96	0·72	4° 31′	3·571429
13	1·20	1·050	0·90	0·64	4° 20′	4·000000
14	1·00	0·860	0·72	0·41	4° 52′	4·347826
15	0·90	0·775	0·65	0·33	4° 56′	4·761905
16	0·79	0·675	0·56	0·25	5° 6′	5·263158
17	0·70	0·600	0·50	0·20	5° 9′	5·882353
18	0·62	0·530	0·44	0·15	5° 9′	6·666667
19	0·54	0·455	0·37	0·11	5° 36′	7·142857
20	0·48	0·410	0·34	0·091	5° 19′	8·333333
21	0·42	0·355	0·29	0·066	5° 38′	9·090909
22	0·37	0·310	0·25	0·049	5° 52′	10·000000

Table B4.

B.A. THREADS

Tolerances in mm.

(C=Close class; N=Normal class)

Values marked with an Asterisk (*) are applied to a basic dimension to which a minus allowance of 0·25 mm. has been made.

BA. No.	Major Diam.		Effective Diam.			Minor Diam.		
	Screws		Screws		Nuts	Screws		Nuts
	C	N	C	N	(All)	C	N	(All)
0	·150	·200*	·100	·125*	·150	·200	·250*	·375
1	·135	·180*	·090	·115*	·140	·185	·230*	·340
2	·120	·160*	·085	·105*	·125	·170	·210*	·305
3	·110	·145*	·080	·100*	·120	·155	·195*	·275
4	·100	·130*	·075	·090*	·110	·145	·180*	·250
5	·090	·120*	·070	·085*	·100	·135	·170*	·220
6	·080	·105*	·060	·080*	·095	·125	·155*	·200
7	·070	·095*	·060	·075*	·090	·115	·145*	·180
8	·065	·085*	·055	·070*	·080	·110	·135*	·160
9	·060	·080*	·050	·065*	·075	·100	·130*	·145
10	·055	·070*	·050	·060*	·070	·095	·120*	·130
11		·080		·055	·065		·110	·115
12		·070		·055	·065		·105	·105
13		·065		·050	·060		·100	·095
14		·060		·050	·060		·095	·085
15		·055		·045	·055		·090	·080
16		·050		·045	·055		·090	·070

Table B5. **B.A. THREADS**

Tolerances in Inches (Unit=0·0001-inch)

(C=Close class; N=Normal class)

Values marked with an Asterisk (*) are applied to a basic dimension to which a minus allowance of 0·00098 in. has been made.

B.A. No.	Major Diam. Screws C	Major Diam. Screws N	Effective Diam. Screws C	Effective Diam. Screws N	Effective Diam. Nuts (All)	Minor Diam. Screws C	Minor Diam. Screws N	Minor Diam. Nuts (All)
0	59	78*	39	49*	59	79	99*	147
1	53	71*	35	45*	55	73	91*	134
2	47	63*	33	42*	50	67	82*	121
3	43	57*	32	39*	47	61	77*	108
4	39	51*	30	35*	43	57	70*	99
5	36	47*	28	33*	39	53	66*	87
6	31	42*	24	32*	38	49	61*	79
7	27	37*	24	29*	36	45	57*	71
8	26	33*	22	28*	31	43	54*	63
9	24	31*	20	26*	29	39	51*	57
10	22	27*	20	24*	27	37	47	51
11		32		22	25		43	45
12		28		22	25		41	41
13		25		19	24		39	38
14		24		20	23		37	34
15		21		18	22		36	31
16		20		18	21		35	28

Table B6. **BRITISH STANDARD CYCLE (B.S.C.)**

Basic Form Dimensions in Inches

Threads per inch	Pitch	Triangular Height	Shortening	Depth of Thread	Depth of Rounding	Radius
56	0·017857	0·01546	0·00297	0·00951	0·00149	0·00297
44	0·022727	0·01968	0·00379	0·01211	0·00189	0·00379
40	0·025000	0·02165	0·00417	0·01332	0·00208	0·00417
32	0·031250	0·02706	0·00521	0·01665	0·00260	0·00521
30	0·033333	0·02887	0·00556	0·01776	0·00278	0·00556
26	0·038462	0·03331	0·00641	0·02049	0·00321	0·00641
24	0·041667	0·03608	0·00694	0·02220	0·00347	0·00694

Table B7. BRITISH STANDARD CYCLE (B.S.C.)
Basic Sizes in Inches

Diam.	Spoke Size	Threads per inch	Major Diam.	Effect. Diam.	Minor Diam.	Application
0·0825	15	56	0·0825	0·0730	0·0635	Spokes and Nipples
0·0905	14	56	0·0905	0·0810	0·0715	,,
0·1025	13	56	0·1025	0·0930	0·0835	,,
0·1145	12	56	0·1145	0·1050	0·0955	,,
$\frac{1}{8}$		40	0·1250	0·1117	0·0984	Bolts and Nuts
0·1291	11	44	0·1291	0·1170	0·1049	Spokes and Nipples
0·1423	10	40	0·1423	0·1290	0·1157	,,
$\frac{5}{32}$		32	0·1563	0·1397	0·1231	Bolts and Nuts
0·1583	9	40	0·1583	0·1450	0·1317	Spokes and Nipples
0·1776	8	32	0·1776	0·1610	0·1444	,,
$\frac{3}{16}$		32	0·1875	0·1709	0·1543	Bolts and Nuts
$\frac{7}{32}$		26	0·2188	0·1983	0·1778	,,
$\frac{1}{4}$		26	0·2500	0·2295	0·2090	,,
$\frac{17}{64}$		26	0·2656	0·2451	0·2246	Cycle and Motor cycle crank cotters
$\frac{9}{32}$		26	0·2813	0·2608	0·2403	Bolts and Nuts
$\frac{5}{16}$		26	0·3125	0·2920	0·2715	,,
$\frac{3}{8}$		26	0·3750	0·3545	0·3340	,,
$\frac{7}{16}$		26	0·4375	0·4170	0·3965	,,
$\frac{1}{2}$		26	0·5000	0·4795	0·4590	,,
$\frac{9}{16}$		26	0·5625	0·5420	0·5215	,,
$\frac{5}{8}$		26	0·6250	0·6045	0·5840	,,
$\frac{11}{16}$		26	0·6875	0·6670	0·6465	,,
$\frac{3}{4}$		26	0·7500	0·7295	0·7090	,,
$\frac{7}{8}$		24	0·8750	0·8528	0·8306	Steering Columns of Juvenile Cycles
$\frac{31}{32}$		30	0·9688	0·9510	0·9332	Steering Columns
1		24	1·0000	0·9778	0·9556	,,
$1\frac{1}{8}$		26	1·1250	1·1045	1·0840	Motor cycle and tandem steering columns
1·290	*	24	1·2900	1·2678	1·2456	Lock rings for sprockets on rear hubs
1·370	†	24	1·3700	1·3478	1·3256	Hub Sprockets and Bottom Bracket Cups
1·450	†	26	1·4500	1·4295	1·4090	Tandem Bottom Bracket Cups
$1\frac{9}{16}$	*	24	1·5625	1·5403	1·5181	Carrier Cycle Sprockets and Lock Rings
$1\frac{5}{8}$		24	1·6250	1·6028	1·5806	,,

* = Left hand; † = Right and Left hand.

NOTE.—The spoke size corresponds to the effective diameter which constitutes the original diameter of wire upon which the thread is rolled.

Table B8. **BRITISH STANDARD CYCLE (B.S.C.)**

Tolerances in Units of 0·0001-inch

(C=Close fit; M=Medium fit; F=Free fit)

Nom. Size	Threads per inch	Major Diam.			Effective Diam.			Minor Diam.			Minor Diam.
		Screws only (—)			Screws (—) & Nuts (+)			Screws only (—)			Nuts (+)
		C	M	F	C	M	F	C	M	F	C, M & F
$\frac{1}{8}$	40	33	41	54	17	25	38	38	57	70	90
$\frac{5}{32}$	32	37	45	60	19	28	42	42	63	77	103
$\frac{3}{16}$	,,	37	47	62	19	29	44	42	65	79	103
$\frac{7}{32}$	26	41	51	68	21	32	48	46	71	87	117
$\frac{1}{4}$	,,	42	53	70	22	33	50	47	72	89	117
$\frac{9}{32}$	,,	43	54	71	23	34	51	48	74	90	117
$\frac{5}{16}$	,,	44	55	74	24	36	54	49	75	93	117
$\frac{3}{8}$	,,	45	57	77	25	38	57	50	77	96	117
$\frac{7}{16}$	,,	47	60	80	27	40	60	52	79	99	117
$\frac{1}{2}$	,,	48	61	83	28	42	63	53	81	102	117
$\frac{9}{16}$	,,	49	63	86	29	44	66	54	83	105	117
$\frac{5}{8}$	,,	50	65	88	30	45	68	55	85	107	117
$\frac{11}{16}$	,,	51	67	91	31	47	71	56	86	110	117
$\frac{3}{4}$	,,	52	68	92	32	48	72	57	88	111	117

Table B9. **BRITISH STANDARD 7°/45° BUTTRESS**

Basic Form Dimensions in Inches

Threads per inch	Pitch	Triangular Height	Shortening at Tip	Shortening at Root	Depth of Thread	Radius at Root	Major Diam. minus Effect. Diam.
20	0·05000	0·0445	0·0123	0·0070	0·0253	0·0060	0·0200
16	0·06250	0·0557	0·0153	0·0087	0·0316	0·0075	0·0250
12	0·08333	0·0742	0·0204	0·0116	0·0421	0·0100	0·0333
10	0·10000	0·0891	0·0245	0·0140	0·0506	0·0121	0·0400
8	0·12500	0·1113	0·0307	0·0174	0·0632	0·0151	0·0500
6	0·16667	0·1484	0·0409	0·0233	0·0843	0·0201	0·0667
5	0·20000	0·1781	0·0491	0·0279	0·1012	0·0241	0·0800
4	0·25000	0·2227	0·0613	0·0349	0·1265	0·0301	0·1000
3	0·33333	0·2969	0·0818	0·0465	0·1686	0·0402	0·1333
2·5	0·40000	0·3563	0·0981	0·0558	0·2023	0·0482	0·1600
2	0·50000	0·4453	0·1227	0·0697	0·2529	0·0603	0·2000
1·5	0·66667	0·5938	0·1635	0·0930	0·3372	0·0804	0·2667
1·25	0·80000	0·7125	0·1963	0·1116	0·4047	0·0964	0·3200
1	1·00000	0·8906	0·2453	0·1395	0·5059	0·1206	0·4000

Table B10. BRITISH STANDARD 7°/45° BUTTRESS
Basic Nominal Dimensions: Tolerances and Allowances
(C=Close class; M=Medium class; F=Free class)

Nom. Diam. (ins.)	Threads per inch	Basic Major Diam.	Basic Effective Diam.	Basic Minor Diam.	Tolerances*			Allowance*
					C	M	F	
1	20	1·0000	0·9800	0·9494	40	60	—	40
	16	,,	0·9750	0·9368	43	65	—	43
	12	,,	0·9667	0·9158	48	71	107	47
	10	,,	0·9600	0·8988	51	76	114	51
	8	,,	0·9500	0·8736	55	83	125	55
	6	,,	0·9333	0·8314	61	92	138	61
1⅛	20	1·1250	1·1050	1·0744	40	60	—	40
	16	,,	1·1000	1·0618	43	65	—	43
	12	,,	1·0917	1·0408	48	71	107	47
	10	,,	1·0850	1·0238	51	76	114	51
	8	,,	1·0750	0·9986	55	83	125	55
	6	,,	1·0583	0·9564	61	92	138	61
1¼	20	1·2500	1·2300	1·1994	40	60	—	40
	16	,,	1·2250	1·1868	43	65	—	43
	12	,,	1·2167	1·1658	48	71	107	47
	10	,,	1·2100	1·1488	51	76	114	51
	8	,,	1·2000	1·1236	55	83	125	55
	6	,,	1·1833	1·0814	61	92	138	61
1⅜	20	1·3750	1·3550	1·3244	40	60	—	40
	16	,,	1·3500	1·3118	43	65	—	43
	12	,,	1·3417	1·2908	48	71	107	47
	10	,,	1·3350	1·2738	51	76	114	51
	8	,,	1·3250	1·2486	55	83	125	55
	6	,,	1·3083	1·2064	61	92	138	61
1½	20	1·5000	1·4800	1·4494	40	60	—	40
	16	,,	1·4750	1·4368	43	65	—	43
	12	,,	1·4667	1·4158	48	71	107	47
	10	,,	1·4600	1·3988	51	76	114	51
	8	,,	1·4500	1·3736	55	83	125	55
	6	,,	1·4333	1·3314	61	92	138	61
1¾	20	1·7500	1·7300	1·6994	43	64	—	43
	16	,,	1·7250	1·6868	46	68	—	46
	12	,,	1·7167	1·6658	50	75	113	50
	10	,,	1·7100	1·6488	53	80	120	54
	8	,,	1·7000	1·6236	58	86	130	58
	6	,,	1·6833	1·5814	64	96	144	64
2	20	2·0000	1·9800	1·9494	43	64	—	43
	16	,,	1·9750	1·9368	46	68	—	46
	12	,,	1·9667	1·9158	50	75	113	50
	10	,,	1·9600	1·8988	53	80	120	54
	8	,,	1·9500	1·8736	58	86	130	58
	6	,,	1·9333	1·8314	64	96	144	64

* See footnote on page 63. *(continued on opposite page*

Table B10 BRITISH STANDARD 7°/45° BUTTRESS

continued. **Basic Nominal Dimensions: Tolerances and Allowances**

(C=Close class; M=Medium class; F=Free class)

Nom. Diam. (ins.)	Threads per inch	Basic Major Diam.	Basic Effective Diam.	Basic Minor Diam.	Tolerances* C	Tolerances* M	Tolerances* F	Allowance*
2¼	20	2·2500	2·2300	2·1994	43	64	—	43
	16	,,	2·2250	2·1868	46	68	—	46
	12	,,	2·2167	2·1658	50	75	113	50
	10	,,	2·2100	2·1488	53	80	120	54
	8	,,	2·2000	2·1236	58	86	130	58
	6	,,	2·1833	2·0814	64	96	144	64
2½	20	2·5000	2·4800	2·4494	43	64	—	43
	16	,,	2·4750	2·4368	46	68	—	46
	12	,,	2·4667	2·4158	50	75	113	50
	10	,,	2·4600	2·3988	53	80	120	54
	8	,,	2·4500	2·3736	58	86	130	58
	6	..	2·4333	2·3314	64	96	144	64
2¾	20	2·7500	2·7300	2·6994	46	68	—	46
	16	,,	2·7250	2·6868	49	73	—	49
	12	,,	2·7167	2·6658	53	80	119	53
	10	,,	2·7100	2·6488	56	84	126	57
	8	,,	2·7000	2·6236	61	91	136	61
	6	,,	2·6833	2·5814	67	100	150	67
	4	,,	2·6500	2·4970	77	116	174	78
3	20	3·0000	2·9800	2·9494	46	68	—	46
	16	,,	2·9750	2·9368	49	73	—	49
	12	,,	2·9667	2·9158	53	80	119	53
	10	,,	2·9600	2·8988	56	84	126	57
	8	,,	2·9500	2·8736	61	91	136	61
	6	,,	2·9333	2·8314	67	100	150	67
	4	,,	2·9000	2·7470	77	116	174	78
3½	20	3·5000	3·4800	3·4494	46	68	—	46
	16	,,	3·4750	3·4368	49	73	—	49
	12	,,	3·4667	3·4158	53	80	119	53
	10	,,	3·4600	3·3988	56	84	126	57
	8	,,	3·4500	3·3736	61	91	136	61
	6	,,	3·4333	3·3314	67	100	150	67
	4	,,	3·4000	3·2470	77	116	174	78
4	20	4·0000	3·9800	3·9494	46	68	—	46
	16	,,	3·9750	3·9368	49	73	—	49
	12	,,	3·9667	3·9158	53	80	119	53
	10	,,	3·9600	3·8988	56	84	126	57
	8	,,	3·9500	3·8736	61	91	136	61
	6	,,	3·9333	3·8314	67	100	150	67
	4	,,	3·9000	3·7470	77	116	174	78

*Expressed in units of 0·0001 inch. As given they are for the major diameter of screw, effective diameter of screw and nut, and minor diameter of nut. It is recommended that the tolerances for the major diameter of nut and the minor diameter of screw be double the tolerance figures given.

Table B11 BRITISH STANDARD 0°/52° BUTTRESS

Basic Form Dimensions in Inches

Threads per inch	Pitch	Triangular Height	Shortening		Depth of Thread	Radius at Root	Major Diam. minus Effect. Diam.
			at Tip	at Root			
20	0·05000	0·0391	0·0095	0·0049	0·0246	0·0046	0·0200
16	0·06250	0·0488	0·0119	0·0061	0·0308	0·0058	0·0250
12	0·08333	0·0651	0·0159	0·0081	0·0411	0·0077	0·0333
10	0·10000	0·0781	0·0191	0·0098	0·0493	0·0093	0·0400
8	0·12500	0·0977	0·0238	0·0122	0·0616	0·0116	0·0500
6	0·16667	0·1302	0·0318	0·0163	0·0822	0·0155	0·0667
5	0·20000	0·1563	0·0381	0·0195	0·0986	0·0186	0·0800
4	0·25000	0·1953	0·0477	0·0244	0·1232	0·0232	0·1000
3	0·33333	0·2604	0·0635	0·0326	0·1643	0·0310	0·1333
2½	0·40000	0·3125	0·0763	0·0391	0·1972	0·0372	0·1600
2	0·50000	0·3906	0·0953	0·0488	0·2465	0·0465	0·2000
1½	0·66667	0·5209	0·1271	0·0651	0·3287	0·0620	0·2667
1¼	0·80000	0·6250	0·1525	0·0781	0·3944	0·0744	0·3200
1	1·00000	0·7813	0·1906	0·0977	0·4930	0·0930	0·4000

Table B12. CORDEAUX THREAD

Size (in.)	Threads per inch	Pitch (in.)	Major Diam. (in.)	Depth of Thread (in.)
½	7	0·1429	0·5025 −0·001 +0·000	0·0642
⅝	6	0·1667	0·6275 −0·001 +0·000	0·0832

Table B13. WATER WELL CASING THREADS—SQUARE FORM

Basic Form Dimensions in Inches

Nom. Size (in.)	Threads per inch	Pitch	Depth of Thread (h)	Width of Thread (t_1)	Width of Groove (t_2)	Flank Clearance (a) min.	Radial Clearance (b)
4	4	0·2500	0·062	0·1210	0·1290	0·008	0·004
5	4	,,	0·062	0·1210	0·1290	0·008	0·004
6	4	,,	0·078	0·1210	0·1290	0·008	0·005
8	4	,,	0·078	0·1210	0·1290	0·008	0·008
10	4	,,	0·094	0·1210	0·1290	0·008	0·008
12	4	,,	0·094	0·1210	0·1290	0·008	0·008
13	4	,,	0·094	0·1210	0·1290	0·008	0·008
15	4	,,	0·094	0·1200	0·1300	0·010	0·012
18	4	,,	0·094	0·1200	0·1300	0·010	0·012
21	4	,,	0·094	0·1200	0·1300	0·010	0·012
24	4	,,	0·094	0·1200	0·1300	0·010	0·012
27	$2\frac{2}{3}$	0·3750	0·1250	0·1800	0·1950	0·015	0·015
30	$2\frac{2}{3}$	,,	0·1250	0·1800	0·1950	0·015	0·015
33	$2\frac{2}{3}$	,,	0·1250	0·1800	0·1950	0·015	0·015
36	$2\frac{2}{3}$	,,	0·1250	0·1800	0·1950	0·015	0·015
40	$2\frac{2}{3}$	,,	0·1560	0·1780	0·1970	0·019	0·020
44	$2\frac{2}{3}$	,,	0·1560	0·1780	0·1970	0·019	0·020
48	$2\frac{2}{3}$	,,	0·1560	0·1780	0·1970	0·019	0·020

N.B. The flank and radial clearances may be increased by not more than 0·004 in.

Table B14. WATER WELL CASING—SQUARE FORM

Casing and Thread Diameters in Inches

Nom. Size (ins.)	Outside Diameter of Casing	Major Diameter of Male Thread	Minor Diameter of Female Thread	Nom. Size (ins.)	Outside Diameter of Casing	Major Diameter of Male Thread	Minor Diameter of Female Thread
4	4·5	4·246	4·130	21	22·0	21·582	21·418
5	5·5	5·246	5·130	24	25·0	24·582	24·418
6	6·625	6·323	6·177	27	28·25	27·798	27·578
8	8·625	8·320	8·180	30	31·25	30·798	30·578
10	10·75	10·399	10·227	33	34·25	33·735	33·515
12	12·75	12·399	12·227	36	37·25	36·735	36·515
13	14·0	13·649	13·477	40	41·375	40·824	40·552
15	16·0	15·582	15·418	44	45·375	44·824	44·552
18	19·0	18·582	18·418	48	49·375	48·824	48·552

Table B15. EDISON SCREW THREADS FOR LAMPS

Dimensions in Inches

Size No.	Pitch (p)	Radius (r)	Major Diam.		Minor Diam.	
			Cap (D_s)	Socket (D_n)	Cap (d_s)	Socket (d_n)
E40 (Goliath)	0·2500	0·0728	1·5551 (max.) 1·5374 (min.)	1·5768 (max.) 1·5591 (min.)	1·4134 (max.) 1·3957 (min.)	1·4350 (max.) 1·4173 (min.)
E27 (Medium)	0·1429	0·0404	1·0413 (max.) 1·0295 (min.)	1·0571 (max.) 1·0453 (min.)	0·9551 (max.) 0·9433 (min.)	0·9709 (max.) 0·9591 (min.)
E14 (Small)	0·1111	0·0324	0·5469 (max.) 0·5394 (min.)	0·5575 (max.) 0·5500 (min.)	0·4839 (max.) 0·4764 (min.)	0·4945 (max.) 0·4870 (min.)
E10 (Miniature)	0·0714	0·0209	0·3752 (max.) 0·3685 (min.)	0·3850 (max.) 0·3783 (min.)	0·3350 (max.)	0·3449 (max.) 0·3382 (min.)
E5 (Lilliput)	0·0394	0·0115	0·2098 (max.) 0·2059 (min.)	0·2161 (max.) 0·2122 (min.)	0·1878 (max.)	0·1941 (max.) 0·1902 (min.)

Table B16. EDISON SCREW THREADS FOR LAMPS

Dimensions in mm.

Size No.	Pitch (p)	Radius (r)	Major Diam.		Minor Diam.	
			Cap (D_s)	Socket (D_n)	Cap (d_s)	Socket (d_n)
E40	6·350	1·85	39·50 (max.) 39·05 (min.)	40·05 (max.) 39·60 (min.)	35·90 (max.) 35·45 (min.)	36·45 (max.) 36·00 (min.)
E27	3·629	1·025	26·45 (max.) 26·15 (min.)	26·85 (max.) 26·55 (min.)	24·26 (max.) 23·96 (min.)	24·66 (max.) 24·36 (min.)
E14	2·822	0·822	13·89 (max.) 13·70 (min.)	14·16 (max.) 13·97 (min.)	12·29 (max.) 12·10 (min.)	12·56 (max.) 12·37 (min.)
E10	1·814	0·531	9·53 (max.) 9·36 (min.)	9·78 (max.) 9·61 (min.)	8·51 (max.)	8·76 (max.) 8·59 (min.)
E5	1·000	0·293	5·33 (max.) 5·23 (min).	5·49 (max.) 5·39 (min.)	4·77 (max.)	4·93 (max.) 4·83 (min.)

Table B17. BOTTLE CLOSURE THREADS

Basic Form Dimensions in Inches

Threads per inch	h	r_1	r_2 (max.)	t
12	0·030	0·022	0·012	0·060
8	0·042	0·031	0·016	0·084
6	0·047	0·034	0·025	0·094
5	0·060	0·044	0·030	0·120

Table B18. BOTTLE CLOSURE THREADS

Dimensions in Inches

(R-3/2 Finish)				(R-4 Finish)			
Size No.	Threads per inch	Major Diam. (D)	Major Diam. Tolerance	Size No.	Threads per inch	Major Diam. (D)	Major Diam. Tolerance
18	8	0·694	+0·010 —0·010	13	12	0·506	+0·008 —0·007
20	8	0·773	"	14	12	0·535	"
22	8	0·852	"	15	12	0·573	"
24	8	0·930	"	18	8	0·694	+0·010 —0·010
28	6	1·075	+0·013 —0·012	20	8	0·773	"
30	6	1·114	"	22	8	0·852	"
33	6	1·252	"	24	8	0·930	"
38	6	1·456	+0·018 —0·017	28	6	1·075	+0·013 —0·012
40	6	1·560	"	31	6	1·193	"
43	6	1·634	"				
48	6	1·850	"				
51	6	1·948	"				
53	6	2·047	"				
58	6	2·204	"				
60	6	2·322	"				
63	6	2·441	"				
66	6	2·559	"				
70	6	2·716	"				
77	5	3·015	+0·020 —0·020				
83	5	3·248	"				
89	5	3·491	"				
100	5	3·917	"				
120	5	4·704	"				

'able B19. BRITISH STANDARD METRIC THREAD (S.I.)

(Superseded by British Standard I.S.O. Metric Thread)

Thread Profile Dimensions in mm.

Pitch	Triangular Height H	Shortening at Crest S_1	Shortening at Root S_2	Maximum Radius at Root	Depth of Thread for Flat Roots
1·0	0·8660	0·1083	0·1894	0·1263	0·5683
1·25	1·0825	0·1353	0·2368	0·1579	0·7104
1·5	1·2990	0·1624	0·2842	0·1895	0·8525
1·75	1·5155	0·1894	0·3315	0·2210	0·9946
2·0	1·7320	0·2165	0·3789	0·2526	1·1367
2·5	2·1651	0·2706	0·4736	0·3158	1·4208
3·0	2·5981	0·3248	0·5683	0·3789	1·7050
3·5	3·0311	0·3789	0·6631	0·4421	1·9892
4·0	3·4641	0·4330	0·7578	0·5052	2·2733
4·5	3·8971	0·4871	0·8525	0·5684	2·5575
5·0	4·3301	0·5413	0·9472	0·6315	2·8416
5·5	4·7631	0·5954	1·0419	0·6947	3·1258
6·0	5·1962	0·6495	1·1367	0·7578	3·4100

able B20. BRITISH STANDARD METRIC THREAD (S.I.)

(Superseded by British Standard I.S.O. Metric Thread)

Thread Profile Dimensions in Inches

Pitch (mm.)	Triangular Height H	Shortening at Crest S_1	Shortening at Root S_2	Maximum Radius at Root	Depth of Thread for Flat Roots
1·0	0·03410	0·00426	0·00746	0·00497	0·02238
1·25	0·04262	0·00533	0·00932	0·00622	0·02797
1·5	0·05114	0·00639	0·01119	0·00746	0·03356
1·75	0·05967	0·00746	0·01305	0·00870	0·03916
2·0	0·06819	0·00852	0·01492	0·00994	0·04475
2·5	0·08524	0·01065	0·01865	0·01243	0·05594
3·0	0·10229	0·01279	0·02238	0·01492	0·06713
3·5	0·11933	0·01492	0·02610	0·01740	0·07831
4·0	0·13639	0·01705	0·02983	0·01989	0·08950
4·5	0·15343	0·01918	0·03356	0·02238	0·10069
5·0	0·17048	0·02131	0·03729	0·02486	0·11188
5·5	0·18753	0·02344	0·04102	0·02734	0·12306
6·0	0·20457	0·02557	0·04475	0·02983	0·13425

Table B21. BRITISH STANDARD METRIC THREAD (S.I.)
(Superseded by British Standard I.S.O. Metric Thread)
Maximum Metal Dimensions in mm.

Nominal Diameter	Pitch	Major Diameter	Effective Diameter	Minor Diameter
6	1·0	6·000	5·350	4·863
7	1·0	7·000	6·350	5·863
8	1·25	8·000	7·188	6·579
9	1·25	9·000	8·188	7·579
10	1·5	10·000	9·026	8·295
11	1·5	11·000	10·026	9·295
12	1·75	12·000	10·863	10·011
14	2·0	14·000	12·701	11·727
16	2·0	16·000	14·701	13·727
18	2·5	18·000	16·376	15·158
20	2·5	20·000	18·376	17·158
22	2·5	22·000	20·376	19·158
24	3·0	24·000	22·051	20·590
27	3·0	27·000	25·051	23·590
30	3.5	30·000	27·727	26·022
33	3·5	33·000	30·727	29·022
36	4·0	36·000	33·402	31·453
39	4·0	39·000	36·402	34·453
42	4·5	42·000	39·077	36·885
45	4·5	45·000	42·077	39·885
48	5·0	48·000	44·752	42·317
52	5·0	52·000	48·752	46·317
56	5·5	56·000	52·428	49·748
60	5·5	60·000	56·428	53·748

Table B22. BRITISH STANDARD METRIC THREAD (S.I.)

(Superseded by British Standard I.S.O. Metric Thread)

Maximum Metal Dimensions in Inches

Nominal Diameter (mm.)	Pitch	Major Diameter	Effective Diameter	Minor Diameter
6	0·0394	0·2362	0·2106	0·1915
7	0·0394	0·2756	0·2500	0·2308
8	0·0492	0·3150	0·2830	0·2590
9	0·0492	0·3543	0·3224	0·2984
10	0·0591	0·3937	0·3554	0·3266
11	0·0591	0·4331	0·3947	0·3659
12	0·0689	0·4724	0·4277	0·3941
14	0·0787	0·5512	0·5000	0·4617
16	0·0787	0·6299	0·5788	0·5404
18	0·0984	0·7087	0·6447	0·5968
20	0·0984	0·7874	0·7235	0·6755
22	0·0984	0·8661	0·8022	0·7543
24	0·1181	0·9449	0·8681	0·8106
27	0·1181	1·0630	0·9863	0·9287
30	0·1378	1·1811	1·0916	1·0245
33	0·1378	1·2992	1·2097	1·1426
36	0·1575	1·4173	1·3150	1·2383
39	0·1575	1·5354	1·4331	1·3564
42	0·1772	1·6535	1·5385	1·4522
45	0·1772	1·7717	1·6566	1·5703
48	0·1969	1·8898	1·7619	1·6660
52	0·1969	2·0472	1·9194	1·8235
56	0·2165	2·2047	2·0641	1·9586
60	0·2165	2·3622	2·2216	2·1161

Table B23. BRITISH STANDARD METRIC THREAD (S.I.)

(Superseded by British Standard I.S.O. Metric Thread)

B.S. Tolerance in Units of 0·001 mm.

(C=Close fit; M=Medium fit; F=Free fit)

Nominal Diameter (mm.)	Pitch (mm.)	Major Diam.	Effective Diameter			Minor Diam.
		Screws only (—)	Screws (—) & Nuts (+)			Nuts only (+)
		(C, M, F)	C	M	F	(C, M, F)
6	1·0	158	64	96	144	163
7	1·0	158	67	100	151	163
8	1·25	185	72	108	162	203
9	1·25	185	75	112	168	203
10	1·5	212	79	119	178	244
11	1·5	212	81	122	183	244
12	1·75	239	86	128	192	284
14	2·0	266	91	137	206	324
16	2·0	266	94	142	213	324
18	2·5	320	102	153	229	406
20	2·5	320	105	157	236	406
22	2·5	320	108	162	243	406
24	3·0	374	114	171	256	487
27	3·0	374	118	177	266	487
30	3·5	428	125	187	280	568
33	3·5	428	128	192	288	568
36	4·0	482	134	202	302	650
39	4·0	482	138	206	310	650
42	4·5	536	143	215	322	731
45	4·5	536	146	219	329	731
48	5·0	590	152	227	341	812
52	5·0	590	155	233	349	812
56	5·5	644	161	242	362	893
60	5·5	644	164	247	370	893

Table B24. BRITISH STANDARD METRIC THREAD (S.I.)

(Superseded by British Standard I.S.O. Metric Thread)

B.S. Tolerance in Units of 0·0001-inch

(C=Close fit; M=Medium fit; F=Free fit)

Nominal Diameter (mm.)	Pitch (mm.)	Major Diam. Screws only (—)	Effective Diameter Screws (—) & Nuts (+)			Minor Diam. Nuts only (+)
		(C, M, F)	C	M	F	(C, M, F)
6	1·0	62	25	38	57	64
7	1·0	62	26	39	59	64
8	1·25	73	28	43	64	80
9	1·25	73	30	44	66	80
10	1·5	83	31	47	70	96
11	1·5	83	32	48	72	96
12	1·75	94	34	50	76	112
14	2·0	105	36	54	81	128
16	2·0	105	37	56	84	128
18	2·5	126	40	60	90	160
20	2·5	126	41	62	93	160
22	2·5	126	43	64	96	160
24	3·0	147	45	67	101	192
27	3·0	147	46	70	105	192
30	3·5	169	49	74	110	224
33	3·5	169	50	76	113	224
36	4·0	190	53	80	119	256
39	4·0	190	54	81	122	256
42	4·5	211	56	85	127	288
45	4·5	211	57	86	130	288
48	5·0	232	60	89	134	320
52	5·0	232	61	92	137	320
56	5·5	254	63	95	143	352
60	5·5	254	65	97	146	352

Table B25. **ACME (B.S.1104) Classes 2G, 3G and 4G Maximum Metal Dimensions in Inches**

Screws—

Nominal Diameter *	Threads per Inch	Major Diameter (max.) 2G, 3G, 4G	Effective Diameter (max.) 2G	3G	4G	Minor Diameter (max.) 2G, 3G, 4G
$\frac{3}{8}$	10	0·3750	0·3201	0·3213	0·3226	0·2550
$\frac{7}{16}$	10	0·4375	0·3826	0·3838	0·3851	0·3175
1	6	1·0000	0·9087	0·9107	0·9127	0·8133
$1\frac{1}{8}$	4	1·1250	0·9915	0·9936	0·9958	0·8550
$1\frac{1}{4}$	4	1·2500	1·1160	1·1183	1·1205	0·9800

Nuts—

Nominal Diameter *	Threads per Inch	Major Diameter (min.)	Effective Diameter (minimum)	Minor Diameter (min.)
$\frac{3}{8}$	10	0·3950	0·3250	0·2750
$\frac{7}{16}$	10	0·4575	0·3875	0·3375
1	6	1·0200	0·9167	0·8333
$1\frac{1}{8}$	4	1·1450	1·0000	0·8750
$1\frac{1}{4}$	4	1·2700	1·1250	1·0000

Table B26. **ACME (B.S.1104) Classes 2G, 3G and 4G Tolerances in Units of 0·0001-inch**

Nominal Diameter	t.p.i.	Major Diameter		Effective Diameter			Minor Diameter			
				Screw and Nut			Screw		Nuts	
		Screw	Nut	2G	3G	4G	2G	3G	4G	All
$\frac{3}{8}$	10	50	200	132	61	44	198	91	66	50
$\frac{7}{16}$	10	50	200	135	63	45	202	94	67	50
1	6	83	200	182	85	61	173	127	91	83
$1\frac{1}{8}$	4	125	200	214	100	71	321	150	106	125
$1\frac{1}{4}$	4	125	200	217	101	72	325	151	108	125

* For other diameters refer to pages 112 and 113.

Table B27. ACME (B.S.1104) Class 5G

Maximum Metal Dimensions in Inches

Nominal Diam.	Threads per inch	Screws			Nuts		
		Major Diam. (max.)	Effective Diam. (max.)	Minor Diam. (max.)	Major Diam. (min.)	Effective Diam. (min.)	Minor Diam. (min.)
1/4	16	0·2500	0·2188	0·1775	0·2600	0·2188	0·1875
5/16	14	0·3125	0·2768	0·2311	0·3225	0·2768	0·2411
3/8	10	0·3750	0·3250	0·2550	0·3950	0·3250	0·2750
7/16	10	0·4375	0·3875	0·3175	0·4575	0·3875	0·3375
1/2	10	0·5000	0·4500	0·3800	0·5200	0·4500	0·4000
5/8	8	0·6250	0·5625	0·4800	0·6450	0·5625	0·5000
3/4	6	0·7500	0·6667	0·5633	0·7700	0·6667	0·5833
7/8	6	0·8750	0·7917	0·6883	0·8950	0·7917	0·7083
1	6	1·0000	0·9167	0·8133	1·0200	0·9167	0·8333
1 1/8	4	1·1250	1·0000	0·8550	1·1450	1·0000	0·8750
1 1/4	4	1·2500	1·1250	0·9800	1·2700	1·1250	1·0000
1 3/8	4	1·3750	1·2500	1·1050	1·3950	1·2500	1·1250
1 1/2	4	1·5000	1·3750	1·2300	1·5200	1·3750	1·2500
1 3/4	4	1·7500	1·6250	1·4800	1·7700	1·6250	1·5000
2	4	2·0000	1·8750	1·7300	2·0200	1·8750	1·7500
2 1/4	3	2·2500	2·0833	1·8967	2·2700	2·0833	1·9167
2 1/2	3	2·5000	2·3333	2·1467	2·5200	2·3333	2·1667
2 3/4	3	2·7500	2·5833	2·3967	2·7700	2·5833	2·4167
3	2	3·0000	2·7500	2·4800	3·0200	2·7500	2·5000

Table B28. **ACME (B.S.1104) Class 5G**

Tolerances in Units of 0·0001-inch

Nominal Diam.	Threads per inch	Screws			Nuts		
		Major Diam.	Effective Diam.	Minor Diam.	Major Diam.	Effective Diam.	Minor Diam.
$\frac{1}{4}$	16	50	28	42	100	28	50
$\frac{5}{16}$	14	50	30	45	100	30	50
$\frac{3}{8}$	10	50	35	52	200	35	50
$\frac{7}{16}$	10	50	36	54	200	36	50
$\frac{1}{2}$	10	50	37	55	200	37	50
$\frac{5}{8}$	8	62	41	61	200	41	62
$\frac{3}{4}$	6	83	47	70	200	47	83
$\frac{7}{8}$	6	83	48	71	200	48	83
1	6	83	49	73	200	49	83
$1\frac{1}{8}$	4	125	57	85	200	57	125
$1\frac{1}{4}$	4	125	58	87	200	58	125
$1\frac{3}{8}$	4	125	59	88	200	59	125
$1\frac{1}{2}$	4	125	60	89	200	60	125
$1\frac{3}{4}$	4	125	61	92	200	61	125
2	4	125	63	94	200	63	125
$2\frac{1}{4}$	3	167	70	105	200	70	167
$2\frac{1}{2}$	3	167	71	107	200	71	167
$2\frac{3}{4}$	3	167	73	109	200	73	167
3	2	250	84	126	200	84	250

Section C

Section C

UNIFIED AND AMERICAN THREAD SERIES

UNIFIED AND AMERICAN THREAD SERIES

The range of thread series which can be conveniently classified as th Unified and American thread series, has the thread profile shown in Fig. (for maximum material condition) with the following proportions:

Thread angle (2θ) = 60 deg.; Flank angle (θ) = 30 deg.; Triangula height (H) = 0·86603 p; Height of external thread = $\frac{17}{24}H$ = 0·6134 p; Height of internal thread = $\frac{5}{8}H$ = 0·54127 p; Shortening a crest of screw = $\frac{1}{8}h$ = 0·10825 p; Shortening at root of screw = $\frac{1}{6}H$ = 0·14434 p; Shortening at crest of nut = $\frac{1}{4}H$ = 0·21651 p Shortening at root of nut = $\frac{1}{8}H$ = 0·10825 p.

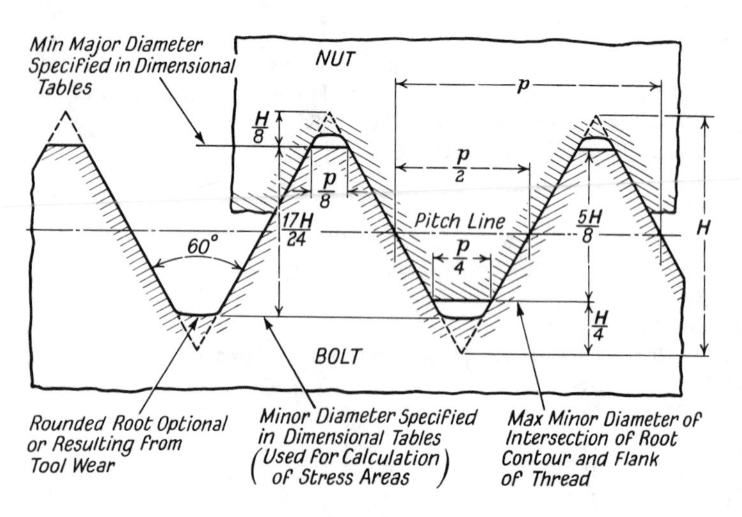

Fig. 1. Unified and American Thread Form

The UNR thread form is identical with that shown except that th root of the thread on the bolt has a radius of 0·14434 p – 0·10825 p.

The series include the following:

(1) Coarse-thread (U.N.C. and N.C.) series
(2) Fine-thread (U.N.F. and N.F.) series
(3) Extra-fine thread (N.E.F.) series

Within each series there are a number of classes of fit. These consis of classes 1A, 2A and 3A for screws; 1B, 2B and 3B for nuts. Not all the classes of fit are applied to all the series, but the corresponding availabl fits may be seen from the following table:

Series	*Fits*
Coarse	1A, 2A, 3A, 1B, 2B, 3B
Fine	1A, 2A, 3A, 1B, 2B, 3B
Extra Fine	2A, 2B, 3A, 3B

With the exception of Classes 1A and 2A, the tolerances to obtain the equired class of fit are applied to basic dimensions in which the basic ıajor diameter is the same as the nominal diameter. In the case of classes A and 2A, the basic dimensions are first reduced by an allowance to btain the maximum metal dimensions.

JNIFIED AND AMERICAN NATIONAL UNIFORM PITCH SERIES

The maximum metal dimensions of the uniform-pitch series may be alculated from the following formulæ, using the allowances given in 'ables C20 to C27 and the values of $\frac{3}{4}H$, $2h_s$ and $2h_n$ from the table below:

crew:

Major Diameter (Maximum) $= D - A$.

Effective Diameter (Maximum) $= D - A - \frac{3}{4}H$.

Minor Diameter (Maximum) $= D - A - 2h_s$.

Vut:

Aajor Diameter (Minimum) $= D$.

Effective Diameter (Minimum) $= D - \frac{3}{4}H$.

Ainor Diameter (Minimum) $= D - 2h_n$.

Series	$\frac{3}{4}H$	$2h_s$	$2h_n$
4-thread	0·16238	0·30672	0·27063
6- ,,	0·10825	0·20448	0·18042
8- ,,	0·08119	0·15336	0·13532
12- ,,	0·05413	0·10224	0·09021
16- ,,	0·04060	0·07668	0·06766
20- ,,	0·03248	0·06134	0·05413
28- ,,	0·02320	0·04382	0·03866
32- ,,	0·02030	0·03834	0·03383

Allowances and tolerances for these series will be found in Tables C20 o C30.

:7-T.P.I. UNIFIED THREAD FOR THIN-WALLED TUBING

This thread is specified only in the classes 2A and 2B, the former for he external thread and the latter for the internal thread. The dimensions or the thread on the specified range of nominal diameters is as follows:

EXTERNAL THREADS (Screws)

Nominal Size	Allowance	Major diameter		Effective diameter			Minor diameter
		Max.	Min.	Max.	Min.	Tol'ance	
$\frac{1}{4}$	0·0010	0·2490	0·2423	0·2249	0·2214	0·0035	0·2036
$\frac{5}{16}$	0·0010	0·3115	0·3048	0·2874	0·2839	0·0035	0·2661
$\frac{3}{8}$	0·0011	0·3739	0·3672	0·3498	0·3462	0·0036	0·3285
0·390	0·0011	0·3889	0·3822	0·3648	0·3612	0·0036	0·3435
$\frac{7}{16}$	0·0011	0·4364	0·4297	0·4123	0·4087	0·0036	0·3910
$\frac{1}{2}$	0·0011	0·4989	0·4922	0·4748	0·4711	0·0037	0·4535
$\frac{9}{16}$	0·0011	0·5614	0·5547	0·5373	0·5336	0·0037	0·5160
$\frac{5}{8}$	0·0011	0·6239	0·6172	0·5998	0·5960	0·0038	0·5785
$\frac{3}{4}$	0·0012	0·7488	0·7421	0·7247	0·7208	0·0039	0·7034
$\frac{7}{8}$	0·0012	0·8738	0·8671	0·8947	0·8458	0·0039	0·8284
1	0·0012	0·9988	0·9921	0·9747	0·9707	0·0040	0·9534

INTERNAL THREADS (Nuts)

Nominal Size	Minor diameter		Effective diameter			Major diameter Min.
	Min.	Max.	Min.	Max.	Tolerance	
$\frac{1}{4}$	0·210	0·219	0·2259	0·2304	0·0045	0·2500
$\frac{5}{16}$	0·272	0·281	0·2884	0·2929	0·0045	0·3125
$\frac{3}{8}$	0·335	0·344	0·3509	0·3556	0·0047	0·3750
0·390	0·350	0·359	0·3659	0·3706	0·0047	0·3900
$\frac{7}{16}$	0·397	0·406	0·4134	0·4181	0·0047	0·4375
$\frac{1}{2}$	0·460	0·469	0·4759	0·4807	0·0048	0·5000
$\frac{9}{16}$	0·522	0·531	0·5384	0·5432	0·0048	0·5625
$\frac{5}{8}$	0·585	0·594	0·6009	0·6059	0·0050	0·6250
$\frac{3}{4}$	0·710	0·719	0·7259	0·7310	0·0051	0·7500
$\frac{7}{8}$	0·835	0·844	0·8509	0·8560	0·0051	0·8750
1	0·960	0·969	0·9759	0·9811	0·0052	1·0000

UNIFIED MINIATURE SERIES (U.N.M.)

Form Proportions: Thread angle (2θ) = 60 deg.; Flank angle (θ) = 30 deg.; Triangular height (H) = 0·86603 p; Depth of engagement (h) = 0·52 p; Shortening at crest of screw (S_1) = 0·108255 p; Shortening a root of screw (S_2) = 0·157775 p; Radius at root of screw (r_2) = 0·157775 p Depth of thread in screw (h_s) = 0·60000 p; Shortening at crest of nut (S_3 = 0·237775 p; Shortening at root of nut (S_4) = 0·072255 p; Radius o root of nut (r_4) = 0·072255 p; Depth of thread in nut (h_n) = 0·556 p The crests of both screw and nut are flat.

The form dimensions in inch and mm. units are given in Table C17 The basic dimensions of the series are given in Table C18, and the corres ponding tolerances in Table C19.

Table C1. UNIFIED AND AMERICAN THREAD FORM

Basic Dimensions (Inches)

Threads per inch	Pitch	Tri-angular Height (H)	Depth of Thread		$\frac{1}{8}H$	$\frac{1}{6}H$
			Screw	Nut		
80	0·012500	0·010825	0·007668	0·006766	0·001353	0·001804
72	0·013889	0·012028	0·008520	0·007518	0·001504	0·002005
64	0·015625	0·013532	0·009585	0·008457	0·001691	0·002255
56	0·017857	0·015465	0·010954	0·009665	0·001933	0·002577
48	0·020833	0·018042	0·012780	0·011276	0·002255	0·003007
44	0·022727	0·019682	0·013942	0·012301	0·002460	0·003280
40	0·025000	0·021651	0·015336	0·013532	0·002706	0·003608
36	0·027778	0·024056	0·017040	0·015035	0·003007	0·004009
32	0·03125	0·027063	0·019170	0·016915	0·003383	0·004511
28	0·035714	0·030929	0·021908	0·019331	0·003866	0·005155
24	0·041667	0·036084	0·025560	0·022553	0·004511	0·006014
20	0·050000	0·043301	0·030672	0·027063	0·005413	0·007217
18	0·055556	0·048113	0·034080	0·030071	0·006014	0·008019
16	0·062500	0·054127	0·038340	0·033829	0·006766	0·009021
14	0·071429	0·061859	0·043817	0·038662	0·007732	0·010310
13	0·076923	0·066617	0·047187	0·041636	0·008327	0·011103
12	0·083333	0·072169	0·051119	0·045105	0·009021	0·012028
11½	0·086957	0·075307	0·053342	0·047067	0·009413	0·012551
11	0·090909	0·078730	0·055767	0·049206	0·009841	0·013122
10	0·100000	0·086603	0·061343	0·054127	0·010825	0·014434
9	0·111111	0·096225	0·068159	0·060141	0·012028	0·016037
8	0·125000	0·108253	0·076679	0·067658	0·013532	0·018042
7	0·142857	0·123718	0·087633	0·077324	0·015465	0·020620
6	0·166667	0·144338	0·102239	0·090211	0·018042	0·024056
5	0·200000	0·173205	0·122687	0·108253	0·021651	0·028868
4½	0·222222	0·192450	0·136319	0·120281	0·024056	0·032075
4	0·250000	0·216506	0·153359	0·135316	0·027063	0·036084

Table C2.

U.N.C. UNIFIED AND N.C. AMERICAN THREADS

(CLASSES 1A, 1B, 2A and 2B)

Maximum Metal Dimensions (Inches)

Size	Threads per inch	Screws (1A and 2A)			Nuts (1B and 2B)		
		Major Dia. Max.	Effective Dia. Max.	Minor Dia. Max.	Major Dia. Min.	Effective Dia. Min.	Minor Dia. Min
1/4	20	0·2489	0·2164	0·1876	0·2500	0·2175	0·1959
5/16	18	0·3113	0·2752	0·2431	0·3125	0·2764	0·2524
3/8	16	0·3737	0·3331	0·2970	0·3750	0·3344	0·3073
7/16	14	0·4361	0·3897	0·3485	0·4375	0·3911	0·3602
1/2	13	0·4985	0·4485	0·4041	0·5000	0·4500	0·4167
9/16	12	0·5609	0·5068	0·4587	0·5625	0·5084	0·4723
5/8	11	0·6234	0·5644	0·5119	0·6250	0·5660	0·5266
3/4	10	0·7482	0·6832	0·6255	0·7500	0·6850	0·6417
7/8	9	0·8731	0·8009	0·7368	0·8750	0·8028	0·7547
1	8	0·9980	0·9168	0·8446	1·0000	0·9188	0·8647
1 1/8	7	1·1228	1·0300	0·9475	1·1250	1·0322	0·9704
1 1/4	7	1·2478	1·1550	1·0725	1·2500	1·1572	1·0954
1 3/8	6	1·3726	1·2643	1·1681	1·3750	1·2667	1·1946
1 1/2	6	1·4976	1·3893	1·2931	1·5000	1·3917	1·3196
1 3/4	5	1·7473	1·6174	1·5019	1·7500	1·6201	1·5335
2	4 1/2	1·9971	1·8528	1·7245	2·0000	1·8557	1·7594
2 1/4	4 1/2	2·2471	2·1028	1·9745	2·2500	2·1057	2·0094
2 1/2	4	2·4969	2·3345	2·1902	2·5000	2·3376	2·2294
2 3/4	4	2·7468	2·5844	2·4401	2·7500	2·5876	2·4794
3	4	2·9968	2·8344	2·6901	3·0000	2·8376	2·7294
3 1/4	4	3·2467	3·0843	2·9400	3·2500	3·0876	2·9794
3 1/2	4	3·4967	3·3343	3·1900	3·5000	3·3376	3·2294
3 3/4	4	3·7466	3·5842	3·4399	3·7500	3·5876	3·4794
4	4	3·9966	3·8342	3·6899	4·0000	3·8376	3·7294

Table C3.

U.N.C. UNIFIED AND N.C. AMERICAN THREADS

(CLASSES 2*, 3*, 3A and 3B)

Maximum Metal Dimensions (Inches)

Size	Threads per inch	Screw (2, 3 and 3A)			Nut (2, 3 and 3B)		
		Major Dia. Max.	Effective Dia. Max.	Minor Dia. Max.	Major Dia. Min.	Effective Dia. Min.	Minor Dia. Min.
1/4	20	0·2500	0·2175	0·1887	0·2500	0·2175	0·1959
5/16	18	0·3125	0·2764	0·2443	0·3125	0·2764	0·2524
3/8	16	0·3750	0·3344	0·2983	0·3750	0·3344	0·3073
7/16	14	0·4375	0·3911	0·3499	0·4375	0·3911	0·3602
1/2	13	0·5000	0·4500	0·4056	0·5000	0·4500	0·4167
9/16	12	0·5625	0·5084	0·4603	0·5625	0·5084	0·4723
5/8	11	0·6250	0·5660	0·5135	0·6250	0·5660	0·5266
3/4	10	0·7500	0·6850	0·6273	0·7500	0·6850	0·6417
7/8	9	0·8750	0·8028	0·7387	0·8750	0·8028	0·7547
1	8	1·0000	0·9188	0·8466	1·0000	0·9188	0·8647
1 1/8	7	1·1250	1·0322	0·9497	1·1250	1·0322	0·9704
1 1/4	7	1·2500	1·1572	1·0747	1·2500	1·1572	1·0954
1 3/8	6	1·3750	1·2667	1·1705	1·3750	1·2667	1·1946
1 1/2	6	1·5000	1·3917	1·2955	1·5000	1·3917	1·3196
1 3/4	5	1·7500	1·6201	1·5046	1·7500	1·6201	1·5335
2	4 1/2	2·0000	1·8557	1·7274	2·0000	1·8557	1·7594
2 1/4	4 1/2	2·2500	2·1057	1·9774	2·2500	2·1057	2·0094
2 1/2	4	2·5000	2·3376	2·1933	2·5000	2·3376	2·2294
2 3/4	4	2·7500	2·5876	2·4433	2·7500	2·5876	2·4794
3	4	3·0000	2·8376	2·6933	3·0000	2·8376	2·7294
3 1/4	4	3·2500	3·0876	2·9433	3·2500	3·0876	2·9794
3 1/2	4	3·5000	3·3376	3·1933	3·5000	3·3376	3·2294
3 3/4	4	3·7500	3·5876	3·4433	3·7500	3·5876	3·4794
4	4	4·0000	3·8376	3·6933	4·0000	3·8376	3·7294

* Classes 2 and 3 are not Unified Classes.

Table C4.

U.N.C. UNIFIED AND N.C. AMERICAN THREADS
Effective Diameter Tolerances (Unit = 0·0001-inch)

Size	1A	1B	2A	2B	3A	3B	2	3
No.								
1			20	26	15	19	19	14
2			21	28	16	21	20	15
3			23	30	17	22	22	16
4			25	33	19	24	24	17
5			26	33	19	25	24	17
6			28	37	21	27	27	19
8			29	38	22	28	27	19
10			33	43	25	32	33	24
12			34	44	26	33	33	24
$\frac{1}{4}$	56	73	37	48	28	36	36	26
$\frac{5}{16}$	61	79	40	53	30	39	41	30
$\frac{3}{8}$	65	85	44	57	33	43	45	32
$\frac{7}{16}$	71	92	47	61	35	46	49	36
$\frac{1}{2}$	74	97	50	65	37	48	52	37
$\frac{9}{16}$	78	102	52	68	39	51	56	40
$\frac{5}{8}$	83	107	55	72	41	54	59	42
$\frac{3}{4}$	88	115	59	77	44	57	64	45
$\frac{7}{8}$	95	123	63	82	47	61	70	49
1	101	132	68	88	51	66	76	54
$1\frac{1}{8}$	109	141	72	94	54	71	85	59
$1\frac{1}{4}$	111	144	74	96	55	72	85	59
$1\frac{3}{8}$	120	155	80	104	60	78	101	71
$1\frac{1}{2}$	121	158	81	105	61	79	101	71
$1\frac{3}{4}$	134	174	89	116	67	87	116	82
2	143	186	95	124	71	93	127	89
$2\frac{1}{4}$	146	190	97	126	73	95	127	89
$2\frac{1}{2}$	155	202	104	135	78	101	140	97
$2\frac{3}{4}$	158	206	105	137	79	103	140	97
3	161	209	107	139	80	104	140	97
$3\frac{1}{4}$	163	212	109	141	82	106	140	97
$3\frac{1}{2}$	166	215	110	143	83	108	140	97
$3\frac{3}{4}$	168	218	112	145	84	109	140	97
4	170	221	113	147	85	111	140	97

'able C5.

U.N.C. UNIFIED AND N.C. AMERICAN THREADS

Major and Minor Diameter Tolerances (Unit = 0·0001-inch)

Sizes	Major Diameter (Screws)			Minor Diameter (Nuts)		
	1A	2A and 3A	2 and 3	1B, 2B	3B	2 and 3
No. 1		38	38	62*	62	62
2		41	40	70*	70	70
3		45	44	81*	81	77
4		51	48	90*	90	89
5		51	48	83*	83	83
6		60	54	100*	100	103
8		60	54	90*	89	82
10		72	66	110*	105	110
12		72	66	100*	97	92
$\frac{1}{4}$	122	81	72	115	108	101
$\frac{5}{16}$	131	87	82	127	106	106
$\frac{3}{8}$	142	94	90	141	109	111
$\frac{7}{16}$	155	103	98	158	115	119
$\frac{1}{2}$	163	109	104	169	117	123
$\frac{9}{16}$	172	114	112	181	120	127
$\frac{5}{8}$	182	121	118	194	125	131
$\frac{3}{4}$	194	129	128	210	128	136
$\frac{7}{8}$	208	139	140	228	138	142
1	225	150	152	250	150	148
$1\frac{1}{8}$	246	164	170	276	171	154
$1\frac{1}{4}$	246	164	170	276	171	154
$1\frac{3}{8}$	273	182	202	306	200	180
$1\frac{1}{2}$	273	182	202	306	200	180
$1\frac{3}{4}$	308	205	232	340	240	216
2	330	220	254	358	267	241
$2\frac{1}{4}$	330	220	254	358	267	241
$2\frac{1}{2}$	357	238	280	375	300	270
$2\frac{3}{4}$	357	238	280	375	300	270
3	357	238	280	375	300	270
$3\frac{1}{4}$	357	238	280	375	300	270
$3\frac{1}{2}$	357	238	280	375	300	270
$3\frac{3}{4}$	357	238	280	375	300	270
4	357	238	280	375	300	270

*2B tolerance only, 1B not given.

Table C6.

U.N.C. UNIFIED AND N.C. AMERICAN THREADS

SMALL SIZES (CLASSES 2A, 2B, 3A, 3B, 2* and 3*)

Maximum Metal Dimensions (Inches)

Size	Threads per inch	Screw (2A)			Nut (2B)		
		Major Dia. Max.	Effective Dia. Max.	Minor Dia. Max.	Major Dia. Min.	Effective Dia. Min.	Minor Dia. Min.
No. 1 (·073)	64	0·0724	0·0623	0·0532	0·0730	0·0629	0·0561
2 (·086)	56	0·0854	0·0738	0·0635	0·0860	0·0744	0·0667
3 (·099)	48	0·0983	0·0848	0·0727	0·0990	0·0855	0·0764
†4 (·112)	40	0·1112	0·0950	0·0805	0·1120	0·0958	0·0849
5 (·125)	40	0·1242	0·1080	0·0935	0·1250	0·1088	0·0979
†6 (·138)	32	0·1372	0·1169	0·0989	0·1380	0·1177	0·1040
†8 (·164)	32	0·1631	0·1428	0·1248	0·1640	0·1437	0·1300
†10 (·190)	24	0·1890	0·1619	0·1379	0·1900	0·1629	0·1450
12 (·216)	24	0·2150	0·1879	0·1639	0·2160	0·1889	0·1710
		Screws (2, 3 & 3A)			Nuts (2, 3 & 3B)		
No. 1 (·073)	64	0·0730	0·0629	0·0538	0·0730	0·0629	0·0561
2 (·086)	56	0·0860	0·0744	0·0641	0·0860	0·0744	0·0667
3 (·099)	48	0·0990	0·0855	0·0734	0·0990	0·0855	0·0764
4 (·112)	40	0·1120	0·0958	0·0813	0·1120	0·0958	0·0849
5 (·125)	40	0·1250	0·1088	0·0943	0·1250	0·1088	0·0979
6 (·138)	32	0·1380	0·1177	0·0997	0·1380	0·1177	0·1040
8 (·164)	32	0·1640	0·1437	0·1257	0·1640	0·1437	0·1300
10 (·190)	24	0·1900	0·1629	0·1389	0·1900	0·1629	0·1450
12 (·216)	24	0·2160	0·1889	0·1649	0·2160	0·1889	0·1710

*Classes 2 and 3 are not Unified Classes.

†Only the sizes indicated thus have been adopted as part of the British Unified Screw Thread System, and these only in classes 2A and 2B.

Table C7.

U.N.F. UNIFIED AND N.F. AMERICAN THREADS

(CLASSES 1A, 1B, 2A and 2B)

Maximum Metal Dimensions (Inches)

Size	Threads per inch	Screws (1A and 2A)			Nuts (1B and 2B)		
		Major Dia. Max.	Effective Dia. Max.	Minor Dia. Max.	Major Dia. Min.	Effective Dia. Min.	Minor Dia. Min.
$\frac{1}{4}$	28	0·2490	0·2258	0·2052	0·2500	0·2268	0·2113
$\frac{5}{16}$	24	0·3114	0·2843	0·2603	0·3125	0·2854	0·2674
$\frac{3}{8}$	24	0·3739	0·3468	0·3228	0·3750	0·3479	0·3299
$\frac{7}{16}$	20	0·4362	0·4037	0·3749	0·4375	0·4050	0·3834
$\frac{1}{2}$	20	0·4987	0·4662	0·4374	0·5000	0·4675	0·4459
$\frac{9}{16}$	18	0·5611	0·5250	0·4929	0·5625	0·5264	0·5024
$\frac{5}{8}$	18	0·6236	0·5875	0·5554	0·6250	0·5889	0·5649
$\frac{3}{4}$	16	0·7485	0·7079	0·6718	0·7500	0·7094	0·6823
$\frac{7}{8}$	14	0·8734	0·8270	0·7858	0·8750	0·8286	0·7977
*1	14	0·9983	0·9519	0·9107	1·0000	0·9536	0·9227
†1	12	0·9982	0·9441	0·8960	1·0000	0·9459	0·9098
$1\frac{1}{8}$	12	1·1232	1·0691	1·0210	1·1250	1·0709	1·0348
$1\frac{1}{4}$	12	1·2482	1·1941	1·1460	1·2500	1·1959	1·1598
$1\frac{3}{8}$	12	1·3731	1·3190	1·2709	1·3750	1·3209	1·2848
$1\frac{1}{2}$	12	1·4981	1·4440	1·3959	1·5000	1·4459	1·4098

*N.F. †U.N.F.

Table C8.

U.N.F. UNIFIED AND N.F. AMERICAN THREADS

(CLASSES 2*, 3*, 3A and 3B)

Maximum Metal Dimensions (Inches)

Size	Threads per inch	Screw (2, 3 and 3A)			Nut (2, 3 and 3B)		
		Major Dia. Max.	Effective Dia. Max.	Minor Dia. Max.	Major Dia. Min.	Effective Dia. Min.	Minor Dia. Min.
1/4	28	0·2500	0·2268	0·2062	0·2500	0·2268	0·2113
5/16	24	0·3125	0·2854	0·2614	0·3125	0·2854	0·2674
3/8	24	0·3750	0·3479	0·3239	0·3750	0·3479	0·3299
7/16	20	0·4375	0·4050	0·3762	0·4375	0·4050	0·3834
1/2	20	0·5000	0·4675	0·4387	0·5000	0·4675	0·4459
9/16	18	0·5625	0·5264	0·4943	0·5625	0·5264	0·5024
5/8	18	0·6250	0·5889	0·5568	0·6250	0·5889	0·5649
3/4	16	0·7500	0·7094	0·6733	0·7500	0·7094	0·6823
7/8	14	0·8750	0·8286	0·7874	0·8750	0·8286	0·7977
†1	14	1·0000	0·9536	0·9124	1·0000	0·9536	0·9227
‡1	12	1·0000	0·9459	0·8978	1·0000	0·9459	0·9098
1 1/8	12	1·1250	1·0709	1·0228	1·1250	1·0709	1·0348
1 1/4	12	1·2500	1·1959	1·1478	1·2500	1·1959	1·1598
1 3/8	12	1·3750	1·3209	1·2728	1·3750	1·3209	1·2848
1 1/2	12	1·5000	1·4459	1·3978	1·5000	1·4459	1·4098

*Classes 2 and 3 are not Unified Classes. †N.F. ‡U.N.F.

able C9.

U.N.F. UNIFIED AND N.F. AMERICAN THREADS

SMALL SIZES (CLASSES 2A, 2B, 3A, 3B, 2* and 3*)

Maximum Metal Dimensions (Inches)

Size	Threads per inch	Screw (2A)			Nut (2B)		
		Major Dia. Max.	Effective Dia. Max.	Minor Dia. Max.	Major Dia. Min.	Effective Dia. Min.	Minor Dia. Min.
o. 0 (·060)	80	0·0595	0·0514	0·0442	0·0600	0·0519	0·0465
1 (·073)	72	0·0724	0·0634	0·0554	0·0730	0·0640	0·0580
2 (·086)	64	0·0854	0·0753	0·0662	0·0860	0·0759	0·0691
3 (·099)	56	0·0983	0·0867	0·0764	0·0990	0·0874	0·0797
4 (·112)	48	0·1113	0·0978	0·0857	0·1120	0·0985	0·0894
5 (·125)	44	0·1243	0·1095	0·0964	0·1250	0·1102	0·1004
6 (·138)	40	0·1372	0·1210	0·1065	0·1380	0·1218	0·1110
8 (·164)	36	0·1632	0·1452	0·1291	0·1640	0·1460	0·1340
†10 (·190)	32	0·1891	0·1688	0·1508	0·1900	0·1697	0·1560
12 (·216)	28	0·2150	0·1918	0·1712	0·2160	0·1928	0·1770
		Screws (2, 3 & 3A)			Nuts (2, 3 & 3B)		
o. 0 (·060)	80	0·0600	0·0519	0·0447	0·0600	0·0519	0·0465
1 (·073)	72	0·0730	0·0640	0·0560	0·0730	0·0640	0·0580
2 (·086)	64	0·0860	0·0759	0·0668	0·0860	0·0759	0·0691
3 (·099)	56	0·0990	0·0874	0·0771	0·0990	0·0874	0·0797
4 (·112)	48	0·1120	0·0985	0·0864	0·1120	0·0985	0·0894
5 (·125)	44	0·1250	0·1102	0·0971	0·1250	0·1102	0·1004
6 (·138)	40	0·1380	0·1218	0·1073	0·1380	0·1218	0·1110
8 (·164)	36	0·1640	0·1460	0·1299	0·1640	0·1460	0·1340
10 (·190)	32	0·1900	0·1697	0·1517	0·1900	0·1697	0·1560
12 (·216)	28	0·2160	0·1928	0·1722	0·2160	0·1928	0·1770

*Classes 2 and 3 are not Unified Classes.

†Only the size indicated thus has been adopted as part of the British Unified Screw Thread System, and this only in the classes 2A and 2B.

Table C10.

U.N.F. UNIFIED AND N.F. AMERICAN THREADS
Effective Diameter Tolerances (Unit = 0·0001-Inch)

Size	1A	1B	2A	2B	3A	3B	2	3
No.								
0			18	23	13	17	17	13
1			19	25	14	19	18	13
2			20	27	15	20	19	14
3			22	28	16	21	20	15
4			24	31	18	23	22	16
5			25	32	19	24	23	16
6			26	34	20	25	24	17
8			28	36	21	27	25	18
10			30	39	23	29	27	19
12			32	42	24	31	31	22
$\frac{1}{4}$	50	65	33	43	25	32	31	22
$\frac{5}{16}$	55	71	37	48	27	36	33	24
$\frac{3}{8}$	57	74	38	49	29	37	33	24
$\frac{7}{16}$	62	81	42	54	31	41	36	26
$\frac{1}{2}$	64	84	43	56	32	42	36	26
$\frac{9}{16}$	68	89	45	59	34	44	41	30
$\frac{5}{8}$	70	91	47	60	35	45	41	30
$\frac{3}{4}$	75	98	50	65	38	49	45	32
$\frac{7}{8}$	81	106	54	70	41	53	49	36
*1	84	109	56	73	42	54	49	36
†1	88	114	59	76	44	57	—	—
$1\frac{1}{8}$	90	117	60	78	45	59	56	40
$1\frac{1}{4}$	92	120	62	80	46	60	56	40
$1\frac{3}{8}$	94	123	63	82	47	61	56	40
$1\frac{1}{2}$	96	125	64	83	48	63	56	40

*N.F. †U.N.F.

able C11.

U.N.F. UNIFIED AND N.F. AMERICAN THREADS

Major and Minor Diameter Tolerances (Unit = 0·0001-Inch)

Sizes	Major Diameter (Screws)			Minor Diameter (Nuts)		
	1A	2A and 3A	2 and 3	1B, 2B	3B	2 and 3
No. 0		32	34	49*	49	49
1		35	36	55*	55	54
2		38	38	62*	62	55
3		41	40	68*	68	59
4		45	44	74*	74	66
5		48	46	75*	75	64
6		51	48	80*	76	70
8		55	50	80*	76	63
10		60	54	80*	81	62
12		65	62	90*	87	62
$\frac{1}{4}$	98	65	62	84	77	60
$\frac{5}{16}$	108	72	66	97	80	65
$\frac{3}{8}$	108	72	66	97	73	65
$\frac{7}{16}$	122	81	72	115	82	72
$\frac{1}{2}$	122	81	72	115	78	72
$\frac{9}{16}$	131	87	82	127	82	76
$\frac{5}{8}$	131	87	82	127	81	76
$\frac{3}{4}$	142	94	90	141	85	80
$\frac{7}{8}$	155	103	98	158	91	85
‡1	155	103	98	88	88	85
§1	172	114	—	181	100	—
$1\frac{1}{8}$	172	114	112	181	100	90
$1\frac{1}{4}$	172	114	112	181	100	90
$1\frac{3}{8}$	172	114	112	181	100	90
$1\frac{1}{2}$	172	114	112	181	100	90

*2B tolerance only, 1B not given.
‡N.F.
§U.N.F.

Table C12. AMERICAN EXTRA FINE (N.E.F.) THREAD

Basic Dimensions (Inches)

Nom. Size	Threads per inch	Major Diameter	Effective Diameter	Minor Diameter		Mean Helix Angle
				Screw	Nut	
12 (0·216)	32	0·2160	0·1957	0·1777	0·1822	2° 55′
$\frac{1}{4}$	32	0·2500	0·2297	0·2117	0·2162	2° 29′
$\frac{5}{16}$	32	0·3125	0·2922	0·2742	0·2787	1° 57′
$\frac{3}{8}$	32	0·3750	0·3547	0·3367	0·3412	1° 36′
$\frac{7}{16}$	28	0·4375	0·4143	0·3937	0·3988	1° 34′
$\frac{1}{2}$	28	0·5000	0·4768	0·4562	0·4613	1° 22′
$\frac{9}{16}$	24	0·5625	0·5354	0·5114	0·5174	1° 25′
$\frac{5}{8}$	24	0·6250	0·5979	0·5739	0·5799	1° 16′
$\frac{11}{16}$	24	0·6875	0·6604	0·6364	0·6424	1° 9′
$\frac{3}{4}$	20	0·7500	0·7175	0·6887	0·6959	1° 16′
$\frac{13}{16}$	20	0·8125	0·7800	0·7512	0·7584	1° 10′
$\frac{7}{8}$	20	0·8750	0·8425	0·8137	0·8209	1° 5′
$\frac{15}{16}$	20	0·9375	0·9050	0·8762	0·8834	1° 0′
1	20	1·0000	0·9675	0·9387	0·9459	57′
$1\frac{1}{16}$	18	1·0625	1·0264	0·9943	1·0024	59′
$1\frac{1}{8}$	18	1·1250	1·0889	1·0568	1·0649	56′
$1\frac{3}{16}$	18	1·1875	1·1514	1·1193	1·1274	53′
$1\frac{1}{4}$	18	1·2500	1·2139	1·1818	1·1899	50′
$1\frac{5}{16}$	18	1·3125	1·2764	1·2443	1·2524	48′
$1\frac{3}{8}$	18	1·3750	1·3389	1·3068	1·3149	45′
$1\frac{7}{16}$	18	1·4375	1·4014	1·3693	1·3774	43′
$1\frac{1}{2}$	18	1·5000	1·4639	1·4318	1·4399	42′
$1\frac{9}{16}$	18	1·5625	1·5624	1·4943	1·5024	40′
$1\frac{5}{8}$	18	1·6250	1·5889	1·5568	1·5649	38′
$1\frac{11}{16}$	18	1·6875	1·6514	1·6193	1·6274	37′
$1\frac{3}{4}$	16	1·7500	1·7094	1·6733	1·6823	40′
2	16	2·0000	1·9594	1·9233	1·9323	35′

able C13. U.N.E.F. AMERICAN THREADS

CLASS 2A (SCREW)

Limits of Size (Inches)

Nominal Size	Threads per inch	Major Diameter		Effective Diameter		Minor Diameter Max.
		Max.	Min.	Max.	Min.	
12	32	0·2151	0·2091	0·1948	0·1917	0·1768
1/4	32	0·2490	0·2430	0·2287	0·2255	0·2107
5/16	32	0·3115	0·3055	0·2912	0·2880	0·2732
3/8	32	0·3740	0·3680	0·3537	0·3503	0·3357
7/16	28	0·4364	0·4299	0·4132	0·4096	0·3926
1/2	28	0·4989	0·4924	0·4757	0·4720	0·4551
9/16	24	0·5613	0·5541	0·5342	0·5303	0·5102
5/8	24	0·6238	0·6166	0·5967	0·5927	0·5727
11/16	24	0·6863	0·6791	0·6592	0·6552	0·6352
3/4	20	0·7487	0·7406	0·7162	0·7118	0·6874
13/16	20	0·8112	0·8031	0·7787	0·7743	0·7498
7/8	20	0·8737	0·8656	0·8412	0·8368	0·8124
15/16	20	0·9361	0·9280	0·9036	0·8991	0·8748
1	20	0·9986	0·9905	0·9661	0·9616	0·9373
1 1/16	18	1·0611	1·0524	1·0250	1·0203	0·9929
1 1/8	18	1·1236	1·1149	1·0875	1·0828	1·0554
1 3/16	18	1·1860	1·1773	1·1499	1·1450	1·1178
1 1/4	18	1·2485	1·2398	1·2124	1·2075	1·1803
1 5/16	18	1·3110	1·3023	1·2749	1·2700	1·2428
1 3/8	18	1·3735	1·3648	1·3374	1·3325	1·3053
1 7/16	18	1·4360	1·4273	1·3999	1·3949	1·3678
1 1/2	18	1·4985	1·4898	1·4624	1·4574	1·4303
1 9/16	18	1·5610	1·5523	1·5249	1·5199	1·4928
1 5/8	18	1·6235	1·6148	1·5874	1·5824	1·5553
1 11/16	18	1·6860	1·6773	1·6499	1·6448	1·6178
1 3/4	16	1·7484	1·7390	1·7078	1·7025	1·6717
2	16	1·9984	1·9890	1·9578	1·9524	1·9217

Table C14. U.N.E.F. AMERICAN THREADS
CLASS 2B (NUT)
Limits of Size (Inches)

Nominal Size	Threads per inch	Major Diameter (min.)	Effective Diameter		Minor Diameter	
			Min.	Max.	Min.	Max.
12	32	0·2160	0·1957	0·1998	0·1820	0·1900
$\frac{1}{4}$	32	0·2500	0·2297	0·2339	0·2160	0·2240
$\frac{5}{16}$	32	0·3125	0·2922	0·2964	0·2790	0·2860
$\frac{3}{8}$	32	0·3750	0·3547	0·3591	0·3410	0·3490
$\frac{7}{16}$	28	0·4375	0·4143	0·4189	0·3990	0·4070
$\frac{1}{2}$	28	0·5000	0·4768	0·4816	0·4610	0·4700
$\frac{9}{16}$	24	0·5625	0·5354	0·5405	0·5170	0·5270
$\frac{5}{8}$	24	0·6250	0·5979	0·6031	0·5800	0·5900
$\frac{11}{16}$	24	0·6875	0·6604	0·6656	0·6420	0·6520
$\frac{3}{4}$	20	0·7500	0·7175	0·7232	0·6960	0·7070
$\frac{13}{16}$	20	0·8125	0·7800	0·7857	0·7580	0·7700
$\frac{7}{8}$	20	0·8750	0·8425	0·8482	0·8210	0·8320
$\frac{15}{16}$	20	0·9375	0·9050	0·9109	0·8830	0·8950
1	20	1·0000	0·9675	0·9734	0·9460	0·9570
$1\frac{1}{16}$	18	1·0625	1·0264	1·0326	1·0020	1·0150
$1\frac{1}{8}$	18	1·1250	1·0889	1·0951	1·0650	1·0780
$1\frac{3}{16}$	18	1·1875	1·1514	1·1577	1·1270	1·1400
$1\frac{1}{4}$	18	1·2500	1·2139	1·2202	1·1900	1·2030
$1\frac{5}{16}$	18	1·3125	1·2764	1·2827	1·2520	1·2650
$1\frac{3}{8}$	18	1·3750	1·3389	1·3452	1·3150	1·3280
$1\frac{7}{16}$	18	1·4375	1·4014	1·4079	1·3770	1·3900
$1\frac{1}{2}$	18	1·5000	1·4639	1·4704	1·4400	1·4520
$1\frac{9}{16}$	18	1·5625	1·5264	1·5329	1·5020	1·5150
$1\frac{5}{8}$	18	1·6250	1·5889	1·5954	1·5650	1·5780
$1\frac{11}{16}$	18	1·6875	1·6514	1·6580	1·6270	1·6400
$1\frac{3}{4}$	16	1·7500	1·7094	1·7163	1·6820	1·6960
2	16	2·0000	1·9594	1·9664	1·9320	1·9460

able C15. **U.N.E.F. AMERICAN THREADS**
CLASS 3A (SCREW)
Limits of Size (Inches)

Nominal Size	Threads per inch	Major Diameter		Effective Diameter		Minor Diameter Max.
		Max.	Min.	Max.	Min.	
12	32	0·2160	0·2100	0·1957	0·1933	0·1777
$\frac{1}{4}$	32	0·2500	0·2440	0·2297	0·2273	0·2117
$\frac{5}{16}$	32	0·3125	0·3065	0·2922	0·2898	0·2742
$\frac{3}{8}$	32	0·3750	0·3690	0·3547	0·3522	0·3367
$\frac{7}{16}$	28	0·4375	0·4310	0·4143	0·4116	0·3937
$\frac{1}{2}$	28	0·5000	0·4935	0·4768	0·4740	0·4562
$\frac{9}{16}$	24	0·5625	0·5553	0·5354	0·5325	0·5114
$\frac{5}{8}$	24	0·6250	0·6178	0·5979	0·5949	0·5739
$\frac{11}{16}$	24	0·6875	0·6803	0·6604	0·6574	0·6364
$\frac{3}{4}$	20	0·7500	0·7419	0·7175	0·7142	0·6887
$\frac{13}{16}$	20	0·8125	0·8044	0·7800	0·7767	0·7512
$\frac{7}{8}$	20	0·8750	0·8669	0·8425	0·8392	0·8137
$\frac{15}{16}$	20	0·9375	0·9294	0·9050	0·9016	0·8762
1	20	1·0000	0·9919	0·9675	0·9641	0·9387
$1\frac{1}{16}$	18	1·0625	1·0538	1·0264	1·0228	0·9943
$1\frac{1}{8}$	18	1·1250	1·1163	1·0889	1·0853	1·0568
$1\frac{3}{16}$	18	1·1875	1·1788	1·1514	1·1478	1·1193
$1\frac{1}{4}$	18	1·2500	1·2413	1·2139	1·2103	1·1818
$1\frac{5}{16}$	18	1·3125	1·3038	1·2764	1·2728	1·2443
$1\frac{3}{8}$	18	1·3750	1·3663	1·3389	1·3353	1·3068
$1\frac{7}{16}$	18	1·4375	1·4288	1·4014	1·3977	1·3693
$1\frac{1}{2}$	18	1·5000	1·4913	1·4639	1·4602	1·4318
$1\frac{9}{16}$	18	1·5625	1·5538	1·5264	1·5227	1·4943
$1\frac{5}{8}$	18	1·6250	1·6163	1·5889	1·5852	1·5568
$1\frac{11}{16}$	18	1·6875	1·6788	1·6514	1·6476	1·6193
$1\frac{3}{4}$	16	1·7500	1·7406	1·7094	1·7054	1·6733
2	16	2·0000	1·9906	1·9594	1·9554	1·9233

Table C16. U.N.E.F. AMERICAN THREADS
CLASS 3B (NUT)
Limits of Size (Inches)

Nominal Size	Threads per inch	Major Diameter Min.	Effective Diameter		Minor Diameter	
			Min.	Max.	Min.	Max.
12	32	0·2160	0·1957	0·1988	0·1820	0·1895
1/4	32	0·2500	0·2297	0·2328	0·2160	0·2229
5/16	32	0·3125	0·2922	0·2953	0·2790	0·2847
3/8	32	0·3750	0·3547	0·3580	0·3410	0·3469
7/16	28	0·4375	0·4143	0·4178	0·3990	0·4051
1/2	28	0·5000	0·4768	0·4804	0·4610	0·4676
9/16	24	0·5625	0·5354	0·5392	0·5170	0·5244
5/8	24	0·6250	0·5979	0·6018	0·5800	0·5869
11/16	24	0·6875	0·6604	0·6643	0·6420	0·6494
3/4	20	0·7500	0·7175	0·7218	0·6960	0·7037
13/16	20	0·8125	0·7800	0·7843	0·7580	0·7662
7/8	20	0·8750	0·8425	0·8468	0·8210	0·8287
15/16	20	0·9375	0·9050	0·9094	0·8830	0·8912
1	20	1·0000	0·9675	0·9719	0·9460	0·9537
1 1/16	18	1·0625	1·0264	1·0310	1·0020	1·0105
1 1/8	18	1·1250	1·0889	1·0935	1·0650	1·0730
1 3/16	18	1·1875	1·1514	1·1561	1·1270	1·1355
1 1/4	18	1·2500	1·2139	1·2186	1·1900	1·1980
1 5/16	18	1·3125	1·2764	1·2811	1·2520	1·2605
1 3/8	18	1·3750	1·3389	1·3436	1·3150	1·3230
1 7/16	18	1·4375	1·4014	1·4062	1·3770	1·3855
1 1/2	18	1·5000	1·4639	1·4687	1·4400	1·4480
1 9/16	18	1·5625	1·5264	1·5312	1·5020	1·5105
1 5/8	18	1·6250	1·5889	1·5937	1·5650	1·5730
1 11/16	18	1·6875	1·6514	1·6563	1·6270	1·6355
1 3/4	16	1·7500	1·7094	1·7146	1·6820	1·6908
2	16	2·0000	1·9594	1·9646	1·9320	1·9408

Basic Dimensions (Inches)

p	t.p.i.	H	S_1	S_2	S_3	S_4	r_2	r_4	h	h_s	h_n
0·003150	317·500	0·00273	0·00034	0·00050	0·00075	0·00023	0·00050	0·00023	0·00164	0·00189	0·00175
0·003543	282·222	0·00307	0·00038	0·00056	0·00084	0·00026	0·00056	0·00026	0·00184	0·00213	0·00197
0·003937	254·000	0·00341	0·00043	0·00062	0·00094	0·00028	0·00062	0·00028	0·00205	0·00236	0·00219
0·004921	203·200	0·00426	0·00053	0·00078	0·00117	0·00036	0·00078	0·00036	0·00256	0·00295	0·00274
0·005906	169·333	0·00511	0·00064	0·00093	0·00140	0·00043	0·00093	0·00043	0·00307	0·00354	0·00328
0·006890	145·143	0·00597	0·00075	0·00109	0·00164	0·00050	0·00109	0·00050	0·00358	0·00413	0·00383
0·007874	127·000	0·00682	0·00085	0·00124	0·00187	0·00057	0·00124	0·00057	0·00409	0·00472	0·00438
0·008858	112·889	0·00767	0·00096	0·00140	0·00211	0·00064	0·00140	0·00064	0·00461	0·00531	0·00493
0·009843	101·600	0·00852	0·00107	0·00155	0·00234	0·00071	0·00155	0·00071	0·00512	0·00591	0·00547
0·011811	84·667	0·01023	0·00128	0·00186	0·00281	0·00085	0·00186	0·00085	0·00614	0·00709	0·00657

Basic Dimensions (mm.)

p	t.p.i.	H	S_1	S_2	S_3	S_4	r_2	r_4	h	h_s	h_n
0·080	317·500	0·0693	0·0087	0·0126	0·0190	0·0058	0·0126	0·0058	0·0416	0·0479	0·0445
0·090	282·222	0·0779	0·0097	0·0142	0·0214	0·0065	0·0142	0·0065	0·0468	0·0539	0·0500
0·100	254·000	0·0866	0·0108	0·0158	0·0238	0·0072	0·0158	0·0072	0·0520	0·0599	0·0556
0·125	203·200	0·1083	0·0135	0·0197	0·0297	0·0090	0·0197	0·0090	0·0650	0·0749	0·0695
0·150	169·333	0·1299	0·0162	0·0237	0·0356	0·0108	0·0237	0·0108	0·0780	0·0899	0·0834
0·175	145·143	0·1516	0·0189	0·0276	0·0416	0·0126	0·0276	0·0126	0·0910	0·1049	0·0973
0·200	127·000	0·1732	0·0217	0·0316	0·0476	0·0144	0·0316	0·0144	0·1040	0·1199	0·1112
0·225	112·889	0·1949	0·0244	0·0355	0·0535	0·0162	0·0355	0·0162	0·1170	0·1348	0·1251
0·250	101·600	0·2165	0·0271	0·0394	0·0594	0·0180	0·0394	0·0180	0·1300	0·1498	0·1390
0·300	84·667	0·2598	0·0325	0·0473	0·0713	0·0217	0·0473	0·0217	0·1560	0·1798	0·1668

Table C18. **UNIFIED MINIATURE SERIES**

Basic Dimensions *(mm.)

Nominal Size	Pitch	Screw			Nut		
		Major Dia. (max.)	Effective Dia. (max.)	Minor Dia. (max.)	Major Dia. (min.))	Effective Dia. (min.)	Mino Dia. (min.
0·30 UNM	0·080	0·300	0·248	0·204	0·306	0·248	0·217
0·35 ,,	0·090	0·350	0·292	0·242	0·356	0·292	0·256
0·40 ,,	0·100	0·400	0·335	0·280	0·407	0·335	0·296
0·45 ,,	0·100	0·450	0·385	0·330	0·457	0·385	0·346
0·50 ,,	0·125	0·500	0·419	0·350	0·509	0·419	0·370
0·55 ,,	0·125	0·550	0·469	0·400	0·559	0·469	0·420
0·60 ,,	0·150	0·600	0·503	0·420	0·611	0·503	0·444
0·70 ,,	0·175	0·700	0·586	0·490	0·713	0·586	0·518
0·80 ,,	0·200	0·800	0·670	0·560	0·814	0·670	0·592
0·90 ,,	0·225	0·900	0·754	0·630	0·916	0·754	0·666
1·00 ,,	0·250	1·000	0·838	0·700	1·018	0·838	0·740
1·10 ,,	0·250	1·100	0·938	0·800	1·118	0·938	0·840
1·20 ,,	0·250	1·200	1·038	0·900	1·218	1·038	0·940
1·40 ,,	0·300	1·400	1·205	1·040	1·422	1·205	1·088

Basic Dimensions *(Inches)

Nominal Size	Pitch	Screw Major Dia. (max.)	Screw Effective Dia. (max.)	Screw Minor Dia. (max.)	Nut Major Dia. (min.)	Nut Effective Dia. (min.)	Nut Mino Dia. (min.
0·30 UNM	0·003150	0·0118	0·0098	0·0080	0·0120	0·0098	0·008
0·35 ,,	0·003543	0·0138	0·0115	0·0095	0·0140	0·0115	0·010
0·40 ,,	0·003937	0·0157	0·0132	0·0110	0·0160	0·0132	0·011
0·45 ,,	0·003937	0·0177	0·0152	0·0130	0·0180	0·0152	0·013
0·50 ,,	0·004921	0·0197	0·0165	0·0138	0·0200	0·0165	0·014
0·55 ,,	0·004921	0·0217	0·0185	0·0157	0·0220	0·0185	0·016
0·60 ,,	0·005906	0·0236	0·0198	0·0165	0·0240	0·0198	0·017
0·70 ,,	0·006890	0·0276	0·0231	0·0193	0·0281	0·0231	0·020
0·80 ,,	0·007874	0·0315	0·0264	0·0220	0·0321	0·0264	0·023
0·90 ,,	0·008858	0·0354	0·0297	0·0248	0·0361	0·0297	0·026
1·00 ,,	0·009843	0·0394	0·0330	0·0276	0·0401	0·0330	0·029
1·10 ,,	0·009843	0·0433	0·0369	0·0315	0·0440	0·0369	0·033
1·20 ,,	0·009843	0·0472	0·0409	0·0354	0·0480	0·0409	0·037
1·40 ,,	0·011811	0·0551	0·0474	0·0409	0·0560	0·0474	0·042

*These dimensions are the maximum-metal dimensions to which the tolerances giver in Table C19 are applied.

ble C19. UNIFIED MINIATURE SERIES

Tolerances

LERANCES IN UNITS OF 0·001 mm.

Nominal ize (mm.)	Screw			Nut		
	Major Diameter	Effective Diameter	Minor Diameter	Major Diameter	Effective Diameter	Minor Diameter
·30 UNM	16	14	21	21	14	37
·35 ,,	17	15	22	24	15	41
·40 ,,	18	16	24	25	16	44
·45 ,,	18	16	24	25	16	44
·50 ,,	21	18	28	29	18	52
·55 ,,	21	18	28	29	18	52
·60 ,,	24	20	32	33	20	60
·70 ,,	27	22	36	37	22	68
·80 ,,	30	24	40	42	24	76
·90 ,,	33	26	44	46	26	84
·00 ,,	36	28	48	50	28	92
·10 ,,	36	28	48	50	28	92
·20 ,,	36	28	48	50	28	92
·40 ,,	42	32	56	58	32	108

LERANCES IN UNITS OF 0·0001 inch

Nominal ize (mm.)	Screw Major Diameter	Screw Effective Diameter	Screw Minor Diameter	Nut Major Diameter	Nut Effective Diameter	Nut Minor Diameter
·30 UNM	6	6	8	9	6	15
·35 ,,	7	6	9	9	6	16
·40 ,,	7	6	9	10	6	17
·45 ,,	7	7	10	10	6	18
·50 ,,	8	7	11	12	7	20
·55 ,,	9	8	11	11	7	20
·60 ,,	9	8	12	14	8	23
·70 ,,	9	9	14	14	9	27
·80 ,,	12	10	15	16	9	30
·90 ,,	13	10	17	18	10	33
·00 ,,	14	11	19	19	11	36
·10 ,,	14	11	19	20	11	36
·20 ,,	14	12	19	19	11	36
·40 ,,	16	12	21	23	13	43

Table C20. UNIFIED 4-THREAD SERIES (4 U.N.)
Tolerances and Allowances (Unit = 0·0001-Inch)

Nom. Size (ins.)	Screw					Nut			
	Allowance Class 2A only	Tolerances				Tolerances			
		Major Diam.		Effective Diam.		Effective Diam.		Minor Diam.	
		2A	3A	2A	3A	2B	3B	2B	3B
2½	31	238	238	104	78	135	101	375	300
2¾	32	238	238	105	79	137	103	375	300
3	32	238	238	107	80	139	104	375	300
3¼	33	238	238	109	82	141	106	375	300
3½	33	238	238	110	83	143	108	375	300
3¾	34	238	238	112	84	145	109	375	300
4	34	238	238	113	85	147	111	375	300
4¼	34	238	238	115	86	149	112	375	300
4½	35	238	238	116	87	151	113	375	300
4¾	35	238	238	117	88	153	114	375	300
5	36	238	238	119	89	154	116	375	300
5¼	36	238	238	120	90	156	117	375	300
5½	36	238	238	121	91	158	118	375	300
5¾	37	238	238	122	92	159	119	375	300
6	37	238	238	124	93	161	120	375	300

Table C21. UNIFIED 6-THREAD SERIES (6 U.N.)
Tolerances and Allowances (Unit = 0·0001-Inch)

Nom. Size (ins.)	Screw					Nut			
	Allowance Class 2A only	Tolerances				Tolerances			
		Major Diam.		Effective Diam.		Effective Diam.		Minor Diam.	
		2A	3A	2A	3A	2B	3B	2B	3B
1⅜	24	182	182	80	60	104	78	306	200
1½	24	182	182	81	61	105	79	306	200
1⅝	25	182	182	82	62	107	80	306	200
1¾	25	182	182	83	63	108	81	306	200
1⅞	25	182	182	84	63	110	82	306	200
2	26	182	182	86	64	111	83	306	200
2¼	26	182	182	88	66	114	85	306	200
2½	27	182	182	90	67	116	87	306	200
2¾	27	182	182	91	68	119	89	306	200
3	28	182	182	93	70	121	91	306	300
3¼	28	182	182	95	71	123	92	306	200
3½	29	182	182	96	72	125	94	306	200
3¾	29	182	182	98	73	127	95	306	200
4	30	182	182	99	74	129	97	306	200
4¼	30	182	182	101	75	131	98	306	200
4½	31	182	182	102	77	133	99	306	200
4¾	31	182	182	103	77	134	101	306	200
5	31	182	182	105	78	136	102	306	200

ble C22. UNIFIED 8-THREAD SERIES (8 U.N.)
Tolerances and Allowances (Unit = 0·0001-Inch)

om. ize ns.)	Screw					Nut			
	Allow-ance Class 2A only	Tolerances				Tolerances			
		Major Diam.		Effective Diam.		Effective Diam.		Minor Diam.	
		2A	3A	2A	3A	2B	3B	2B	3B
1	20	150	150	68	51	88	66	250	150
1⅛	21	150	150	69	52	90	67	250	150
1¼	21	150	150	70	53	92	69	250	150
1⅜	22	150	150	72	54	93	70	250	150
1½	22	150	150	73	55	95	71	250	150
1⅝	22	150	150	74	56	97	72	250	150
1¾	23	150	150	75	56	98	74	250	150
1⅞	23	150	150	77	57	100	75	250	150
2	23	150	150	78	58	101	76	250	150
2¼	24	150	150	80	60	104	78	250	150
2½	24	150	150	82	61	106	80	250	150
2¾	25	150	150	83	63	108	81	250	150
3	26	150	150	85	64	111	83	250	150
3¼	26	150	150	87	65	113	84	250	150
3½	26	150	150	88	66	115	86	250	150
3¾	27	150	150	90	67	117	88	250	150
4	27	150	150	91	68	119	89	250	150

ble C23. UNIFIED 12-THREAD SERIES (12 U.N.)
Tolerances and Allowances (Unit = 0·0001-Inch)

om. ize ns.)	Screw					Nut			
	Allow-ance Class 2A only	Tolerances				Tolerances			
		Major Diam.		Effective Diam.		Effective Diam.		Minor Diam.	
		2A	3A	2A	3A	2B	3B	2B	3B
9/16	16	114	114	52	39	68	51	181	120
⅝	16	114	114	54	41	71	53	181	115
¾	17	114	114	55	41	72	54	181	109
⅞	17	114	114	55	41	72	54	181	104
1	18	114	114	59	44	76	57	181	100
1⅛	18	114	114	60	45	78	59	181	100
1¼	18	114	114	62	46	80	60	181	100
1⅜	19	114	114	63	47	82	61	181	100
1½	19	114	114	64	48	83	63	181	100
1⅝	18	114	114	59	44	76	57	181	100
1¾	18	114	114	60	45	78	58	181	100
1⅞	18	114	114	60	45	78	58	181	100
2	18	114	114	61	45	79	59	181	100
2¼	18	114	114	61	45	79	59	181	100
2½	19	114	114	62	46	81	60	181	100
2¾	19	114	114	62	46	81	60	181	100
3	19	114	114	63	47	82	62	181	100
3¼	19	114	114	63	47	82	62	181	100
3½	19	114	114	64	48	84	63	181	100
3¾	19	114	114	64	48	84	63	181	100
4	20	114	114	65	49	85	64	181	100

Table C24. UNIFIED 16-THREAD SERIES (16 U.N.)
Tolerances and Allowances (Unit = 0·0001-Inch)

Nom. Size (ins.)	Screw					Nut			
	Allowance Class 2A only	Tolerances				Tolerances			
		Major Diam.		Effective Diam.		Effective Diam.		Minor Diam.	
		2A	3A	2A	3A	2B	3B	2B	3B
$\frac{3}{8}$	13	94	94	44	33	57	43	141	109
$\frac{7}{16}$	14	94	94	46	34	59	40	141	102
$\frac{1}{2}$	14	94	94	47	35	61	46	141	96
$\frac{9}{16}$	14	94	94	47	35	61	46	141	92
$\frac{5}{8}$	14	94	94	48	36	62	46	141	89
$\frac{3}{4}$	15	94	94	50	38	65	49	141	85
$\frac{7}{8}$	15	94	94	49	36	63	47	141	85
1	15	94	94	50	37	65	49	141	85
$1\frac{1}{8}$	15	94	94	50	37	65	49	141	85
$1\frac{1}{4}$	15	94	94	51	38	66	50	141	85
$1\frac{3}{8}$	15	94	94	51	38	66	50	141	85
$1\frac{1}{2}$	16	94	94	52	39	68	51	141	85
$1\frac{5}{8}$	16	94	94	52	39	68	51	141	85
$1\frac{3}{4}$	16	94	94	53	40	69	52	141	85
$1\frac{7}{8}$	16	94	94	53	40	69	52	141	85
2	16	94	94	54	40	70	52	141	85
$2\frac{1}{4}$	16	94	94	54	41	70	52	141	85
$2\frac{1}{2}$	17	94	94	55	41	72	54	141	85
$2\frac{3}{4}$	17	94	94	55	41	72	54	141	85
3	17	94	94	56	42	73	55	141	85

Table C25. UNIFIED 20-THREAD SERIES (20 U.N.)
Tolerances and Allowances (Unit = 0·0001-Inch)

Nom. Size (ins.)	Screw					Nut			
	Allowance Class 2A only	Tolerances				Tolerances			
		Major Diam.		Effective Diam.		Effective Diam.		Minor Diam.	
		2A	3A	2A	3A	2B	3B	2B	3B
$\frac{1}{4}$	11	81	81	37	28	48	36	115	108
$\frac{5}{16}$	12	81	81	40	30	52	39	115	96
$\frac{3}{8}$	12	81	81	41	31	54	40	115	88
$\frac{7}{16}$	13	81	81	42	31	54	41	115	82
$\frac{1}{2}$	13	81	81	43	32	56	42	115	78
$\frac{9}{16}$	13	81	81	42	32	55	41	115	78
$\frac{5}{8}$	13	81	81	43	32	56	42	115	78
$\frac{3}{4}$	13	81	81	44	33	57	43	115	78
$\frac{7}{8}$	13	81	81	44	33	57	43	115	78
1	14	81	81	45	34	59	44	115	78
$1\frac{1}{8}$	14	81	81	45	34	59	44	115	78
$1\frac{1}{4}$	14	81	81	47	35	61	45	115	78
$1\frac{3}{8}$	14	81	81	47	35	61	45	115	78
$1\frac{1}{2}$	14	81	81	48	36	62	46	115	78
$1\frac{5}{8}$	14	81	81	48	36	62	46	115	78
$1\frac{3}{4}$	15	81	81	48	36	63	47	115	78
$1\frac{7}{8}$	15	81	81	48	36	63	47	115	78
2	15	81	81	49	37	64	48	115	78
$2\frac{1}{4}$	15	81	81	49	38	64	48	115	78
$2\frac{1}{2}$	15	81	81	51	38	66	50	115	78
$2\frac{3}{4}$	15	81	81	51	38	66	50	115	78
3	16	81	81	52	39	68	51	115	78

able C26. UNIFIED 28-THREAD SERIES (28 U.N.)

Tolerances and Allowances (Unit = 0·0001-Inch)

Nom. Size (ins.)	Screw					Nut			
	Allowance Class 2A only	Tolerances				Tolerances			
		Major Diam.		Effective Diam.		Effective Diam.		Minor Diam.	
		2A	3A	2A	3A	2B	3B	2B	3B
$\frac{1}{4}$	10	65	65	33	25	43	32	84	77
$\frac{5}{16}$	10	65	65	34	26	44	33	84	69
$\frac{3}{8}$	11	65	65	36	27	46	35	84	63
$\frac{7}{16}$	11	65	65	36	27	46	35	84	63
$\frac{1}{2}$	11	65	65	37	28	48	36	84	63
$\frac{9}{16}$	11	65	65	37	28	48	36	84	63
$\frac{5}{8}$	11	65	65	38	28	49	37	84	63
$\frac{3}{4}$	12	65	65	38	29	50	37	84	63
$\frac{7}{8}$	12	65	65	38	29	50	37	84	63
1	12	65	65	40	30	52	39	84	63
$1\frac{1}{8}$	12	65	65	40	30	52	39	84	63
$1\frac{1}{4}$	12	65	65	41	31	53	40	84	63
$1\frac{3}{8}$	12	65	65	41	31	53	40	84	63
$1\frac{1}{2}$	13	65	65	42	31	55	41	84	63

able C27. UNIFIED 32-THREAD SERIES (32 U.N.)

Tolerances and Allowances (Unit = 0·0001-Inch)

Nom. Size (ins.)	Screw					Nut			
	Allowance Class 2A only	Tolerances				Tolerances			
		Major Diam.		Effective Diam.		Effective Diam.		Minor Diam.	
		2A	3A	2A	3A	2B	3B	2B	3B
$\frac{1}{4}$	10	60	60	32	24	42	31	74	67
$\frac{5}{16}$	10	60	60	32	24	42	31	74	60
$\frac{3}{8}$	10	60	60	34	25	44	33	74	57
$\frac{7}{16}$	10	60	60	34	25	44	33	74	57
$\frac{1}{2}$	10	60	60	35	26	45	34	74	57
$\frac{9}{16}$	10	60	60	35	26	45	34	74	57
$\frac{5}{8}$	11	60	60	36	27	46	35	74	57
$\frac{3}{4}$	11	60	60	36	27	47	36	74	57
$\frac{7}{8}$	11	60	60	36	27	47	36	74	57
1	11	60	60	38	28	49	37	74	57

Table C28. AMERICAN NATIONAL 8-THREAD SERIES (8 N)

(CLASSES 2 and 3)

Tolerances (Unit = 0·0001-Inch)

Nominal Size (ins.)	Screw			Nut		
	Major Diameter 2 and 3	Effective Diameter 2	Effective Diameter 3	Effective Diameter 2	Effective Diameter 3	Minor Diameter 2 and 3
1	152	76	54	76	54	148
1⅛	152	79	55	79	55	148
1¼	152	83	58	83	58	148
1⅜	152	86	61	86	61	148
1½	152	90	63	90	63	148
1⅝	152	93	65	93	65	148
1¾	152	97	68	97	68	148
1⅞	152	100	70	100	70	148
2	152	104	73	104	73	148
2⅛	152	107	75	107	75	148
2¼	152	110	77	110	77	148
2½	152	117	82	117	82	148
2¾	152	124	87	124	87	148
3	152	130	92	130	92	148
3¼	152	132	93	132	93	148
3½	152	133	93	133	93	148
3¾	152	134	94	134	94	148
4	152	135	95	135	95	148
4¼	152	137	96	137	96	148
4½	152	138	97	138	97	148
4¾	152	139	98	139	98	148
5	152	140	99	140	99	148
5¼	152	141	99	141	99	148
5½	152	142	100	142	100	148
5¾	152	143	101	143	101	148
6	152	144	102	144	102	148

able C29. AMERICAN NATIONAL 12-THREAD SERIES (12 N)

(CLASSES 2 and 3)

Tolerances (Unit = 0·0001-Inch)

Nominal Size (ins.)	Screw			Nut		
	Major Diameter 2 and 3	Effective Diameter 2	Effective Diameter 3	Effective Diameter 2	Effective Diameter 3	Minor Diameter 2 and 3
$\frac{1}{2}$	112	56	40	56	40	127
$\frac{9}{16}$	112	56	40	56	40	127
$\frac{5}{8}$	112	56	40	56	40	90
$\frac{11}{16}$	112	56	40	56	40	90
$\frac{3}{4}$	112	56	40	56	40	90
$\frac{13}{16}$	112	56	40	56	40	90
$\frac{7}{8}$	112	56	40	56	40	90
$\frac{15}{16}$	112	56	40	56	40	90
1	112	56	40	56	40	90
$1\frac{1}{16}$	112	56	40	56	40	90
$1\frac{1}{8}$	112	56	40	56	40	90
$1\frac{3}{16}$	112	56	40	56	40	90
$1\frac{1}{4}$	112	56	40	56	40	90
$1\frac{5}{16}$	112	56	40	56	40	90
$1\frac{3}{8}$	112	56	40	56	40	90
$1\frac{7}{16}$	112	56	40	56	40	90
$1\frac{1}{2}$	112	56	40	56	40	90
$1\frac{5}{8}$	112	64	45	64	45	90
$1\frac{3}{4}$	112	65	46	65	46	90
$1\frac{7}{8}$	112	66	46	66	46	90
2	112	67	47	67	47	90
$2\frac{1}{8}$	112	68	48	68	48	90
$2\frac{1}{4}$	112	69	48	69	48	90
$2\frac{3}{8}$	112	70	49	70	49	90
$2\frac{1}{2}$	112	71	49	71	49	90
$2\frac{5}{8}$	112	71	50	71	50	90
$2\frac{3}{4}$	112	72	50	72	50	90
$2\frac{7}{8}$	112	73	51	73	51	90
3	112	74	51	74	51	90
$3\frac{1}{8}$	112	74	52	74	52	90
$3\frac{1}{4}$	112	75	52	75	52	90
$3\frac{3}{8}$	112	76	53	76	53	90
$3\frac{1}{2}$	112	76	53	76	53	90
$3\frac{5}{8}$	112	77	54	77	54	90
$3\frac{3}{4}$	112	78	54	78	54	90
$3\frac{7}{8}$	112	78	55	78	55	90
4	112	79	55	79	55	90
$4\frac{1}{4}$	112	80	56	80	56	90
$4\frac{1}{2}$	112	81	57	81	57	90
$4\frac{3}{4}$	112	83	58	83	58	90
5	112	84	59	84	59	90
$5\frac{1}{4}$	112	85	59	85	59	90
$5\frac{1}{2}$	112	86	60	86	60	90
$5\frac{3}{4}$	112	87	61	87	61	90
6	112	88	62	88	62	90

Table C30. AMERICAN NATIONAL 16-THREAD SERIES (16 N)
(CLASSES 2 and 3)
Tolerances (Unit = 0·0001-Inch)

Nominal Size (ins.)	Screw			Nut		
	Major Diameter 2 and 3	Effective Diameter 2	Effective Diameter 3	Effective Diameter 2	Effective Diameter 3	Minor Diameter 2 and 3
$\frac{3}{4}$	90	45	32	45	32	80
$\frac{13}{16}$	90	51	35	51	35	80
$\frac{7}{8}$	90	51	36	51	36	80
$\frac{15}{16}$	90	52	36	52	36	80
1	90	52	37	52	37	80
$1\frac{1}{16}$	90	53	37	53	37	80
$1\frac{1}{8}$	90	54	38	54	38	80
$1\frac{3}{16}$	90	54	38	54	38	80
$1\frac{1}{4}$	90	55	38	55	38	80
$1\frac{5}{16}$	90	55	39	55	39	80
$1\frac{3}{8}$	90	56	39	56	39	80
$1\frac{7}{16}$	90	56	40	56	40	80
$1\frac{1}{2}$	90	57	40	57	40	80
$1\frac{9}{16}$	90	58	40	58	40	80
$1\frac{5}{8}$	90	58	41	58	41	80
$1\frac{11}{16}$	90	58	41	58	41	80
$1\frac{3}{4}$	90	59	41	59	41	80
$1\frac{13}{16}$	90	59	42	59	42	80
$1\frac{7}{8}$	90	60	42	60	42	80
$1\frac{15}{16}$	90	60	42	60	42	80
2	90	61	43	61	43	80
$2\frac{1}{16}$	90	61	43	61	43	80
$2\frac{1}{8}$	90	62	43	62	43	80
$2\frac{3}{16}$	90	62	43	62	43	80
$2\frac{1}{4}$	90	62	44	62	44	80
$2\frac{5}{16}$	90	63	44	63	44	80
$2\frac{3}{8}$	90	63	44	63	44	80
$2\frac{7}{16}$	90	64	45	64	45	80
$2\frac{1}{2}$	90	64	45	64	45	80
$2\frac{5}{8}$	90	65	45	65	45	80
$2\frac{3}{4}$	90	66	46	66	46	80
$2\frac{7}{8}$	90	66	46	66	46	80
3	90	67	47	67	47	80
$3\frac{1}{8}$	90	68	47	68	47	80
$3\frac{1}{4}$	90	69	48	69	48	80
$3\frac{3}{8}$	90	69	48	69	48	80
$3\frac{1}{2}$	90	70	49	70	49	80
$3\frac{5}{8}$	90	71	49	71	49	80
$3\frac{3}{4}$	90	71	50	71	50	80
$3\frac{7}{8}$	90	72	50	72	50	80
4	90	72	51	72	51	80

Section D

Section D

AMERICAN TRANSLATIONAL THREADS

American Acme

The basic form of this thread is shown in Fig. 1. The proportions are

$\theta = 14$ deg. 30 mins; $2\theta = 29$ deg.; $H = 1{\cdot}9333566\,p$.
$F = 0{\cdot}3707\,p$; $F_c = 0{\cdot}3707\,p - 0{\cdot}256 \times$ Major diameter allowanc
$F_r = 0{\cdot}3707\,p - 0{\cdot}259 \times$ (Minor diameter allowance – effective diameter allowance). Allowance $= 0{\cdot}020$ for 10 tpi and coarser, and 0·01 for finer pitches.

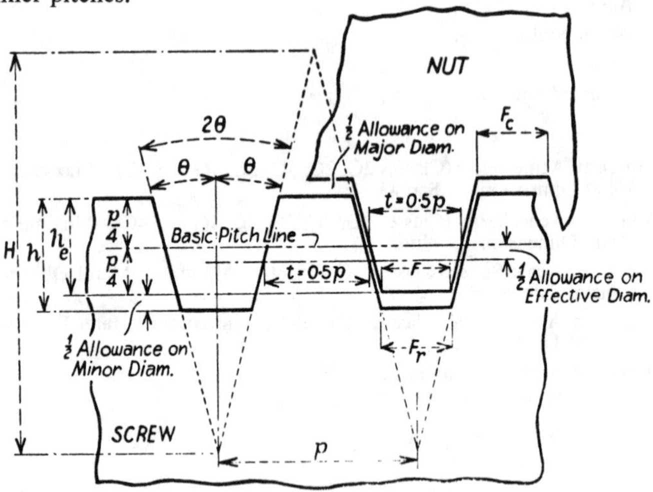

Fig. 1. Acme Basic Form

Maximum Metal Dimensions

The maximum metal dimensions to which tolerances are applied fo Acme nuts and screws for the various classes* are based on the followin formulæ, where N = nominal diameter of screw.

Screw—Maximum Major Diameter (D_s)

Classes

2G, 3G, 4G, 2C, 3C, 4C: $D_s = N$

5C and 6C: $D_s = N - 0{\cdot}025\sqrt{N}$

Screw—Maximum Effective Diameter (E_s)

Classes

2G and 2C: $E_s = N - \frac{1}{2}p - 0{\cdot}008\sqrt{N}$

3G and 3C: $E_s = N - \frac{1}{2}p - 0{\cdot}006\sqrt{N}$

4G and 4C: $E_s = N - \frac{1}{2}p - 0{\cdot}004\sqrt{N}$

5C: $E_s = N - 0{\cdot}025\sqrt{N} - 0{\cdot}008\sqrt{N} - \frac{1}{2}p$

6C: $E_s = N - 0{\cdot}025\sqrt{N} - 0{\cdot}006\sqrt{N} - \frac{1}{2}p$

*For Class 5G (B.S. 1104), see page 54.

rew—*Maximum Minor Diameter* (d_s)

Classes

2G, 3G, 4G } $d_s = N - p - 0{\cdot}020$ in. (10 t.p.i. and coarser)
2C, 3C, 4C } $d_s = N - p - 0{\cdot}010$ in. (finer than 10 t.p.i.)
5C and 6C $d_s = N - 0{\cdot}025\sqrt{N} - p - 0{\cdot}020$ in. (10 t.p.i. and coarser)
$d_s = N - 0{\cdot}025\sqrt{N} - p - 0{\cdot}010$ in. (finer than 10 t.p.i.).

Nut—Minimum Major Diameter (D_n)

Classes

2G, 3G, 4G, { $D_n = N + 0{\cdot}020$ in. (10 t.p.i. and coarser)
$D_n = N + 0{\cdot}010$ in. (finer than 10 t.p.i.)
2C, 3C, 4C: $D_n = N + 0{\cdot}001\ \sqrt{N}$
5C and 6C: $D_n = N - 0{\cdot}024\ \sqrt{N}$

Nut—Minimum Effective Diameter (E_n)

Classes

2G, 3G, 4G }
2C, 3C, 4C } $E_n = N - \frac{1}{2}p$
5C and 6C: $E_n = N - 0{\cdot}025\ \sqrt{N} - \frac{1}{2}p$

Nut—Minimum Minor Diameter (d_n)

Classes

2G, 3G, 4G: $d_n = N - p$
2C, 3C, 4C: $d_n = N - 0{\cdot}9\ p$
5C and 6C: $d_n = N - 0{\cdot}025\ \sqrt{N} - 0{\cdot}9\ p$

olerances. The tolerances of American Acme are based on the following rmulæ:

Effective Diameter (*Screw and Nut*)

Classes

4 and 6: $0{\cdot}010\ \sqrt{p} + 0{\cdot}002\ \sqrt{N}$
2: $0{\cdot}030\ \sqrt{p} + 0{\cdot}006\ \sqrt{N}$
3 and 5: $0{\cdot}014\ \sqrt{p} + 0{\cdot}0028\ \sqrt{N}$

Major Diameter—Screw

Classes

2G, 3G, 4G: $0{\cdot}05\ p$, with a minimum of 0·005 in.
2C: $0{\cdot}0035\ \sqrt{N}$
3C, 5C: $0{\cdot}0015\ \sqrt{N}$
4C, 6C: $0{\cdot}0010\ \sqrt{N}$

Major Diameter—Nut

Classes

2G, 3G, 4G: { 0·020 in. (10 t.p.i. and coarser)
0·010 in. (finer than 10 t.p.i.)
2C: $0{\cdot}0035\ \sqrt{N}$
3C, 5C: $0{\cdot}0035\ \sqrt{N}$
4C, 6C: $0{\cdot}0020\ \sqrt{N}$

Minor Diameter—Screw

Classes

2C, 2G: $0{\cdot}045\ \sqrt{p} + 0{\cdot}009\ \sqrt{N}$
3G, 3C, 5C: $0{\cdot}021\ \sqrt{p} + 0{\cdot}0042\ \sqrt{N}$
4G, 4C, 6C: $0{\cdot}015\ \sqrt{p} + 0{\cdot}003\ \sqrt{N}$

Minor Diameter—Nut

All Classes: 0·05 p, with a minimum of 0·005 in.

AMERICAN 29-DEG. STUB THREAD

This thread has a very strong section and is suitable where space limitations or other considerations make a shallow translating thread desirable

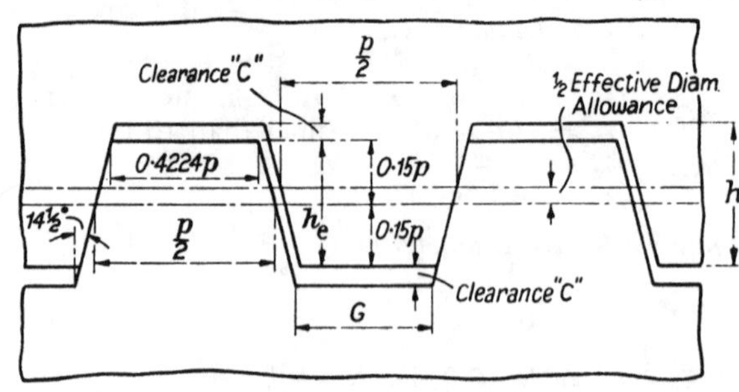

Fig. 2. American 29-deg. Stub Form

The profile is shown in Fig. 2 and has the following proportions:

Thread angle (2θ) = 29 deg.; Flank angle (θ) = 14 deg. 30 min. Depth of engagement (h_e) = 0·3 p; Depth of thread (h) = h_e + where the clearance "c" is 0·010-inch for 10 t.p.i. or coarser, and 0·005-inch for finer pitches. Flat at crest of screw = 0·4224 p; Flat at root of screw (G) = 0·4224 p – 0·52 c; To provide a clearance between the unloaded flanks, an allowance is made to the effective diameter as shown.

AMERICAN 60-DEG. STUB THREAD

The 60-deg. stub thread has the proportions shown in Fig. 3 which may be stated as follows:

Thread angle (2θ) = 60 deg.; Flank angle (θ) = 30 deg.; Depth of engagement (h_e) = 0·433 p; Depth of thread (h) = h_e + 0·02 p = 0·453 p; Flat at crest of screw = 0·250 p; Flat at root of screw = 0·227 p; To provide an axial clearance an allowance is made to the effective diameter as shown.

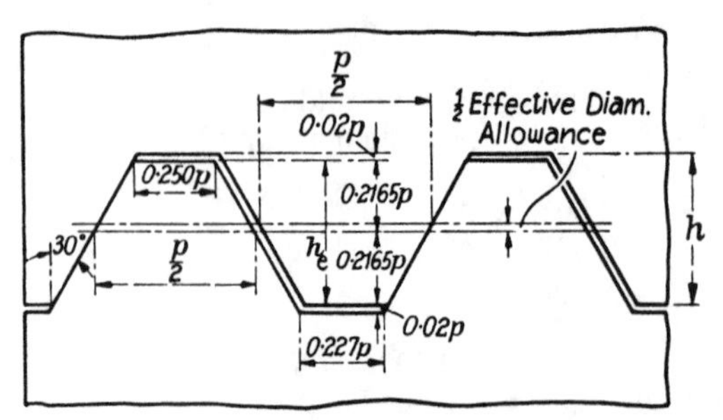

Fig. 3. American 60-deg. Stub Form

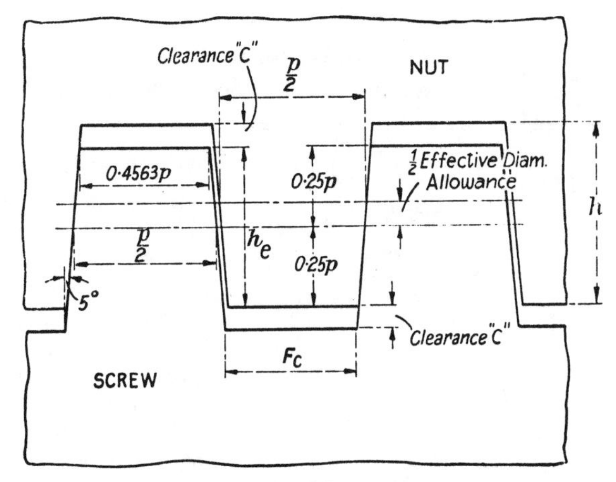

Fig. 4. American Modified Square Form

AMERICAN MODIFIED SQUARE THREAD

This thread is considered to be practically equivalent of the square thread for most applications, yet is capable of economic production. Its profile, which is shown in Fig. 4, has the following proportions:

Thread angle (2θ) = 10 deg.; Flank angle (θ) = 5 deg.; Depth of engagement (h_e) = 0·5 p; Depth of thread (h) = $h_e + c = 0{\cdot}5\ p + c$; where the clearance "$c$" is determined to suit the application of the thread assembly. Flat at crest of screw = 0·4563 p; Flat at root of screw (F_c) = 0·4563 p − 0·17 c. To provide a clearance between the unloaded flanks, an allowance is made to the effective diameter as shown.

Table D1. **AMERICAN ACME**

(CLASSES 2G, 2C, 3G, 3C, 4G, 4C)

Screws—Maximum Metal Dimensions in Inches

Nom. Diam.	Threads per inch	Major Diam. (max.) 2G, 3G, 4G 2C, 3C, 4C	Effective Diameter (max.)			Minor Diam. (max.) 2G, 3G, 4G 2C, 3C, 4C
			2G, 2C	3G, 3C	4G, 4C	
*$\frac{1}{4}$	16	0·2500	0·2148	0·2158	0·2168	0·1775
*$\frac{5}{16}$	14	0·3125	0·2728	0·2738	0·2748	0·2311
*$\frac{3}{8}$	12	0·3750	0·3284	0·3296	0·3309	0·2817
*$\frac{7}{16}$	12	0·4375	0·3909	0·3921	0·3934	0·3442
$\frac{1}{2}$	10	0·5000	0·4443	0·4458	0·4472	0·3800
$\frac{5}{8}$	8	0·6250	0·5562	0·5578	0·5593	0·4800
$\frac{3}{4}$	6	0·7500	0·6598	0·6615	0·6632	0·5633
$\frac{7}{8}$	6	0·8750	0·7842	0·7861	0·7880	0·6883
1	5	1·0000	0·8920	0·8940	0·8960	0·7800
1$\frac{1}{8}$	5	1·1250	1·0165	1·0186	1·0208	0·9050
1$\frac{1}{4}$	5	1·2500	1·1411	1·1433	1·1455	1·0300
1$\frac{3}{8}$	4	1·3750	1·2406	1·2430	1·2453	1·1050
1$\frac{1}{2}$	4	1·5000	1·3652	1·3677	1·3701	1·2300
1$\frac{3}{4}$	4	1·7500	1·6145	1·6171	1·6198	1·4800
2	4	2·0000	1·8637	1·8665	1·8693	1·7300
2$\frac{1}{4}$	3	2·2500	2·0713	2·0743	2·0773	1·8967
2$\frac{1}{2}$	3	2·5000	2·3207	2·3238	2·3270	2·1467
2$\frac{3}{4}$	3	2·7500	2·5700	2·5734	2·5767	2·3967
3	2	3·0000	2·7360	2·7395	2·7430	2·4800
3$\frac{1}{2}$	2	3·5000	3·2350	3·2388	3·2425	2·9800
4	2	4·0000	3·7340	3·7380	3·7420	3·4800
4$\frac{1}{2}$	2	4·5000	4·2330	4·2373	4·2415	3·9800
5	2	5·0000	4·7319	4·7364	4·7409	4·4800

*These sizes are in the 2G, 3G and 4G Series only.

able D2. AMERICAN ACME

(CLASSES 2G, 2C, 3G, 3C, 4G, 4C)

Nuts—Maximum Metal Dimensions in Inches

Nom. Diam.	Threads per inch	Major Diam. (min.) 2G, 3G, 4G	Effective Diam. (min.) 2G, 3G, 4G 2C, 3C, 4C	Minor Diam. (min.) 2G, 3G, 4G	Major Diam. (min.) 2C, 3C, 4C	Minor Diam. (min.) 2C, 3C, 4C
$\frac{1}{4}$	16	0·2600	0·2188*	0·1875		
$\frac{5}{16}$	14	0·3225	0·2768*	0·2411		
$\frac{3}{8}$	12	0·3850	0·3333*	0·2917		
$\frac{7}{16}$	12	0·4475	0·3958*	0·3542		
$\frac{1}{2}$	10	0·5200	0·4500	0·4000	0·5007	0·4100
$\frac{5}{8}$	8	0·6450	0·5625	0·5000	0·6258	0·5125
$\frac{3}{4}$	6	0·7700	0·6667	0·5833	0·7509	0·6000
$\frac{7}{8}$	6	0·8950	0·7917	0·7083	0·8759	0·7250
1	5	1·0200	0·9000	0·8000	1·0010	0·8200
$1\frac{1}{8}$	5	1·1450	1·0250	0·9250	1·1261	0·9450
$1\frac{1}{4}$	5	1·2700	1·1500	1·0500	1·2511	1·0700
$1\frac{3}{8}$	4	1·3950	1·2500	1·1250	1·3762	1·1500
$1\frac{1}{2}$	4	1·5200	1·3750	1·2500	1·5012	1·2750
$1\frac{3}{4}$	4	1·7700	1·6250	1·5000	1·7513	1·5250
2	4	2·0200	1·8750	1·7500	2·0014	1·7750
$2\frac{1}{4}$	3	2·2700	2·0833	1·9167	2·2515	1·9500
$2\frac{1}{2}$	3	2·5200	2·3333	2·1667	2·5016	2·2000
$2\frac{3}{4}$	3	2·7700	2·5833	2·4167	2·7517	2·4500
3	2	3·0200	2·7500	2·5000	3·0017	2·5500
$3\frac{1}{2}$	2	3·5200	3·2500	3·0000	3·5019	3·0500
4	2	4·0200	3·7500	3·5000	4·0020	3·5500
$4\frac{1}{2}$	2	4·5200	4·2500	4·0000	4·5021	4·0500
5	2	5·0200	4·7500	4·5000	5·0022	4·5500

*2G, 3G and 4G Series only.

Table D3. **AMERICAN ACME**

(CLASSES 5C and 6C)

Screws—Maximum Metal Dimensions in Inches

Nom. Diam.	Threads per inch	Major Diam. (max.) 5C, 6C	Effective Diam. (max.)		Minor Diam. (max.) 5C, 6C
			5C	6C	
$\frac{1}{2}$	10	0·4823	0·4266	0·4281	0·3623
$\frac{5}{8}$	8	0·6052	0·5364	0·5380	0·4602
$\frac{3}{4}$	6	0·7283	0·6381	0·6398	0·5416
$\frac{7}{8}$	6	0·8516	0·7608	0·7627	0·6649
1	5	0·9750	0·8670	0·8690	0·7550
$1\frac{1}{8}$	5	1·0985	0·9900	0·9921	0·8785
$1\frac{1}{4}$	5	1·2220	1·1131	1·1153	1·0020
$1\frac{3}{8}$	4	1·3457	1·2113	1·2137	1·0757
$1\frac{1}{2}$	4	1·4694	1·3346	1·3371	1·1994
$1\frac{3}{4}$	4	1·7169	1·5814	1·5840	1·4469
2	4	1·9646	1·8283	1·8311	1·6946
$2\frac{1}{4}$	3	2·2125	2·0338	2·0368	1·8592
$2\frac{1}{2}$	3	2·4605	2·2812	2·2843	2·1072
$2\frac{3}{4}$	3	2·7085	2·5285	2·5319	2·3552
3	2	2·9567	2·6927	2·6962	2·4367
$3\frac{1}{2}$	2	3·4532	3·1882	3·1920	3·9332
4	2	3·9500	3·6840	3·6880	3·4300
$4\frac{1}{2}$	2	4·4470	4·1800	4·1843	3·9270
5	2	4·9441	4·6760	4·6805	4·4241

Table D4. AMERICAN ACME

(CLASSES 5C and 6C)

Nuts—Maximum Metal Dimensions in Inches

Nom. Diam.	Threads per inch	Major Diam. (min.)	Effective Diam. (min.)	Minor Diam. (min.)
½	10	0·4830	0·4323	0·3923
⅝	8	0·6060	0·5427	0·4927
¾	6	0·7292	0·6450	0·5783
⅞	6	0·8525	0·7683	0·7016
1	5	0·9760	0·8750	0·7950
1⅛	5	1·0996	0·9985	0·9185
1¼	5	1·2231	1·1220	1·0420
1⅜	4	1·3469	1·2207	1·1207
1½	4	1·4706	1·3444	1·2444
1¾	4	1·7182	1·5919	1·4919
2	4	1·9660	1·8396	1·7396
2¼	3	2·2140	2·0458	1·9125
2½	3	2·4621	2·2938	2·1605
2¾	3	2·7102	2·5418	2·4085
3	2	2·9584	2·7067	2·5067
3½	2	3·4551	3·2032	3·0032
4	2	3·9520	3·7000	3·5000
4½	2	4·4491	4·1970	3·9970
5	2	4·9463	4·6941	4·4941

Table D5. AMERICAN ACME

(Tolerances in Units of 0·0001-Inch)

Size	Screw Major Diameter				Screw Minor Diameter			Nut Major Diameter			Nut Minor Diam.	Screw and Nut Effective Diameter		
	2G, 3G, 4G	2C	3C, 5C	4C, 6C	2G, 2C	3G, 3C, 5C	4G, 4C, 6C	2G, 3G, 4G	2C, 3C, 5C	4C, 6C	All	4G, 4C, 6C	2G, 2C	3C, 3G, 5C
*1/4	50				157	73	53	100			50	35	105	49
*5/16	50				171	80	57	100			50	38	114	53
*3/8	50				185	87	62	100			50	41	123	58
*7/16	50				189	88	63	100			50	42	126	59
1/2	50	25	11	7	206	96	69	200	25	14	50	46	137	64
5/8	62	28	12	8	230	107	77	200	28	16	62	51	154	72
3/4	83	30	13	9	262	122	87	200	30	17	83	58	174	81
7/8	83	33	14	9	268	125	89	200	33	19	83	60	179	83
1	100	35	15	10	291	136	97	200	35	20	100	65	194	91
1 1/8	100	37	16	11	297	138	99	200	37	21	100	66	198	92
1 1/4	100	39	17	11	302	141	101	200	39	22	100	67	201	94
1 3/8	125	41	18	12	331	154	110	200	41	23	125	73	220	103
1 1/2	125	43	18	12	335	156	112	200	43	24	125	74	223	104
1 3/4	125	46	20	13	344	160	115	200	46	26	125	76	229	107
2	125	49	21	14	352	164	117	200	49	28	125	78	235	110
2 1/4	167	52	22	15	395	184	132	200	52	30	167	88	263	123
2 1/2	167	55	24	16	402	188	134	200	55	32	167	89	268	125
2 3/4	167	58	25	17	409	191	136	200	58	33	167	91	273	127
3	250	61	26	17	474	221	158	200	61	35	250	105	316	147
3 1/2	250	65	28	19	486	226	162	200	65	37	250	108	324	151
4	250	70	30	20	498	232	166	200	70	40	250	111	332	155
4 1/2	250	74	32	21	509	237	169	200	74	42	250	113	339	158
5	250	78	34	22	519	242	173	200	78	45	250	115	346	162

Table D6. AMERICAN 29-DEG. STUB THREAD

Basic Dimensions in Inches

Threads per inch	Pitch	Depth of Engagement	Depth of Thread	Thread Thickness (Basic)	Flat at Crest of Screw	Flat at Root of Screw
16	0·06250	0·0188	0·0238	0·0313	0·0264	0·0238
14	0·07143	0·0214	0·0264	0·0357	0·0302	0·0276
12	0·08333	0·0250	0·0300	0·0417	0·0352	0·0326
10	0·10000	0·0300	0·0400	0·0500	0·0422	0·0370
9	0·11111	0·0333	0·0433	0·0556	0·0469	0·0417
8	0·12500	0·0375	0·0475	0·0625	0·0528	0·0476
7	0·14286	0·0429	0·0529	0·0714	0·0603	0·0551
6	0·16667	0·0500	0·0600	0·0833	0·0704	0·0652
5	0·20000	0·0600	0·0700	0·1000	0·0845	0·0793
4	0·25000	0·0750	0·0850	0·1250	0·1056	0·1004
3½	0·28571	0·0857	0·0957	0·1429	0·1207	0·1155
3	0·33333	0·1000	0·1100	0·1667	0·1408	0·1356
2½	0·40000	0·1200	0·1300	0·2000	0·1690	0·1638
2	0·50000	0·1500	0·1600	0·2500	0·2112	0·2060

Table D7. AMERICAN 60-DEG. STUB THREAD

Basic Dimensions in Inches

Threads per inch	Pitch	Depth of Engagement	Depth of Thread	Thread Thickness (Basic)	Flat at Crest of Screw	Flat at Root of Screw
16	0·06250	0·0271	0·0283	0·0313	0·0156	0·0142
14	0·07143	0·0309	0·0324	0·0357	0·0179	0·0162
12	0·08333	0·0361	0·0378	0·0417	0·0208	0·0189
10	0·10000	0·0433	0·0453	0·0500	0·0250	0·0227
9	0·11111	0·0481	0·0503	0·0556	0·0278	0·0252
8	0·12500	0·0541	0·0566	0·0625	0·0313	0·0284
7	0·14286	0·0619	0·0647	0·0714	0·0357	0·0324
6	0·16667	0·0722	0·0755	0·0833	0·0417	0·0378
5	0·20000	0·0866	0·0906	0·1000	0·0500	0·0454
4	0·25000	0·1083	0·1133	0·1250	0·0625	0·0567

Section E

Section E

AMERICAN PIPE THREADS

American Taper Pipe Threads

The pipe threads used in the U.S.A. are based on the original Brigg system and designed for American standard dimensions of pipe. The series in use are:

(1) National Taper Pipe Threads
(2) Dryseal Pipe Threads
(3) A.P.I. (American Petroleum Institute) Line Pipe Threads.

Many of the specifying basic dimensions are common to all three, and have been conveniently collected in Table E1, where the existing difference have been indicated. The standard taper is 1 in 16 in. diameter. The thread form has the following proportions:

Thread angle (2θ) = 60 deg.; Flank angle (θ) = 30 deg.; Triangular height (H) = $0{\cdot}8660\,p$. Both flanks are equally inclined to the pipe axis.

National Taper Pipe Thread (N.P.T.)

Basic dimensions are derived from the following formulæ:

$h = 0{\cdot}800\,p$;
$E_0 = D_p - (0{\cdot}050\,D_p + 1{\cdot}1)\,p$;
$E_1 = E_0 + 0{\cdot}0625\,L_1$;
$L_2 = (0{\cdot}8\,D_p + 6{\cdot}8)\,p$.

Where E_0 = Effective diameter of thread at end of pipe;
E_1 = Effective diameter at large end of coupling;
L_1 = Hand-engagement length;
L_2 = Length of effective or full thread on pipe;
D_p = The basic outside diameter of pipe related to the nominal size of thread as shown in Table E2.

The limits of truncation for N.P.T. are given in Table E3 and limits of thread depth in Table E4.

S.A.E. Dryseal Pipe Threads

The amount of truncation applied to the fundamental triangle varies from pitch to pitch as shown in Table E5, hence the depth of thread cannot be expressed as a constant fraction of p. Flat roots and crests are specified, although commercially produced crests and roots may be round, so long as the profile lies between the truncation limits specified.

The Dryseal Series are:

Dryseal Standard Taper Pipe Thread: N.P.T.F.
,, S.A.E. Short Taper Pipe Thread: P.T.F.—S.A.E. Short.
,, Fuel Internal Straight Pipe Thread: N.P.S.F.
,, Intermediate Straight Pipe Thread: N.P.S.I.

N.P.T.F.

The basic dimensions are derived from the same formulæ as the National Taper Pipe Threads (*q.v.*).

.T.F.—S.A.E. SHORT

$E_0 = D_p - (0{\cdot}05\ D_p + 1{\cdot}037)\ p;$
$E_1 = E_0 + 0{\cdot}0625\ L_1;$
$L_2 = (0{\cdot}8\ D_p + 5{\cdot}8)\ p.$

Dimensions for the assembly between the S.A.E. Dryseal series are iven in Tables E6 and E7.

..P.I. LINE PIPE THREAD

The amount of truncation is 0·073 p at the crest and 0·033 p at the ɔots of both internal and external threads, leaving a thread depth of = 0·760 p.

The basic depths of threads for the six pitches used are:

t.p.i. =	27;	18;	14;	11½;	8.
h =	0·0281;	0·0422;	0·0543;	0·0661;	0·0950.

nd the amounts of truncation are:

t.p.i.	27;	18;	14;	11½;	8
crest:	0·0027;	0·0041;	0·0052;	0·0063;	0·0091 (in.)
root:	0·0012;	0·0018;	0·0024;	0·0029;	0·0041 (in.)

Sizes of 1 inch nominal and larger have the following tolerances:

Lead: ± 0·003 inch per inch and ± 0·006 inch accumulative.
Depth of thread: plus 0·002 minus 0·006 inch.
Angle (Included): plus or minus 1·5 deg.
Taper per foot on diameter: plus 0·0625, minus 0·0312 inch.

Gauging for effective diameter is effected at the "Hand-tight plane" ·hich therefore becomes the "Gauge Plane."

.MERICAN PARALLEL PIPE THREAD.

A parallel version of the American Standard Pipe Thread is used for ıechanical joints where fluid-tightness is not normally required. Two rades of fit are specified, viz. free-fitting and loose-fitting with the imensions as given in Tables E8 and E9.

Table E1.

AMERICAN NATIONAL TAPER PIPE: DRYSEAL TAPER PIPE AND A.P.I. LINE PIPE THREAD

Basic Dimensions (Inches)

Nom. Size (ins.)	Threads per inch	Hand Engage-ment Length§	Length of Full Thread		Effective Diameter at	
			Pipe	Coupling ‡	End of Pipe	End of Coupling‖
1/16†	27	0·160	0·2611	0·2711	0·27118	0·28118
1/8 N.P.T.	27	0·1615	0·2639		0·36351	0·37360
1/8 Dryseal & A.P.I.	27	0·1615	0·2639	0·2726	0·36351	0·37360
1/4 N.P.T.	18	0·2278	0·4018		0·47739	0·49163
1/4 Dryseal & A.P.I.	18	0·2278	0·4018	0·3945	0·47739	0·49163
3/8	18	0·240	0·4078	0·4067	0·61201	0·62701
1/2	14	0·320	0·5337	0·5343	0·75843	0·77843
3/4	14	0·339	0·5457	0·5533	0·96768	0·98887
1	11½	0·400	0·6828	0·6609	1·21363	1·23863
1¼	11½	0·420	0·7068	0·6809	1·55713	1·58338
1½	11½	0·420	0·7235	0·6809	1·79609	1·82234
2	11½	0·436	0·7565	0·6969	2·26902	2·29627
2½	8	0·682	1·1375	1·0570	2·71953	2·76216
3	8	0·766	1·2000	1·1410	3·34063	3·38850
3½*	8	0·821	1·2500		3·83750	3·88881
4*	8	0·844	1·3000		4·33438	4·38713
5*	8	0·937	1·4063		5·39073	5·44929
6*	8	0·958	1·5125		6·44609	6·50597
8*	8	1·063	1·7125		8·43359	8·50003
10*	8	1·210	1·9250		10·54531	10·62094
12*	8	1·360	2·1250		12·53281	12·61781
14*	8	1·562	2·2500		13·77500	13·87263
16*	8	1·812	2·4500		15·76250	15·87575
18*	8	2·000	2·6500		17·75000	17·87500
20*	8	2·125	2·8500		19·73750	19·87031
24*†	8	2·375	3·2500		23·71250	23·86094

*Not included in the Dryseal Series. †Not included in the A.P.I. Line Pipe Series. ‡Specified for Dryseal only. §Also known as Length: End of pipe to handtight Plane in A.P.I. specifications. ‖Known as "Pitch diameter at handtight plane" in A.P.I. Specifications.

Table E2. EXTERNAL DIAMETER OF PIPE AND THE CORRESPONDING NOMINAL SIZE OF N.P.T. THREAD

(Dimensions in Inches)

Nom. Size	D_p	Nom. Size	D_p	Nom. Size	D_p	Nom. Size	D_p
$\frac{1}{16}$	0·3125	1	1·315	$3\frac{1}{2}$	4·000	12	12·750
$\frac{1}{8}$	0·405	$1\frac{1}{4}$	1·660	4	4·500	14	14·000
$\frac{1}{4}$	0·540	$1\frac{1}{2}$	1·900	5	5·563	16	16·000
$\frac{3}{8}$	0·675	2	2·375	6	6·625	18	18·000
$\frac{1}{2}$	0·840	$2\frac{1}{2}$	2·875	8	8·625	20	20·000
$\frac{3}{4}$	1·050	3	3·500	10	10·750	24	24·000

Table E3. NATIONAL TAPER PIPE THREAD (N.P.T.) LIMITS OF TRUNCATION

(Dimensions in Inches)

Threads per inch	Truncation at Crest and Root	
	Minimum	Maximum
27	0·0012	0·0036
18	0·0018	0·0049
14	0·0024	0·0056
$11\frac{1}{2}$	0·0029	0·0063
8	0·0041	0·0078

Table E4. NATIONAL TAPER PIPE THREAD LIMITS OF DEPTH OF THREAD

(Dimensions in Inches)

Threads per inch	Depth of Thread (h)	
	Maximum	Minimum
27	0·02963	0·02496
18	0·04444	0·03833
14	0·05714	0·05071
$11\frac{1}{2}$	0·06957	0·06261
8	0·10000	0·09275

Table E5. DRYSEAL THREADS—LIMITS OF TRUNCATION (Dimensions in Inches)

Threads per inch	Triangular Height H (ins.)	Truncation			
		At Crest		At Root	
		Min.	Max.	Min.	Max.
27	0·03208	0·0017	0·0035	0·0035	0·0052
18	0·04811	0·0026	0·0043	0·0043	0·0061
14	0·06186	0·0026	0·0043	0·0043	0·0061
11½	0·07531	0·0035	0·0052	0·0052	0·0078
8	0·10825	0·0052	0·0069	0·0069	0·0095

Table E6. ASSEMBLY DIMENSIONS OF (P.T.F.—S.A.E. SHORT) EXTERNAL THREADS WITH (N.P.S.I.) INTERNAL (Dimensions in Inches)

Nom. Size	Threads per inch	Effective Diameter of Pipe	Hand Engagement Length	Length of Full Thread	
				Pipe	Coupling
1/16	27	0·27349	0·1230	0·2241	0·2711
⅛	27	0·36582	0·1244	0·2268	0·2726
¼	18	0·48086	0·1722	0·3462	0·3945
⅜	18	0·61548	0·1844	0·3522	0·4067
½	14	0·76289	0·2486	0·4623	0·5343
¾	14	0·97214	0·2676	0·4743	0·5533
1	11½	1·21906	0·3130	0·5958	0·6609
1¼	11½	1·56256	0·3330	0·6198	0·6809
1½	11½	1·80152	0·3330	0·6365	0·6809
2	11½	2·27445	0·3490	0·6695	0·6969
2½	8	2·72734	0·5570	1·0125	1·0570
3	8	3·34844	0·6410	1·0750	1·1410

able E7. ASSEMBLY DIMENSIONS OF (P.T.F.—S.A.E.) SHORT ꟷTERNAL THREADS WITH (N.P.T.F.) EXTERNAL THREADS

(Dimensions in Inches)

Nom. Size	Threads per inch	Effective Diameter at End of: Pipe	Effective Diameter at End of: Coupling	Hand Engagement Length	Length of Full Thread Coupling
1/16	27	0·27118	0·27887	0·1230	0·2341
1/8	27	0·36351	0·37129	0·1244	0·2356
1/4	18	0·47739	0·48815	0·1722	0·3389
3/8	18	0·61201	0·62354	0·1844	0·3511
1/2	14	0·75843	0·77397	0·2486	0·4629
3/4	14	0·96768	0·98441	0·2676	0·4819
1	11½	1·21363	1·23320	0·3130	0·5739
1¼	11½	1·55713	1·57795	0·3330	0·5939
1½	11½	1·79609	1·81691	0·3330	0·5939
2	11½	2·26902	2·29084	0·3490	0·6099
2½	8	2·71953	2·75435	0·5570	0·9320
3	8	3·34063	3·38069	0·6410	1·0169

able E8. AMERICAN PARALLEL PIPE THREAD FOR FREE-FITTING MECHANICAL JOINTS

(Dimensions in Inches)

Nom. Size	Pipe				Coupling			
	Major Diameter		Effective Diameter		Minor Diameter		Effective Diameter	
	Max.	Min.	Max.	Min.	Max.	Min.	Max.	Min.
1/8	0·397	0·390	0·3725	0·3689	0·364	0·358	0·3783	0·3736
1/4	0·526	0·517	0·4903	0·4859	0·481	0·468	0·4974	0·4916
3/8	0·662	0·653	0·6256	0·6211	0·612	0·603	0·6329	0·6270
1/2	0·823	0·813	0·7769	0·7718	0·759	0·747	0·7851	0·7784
3/4	1·034	1·024	0·9873	0·9820	0·970	0·958	0·9958	0·9889
1	1·293	1·281	1·2369	1·2311	1·211	1·201	1·2462	1·2386
1¼	1·638	1·626	1·5816	1·5756	1·555	1·546	1·5912	1·5834
1½	1·877	1·865	1·8205	1·8144	1·794	1·785	1·8302	1·8223
2	2·351	2·339	2·2944	2·2882	2·268	2·259	2·3044	2·2963
2½	2·841	2·826	2·7600	2·7526	2·727	2·708	2·7720	2·7622
3	3·467	3·452	3·3862	3·3786	3·353	3·334	3·3984	3·3885
3½	3·968	3·953	3·8865	3·8788	3·848	3·835	3·8988	3·8888
4	4·466	4·451	4·3848	4·3771	4·346	4·333	4·3971	4·3871
5	5·528	5·513	5·4469	5·4390	5·408	5·395	5·4598	5·4493
6	6·585	6·570	6·5036	6·4955	6·464	6·452	6·5165	6·5060

Table E9. AMERICAN PARALLEL PIPE THREAD FOR BACK NUT CONNECTIONS (LOOSE FITTING MECHANICAL JOINTS
(Dimensions in Inches)

Nom. Size	Threads per inch	Pipe			Coupling		
		Major Diam.	Effective Diameter		Minor Diam.	Effective Diameter	
		Max.	Max.	Min.	Min.	Max.	Min.
⅛	27	0·409	0·3840	0·3805	0·362	0·3898	0·3863
¼	18	0·541	0·5038	0·4986	0·470	0·5125	0·5073
⅜	18	0·678	0·6409	0·6357	0·607	0·6496	0·6444
½	14	0·844	0·7963	0·7896	0·753	0·8075	0·8008
¾	14	1·054	1·0067	1·0000	0·964	1·0179	1·0112
1	11½	1·318	1·2604	1·2523	1·208	1·2739	1·2658
1¼	11½	1·663	1·6051	1·5970	1·553	1·6187	1·6106
1½	11½	1·902	1·8441	1·8360	1·792	1·8576	1·8495
2	11½	2·376	2·3180	2·3099	2·265	2·3315	2·3234
2½	8	2·877	2·7934	2·7817	2·718	2·8129	2·8012
3	8	3·503	3·4198	3·4081	3·344	3·4393	3·4276
3½	8	4·003	3·9201	3·9084	3·845	3·9396	3·9279
4	8	4·502	4·4184	4·4067	4·343	4·4379	4·4262
5	8	5·564	5·4805	5·4688	5·405	5·5001	5·4884
6	8	6·620	6·5372	6·5255	6·462	6·5567	6·5450
8	8	8·615	8·5313	8·5196	8·456	8·5508	8·5391
10	8	10·735	10·6522	10·6405	10·577	10·6717	10·6600
12	8	12·732	12·6491	12·6374	12·574	12·6686	12·6569

Section F

Section F

A.P.I. CASING AND STANDARD TUBING

The series of casing and tubing threads specified by the America Petroleum Institute (excluding A.P.I. line pipe which is dealt with Section E) are as follows:

1. Casing, Short Thread
2. ,, Long Thread
3. Tubing —Non-upset
4. ,, —External Upset.

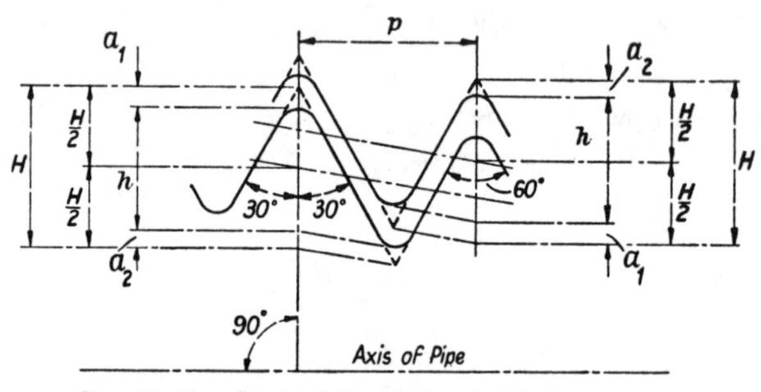

Fig. 1. A.P.I. Casing and Tubing Thread Form

The form of all these threads, is as shown in Fig. 1, which is known the "round thread" with proportions as follows:

Round Thread
$H = 0{\cdot}866\ p$; $h = 0{\cdot}626\ p - 0{\cdot}007$ inch;
a_1 (shortening at crest) $= 0{\cdot}120\ p + 0{\cdot}005$ inch; a_2 (shortening at roo $= 0{\cdot}120\ p + 0{\cdot}002$ inch. These have the values:
8 *t.p.i.*: $H = 0{\cdot}10825$ in.; $h = 0{\cdot}07125$ in.; $a_1 = 0{\cdot}0200$ in $a_2 = 0{\cdot}01700$ in.
10 *t.p.i.*: $H = 0{\cdot}08660$ in.; $h = 0{\cdot}05560$ in.; $a_1 = 0{\cdot}01700$ in $a_2 = 0{\cdot}01400$ in.

For all these threads the taper is 1 in 16 on diameter.

Tolerances
Lead: plus 0·003 in. minus 0·003 in. per inch
,, 0·006 in. ,, 0·006 in. cumulative.
Height: plus 0·002 in., minus 0·004 in.
Thread angle: plus or minus $1\frac{1}{2}$ deg.
Taper on diameter: plus $\frac{1}{16}$ in. minus $\frac{1}{32}$ in.

Table F1. A.P.I. STANDARD CASING—LONG THREAD
8 Threads per Inch all Sizes

Outside Diam. (ins.)	Wall Thickness (ins.)	End of Pipe to Handtight Plane (ins.)	Effect. Diam. at Handtight Plane (ins.)	Total Length to Vanish Point (ins.)
4½	0·250	1·921	4·40337	3·000
4½	0·290	”	”	”
5	0·253	2·296	4·90337	3·375
”	0·296	”	”	”
”	0·362	”	”	”
5½	0·275	2·421	5·40337	3·500
”	0·304	”	”	”
”	0·361	”	”	”
”	0·415	”	”	”
6⅝	0·288	2·796	6·52837	3·875
”	0·352	”	”	”
”	0·417	”	”	”
”	0·475	”	”	”
7	0·317	2·921	6·90337	4·000
”	0·362	”	”	”
”	0·408	”	”	”
”	0·453	”	”	”
”	0·498	”	”	”
”	0·540	”	”	”
7⅝	0·328	2·979	7·52418	4·125
”	0·375	”	”	”
”	0·430	”	”	”
”	0·500	”	”	”
8⅝	0·352	3·354	8·52418	4·500
”	0·400	”	”	”
”	0·450	”	”	”
”	0·500	”	”	”
”	0·557	”	”	”
9⅝	0·352	3·604	9·52418	4·750
”	0·395	”	”	”
”	0·435	”	”	”
”	0·472	”	”	”
”	0·545	”	”	”

Table F2. A.P.I. STANDARD CASING—SHORT THREAD
8 Threads per Inch all Sizes

Outside Diam. (ins.)	Wall Thickness (ins.)	End of Pipe to Handtight Plane (ins.)	Effect. Diam. at Handtight Plane (ins.)	Total Length to Vanish Point (ins.)
4½	0·205	0·921	4·40337	2·000
,,	Others	1·546	,,	2·625
5	0·220	1·421	4·90337	2·500
,,	Others	1·671	,,	2·750
5½	All	1·796	5·40337	2·875
6⅝	All	2·046	6·52837	3·125
7	0·231	1·296	6·90337	2·375
,,	Others	2·046	,,	3·125
7⅝	All	2·104	7·52418	3·250
8⅝	0·264	1·854	8·52418	3·000
,,	Others	2·229	,,	3·375
9⅝	All	,,	9·52418	,,
10¾	0·279	1·604	10·64918	2·750
,,	Others	2·354	,,	3·500
11¾	All	,,	11·64918	,,
13⅜	All	,,	13·27418	,,
16	All	2·854	15·89918	4·000
20	0·438	,,	19·89918	,,

Table F2 *continued.* A.P.I. STANDARD CASING—SHORT THREAD
8 Threads per Inch all Sizes

Outside Diam. (ins.)	Range of Wall Thicknesses (ins.)	Outside Diam. (ins.)	Range of Wall Thicknesses (ins.)
4½	0·205, 0·224, 0·250	9⅝	0·312, 0·352, 0·395
5	0·220, 0·253, 0·296	10¾	0·279, 0·350, 0·400, 0·450, 0·495
5½	0·244, 0·275, 0·304		
6⅝	0·288, 0·352	11¾	0·333, 0·375, 0·435, 0·489
7	0·231, 0·272, 0·317, 0·362	13⅜	0·330, 0·380, 0·430, 0·480, 0·514
7⅝	0·300, 0·328		
8⅝	0·264, 0·304, 0·352, 0·400	16	0·375, 0·438, 0·495
		20	0·438

Table F3. A.P.I. STANDARD TUBING—NON-UPSET

Outside Diam. (ins.)	Threads per inch	Wall Thickness (ins.)	End of Pipe to Hand-tight Plane (ins.)	Effect. Diam. at Hand-tight Plane (ins.)	Total Length to Vanish Point (ins.)
1·900	10	0·145	0·729	1·83826	1·375
$2\frac{3}{8}$	10	0·167	0·979	2·31326	1·625
$2\frac{3}{8}$	10	0·190	0·979	2·31326	1·625
$2\frac{3}{8}$	10	0·254	0·979	2·31326	1·625
$2\frac{7}{8}$	10	0·217	1·417	2·81326	2·063
$2\frac{7}{8}$	10	0·308	1·417	2·81326	2·063
$3\frac{1}{2}$	10	0·216	1·667	3·43826	2·313
$3\frac{1}{2}$	10	0·254	1·667	3·43826	2·313
$3\frac{1}{2}$	10	0·289	1·667	3·43826	2·313
$3\frac{1}{2}$	10	0·375	1·667	3·43826	2·313
4	8	0·226	1·591	3·91395	2·375
$4\frac{1}{2}$	8	0·271	1·779	4·41395	2·563

Table F4. A.P.I. STANDARD TUBING—EXTERNAL UPSET

Outside Diam. (ins.)	Threads per inch	Wall Thickness (ins.)	End of Pipe to Hand-tight Plane (ins.)	Effect. Diam. at Hand-tight Plane (ins.)	Total Length to Vanish Point (ins.)
1·050	10	0·113	0·479	1·25328	1·125
1·315	10	0·133	0·604	1·40706	1·250
1·660	10	0·140	0·729	1·75079	1·375
1·900	10	0·145	0·792	2·03206	1·438
$2\frac{3}{8}$	8	0·190	1·154	2·50775	1·938
$2\frac{3}{8}$	8	0·254	1·154	2·50775	1·938
$2\frac{7}{8}$	8	0·217	1·341	3·00775	2·125
$2\frac{7}{8}$	8	0·308	1·341	3·00775	2·125
$3\frac{1}{2}$	8	0·254	1·591	3·66395	2·375
$3\frac{1}{2}$	8	0·375	1·591	3·66395	2·375
4	8	0·262	1·716	4·16395	2·500
$4\frac{1}{2}$	8	0·271	1·841	4·66395	2·625

Section G

Section G

CONTINENTAL FORMS AND SERIES

French Metric Thread Form

The basic form of thread for screws and nuts is illustrated in Fig. 1. I the screw, the fundamental triangle is truncated by an amount $\frac{1}{8}H$ at tl crest and an amount of $\frac{1}{16}H$ at the root. Similarly, the fundament triangle of the nut is truncated by $\frac{1}{16}H$ at the root of the internal threa and by $\frac{1}{8}H$ at the crest. A radial clearance of $\frac{1}{16}H$ is therefore provide in each instance. The root space of the screw may be rounded to r = 0·054 p.

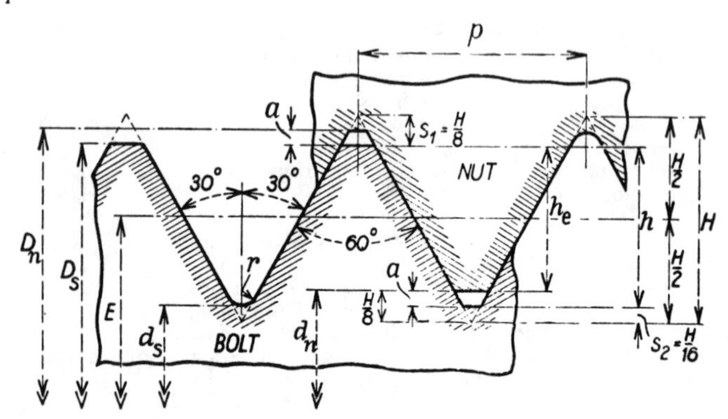

Fig. 1. French Metric Thread—Basic Form

The proportions of the thread are:

H=0·866 p; Depth of Engagement (h_e) = $\frac{3}{4}H$ = 0·650 p: Depth Thread (h) = $h_e + a$ = 0·704 p; Radial Clearance (a) = 0·054 Major Diameter of Screw (D_s) = Nominal Diameter; Effecti Diameter (E) = $D_s - \frac{3}{4}H = D_s - h$; Minor Diameter of Screw (d_s) = $D_s - 2\ (h_e + a)$; Major Diameter of Nut (D_n) = ($D_s + 2a$); Min Diameter of Nut (d_n) = $D_s - 2h_e$.

French Metric Screw Series

These series may be classified as follows:

(1) A series of 30 sizes with diameters ranging from 3·0 to 60 mm (Table G1).

(2) Six uniform pitch series with pitches of 0·75. 1·0, 1·5, 2·0, 3·0, 4 and 6·0 mm. The full comprehensive range of diameters covered b these series is: 6, 7, 8, 9, 10, 11, 12, 14, 16, 18, 20, 22, 24, 27, 30, 3 36, 39, 42, 45, 48, 52, 56, 60, 64, 68, 72, 76, 80, 85 and thenc upwards with diameters ending in the figure "5". The commencin diameters for the various series in this sequence are:

pitch:	0·75;	1·0;	1·5;	2·0;	3·0;	4·0;	6·0.
diameter:	6·0;	6·0;	10·0;	24·0;	36·0;	56·0;	64·0.

RENCH AUTOMOBILE THREADS

These threads are based on a modification of the standard 60-deg ietric form. In the thread of the screw, the fundamental triangle is runcated by $\frac{1}{8}H$ at the crest and the root, the root being rounded by a $\frac{1}{8}H$ adius. In the thread of the nut, the truncation at the root of the major iameter is $\frac{1}{12}H$ and is rounded by a $\frac{1}{12}H$ radius. The truncation of the rest at the minor diameter is $\frac{3}{16}H$. All crests are flat. The proportions re therefore as follows:

$H = 0{\cdot}866\ p$; Depth of engagement $(h) = \frac{11}{16}H = 0{\cdot}595\ p$; Depth of thread of Screw $= \frac{3}{4}H$; Depth of thread of Nut $= \frac{35}{48}H$; Clearance at root of Screw $= H/16$; Clearance at root of Nut $= H/24$,

Major Diameter of Screw (D) = Nominal Diameter.
Effective Diameter $(E) = D - \frac{3}{4}H$.
Minor Diameter of Screw $= D - 1{\cdot}5H$.
Major Diameter of Nut $= D + \frac{1}{12}H$.
Minor Diameter of Nut $= D - 1\frac{3}{8}H = D - 1{\cdot}190\ p$.

Two series of threads are specified "Pas Normal" and "Pas Fins" vith the diameters and pitches as shown in Table G3. Another table, G2, ives the constants for calculating the various basic diameters from the ominal diameter for the range of pitches.

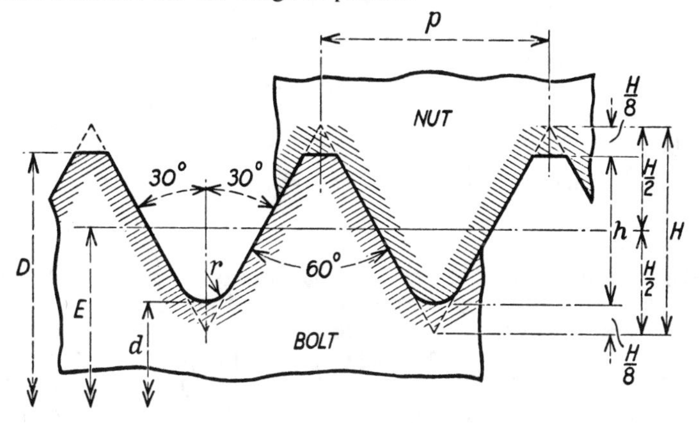

Fig. 2. German Metric Thread—Basic Form

GERMAN METRIC SERIES

The normal series of German metric threads have the basic form hown in Fig. 2, with the specified proportions:

Triangular Height $(H) = 0{\cdot}8660\ p$; Depth of thread $(h) = 0{\cdot}6495\ p$; Effective Diameter $(E) = D - h$; Minor Diameter $(d) = D - 2h$; Radius at Root $(r) = \frac{1}{8}H = 0{\cdot}1082\ p$.

Basic dimensions are given in Table G4. The sizes below 1 mm. are ised for instrument and horological work and may be found in Section H. 'olerances for these threads are given in Table G5.

GERMAN METRIC UNIFORM PITCH SERIES (FEINGEWINDE)

Series of screws of metric thread form and uniform pitch are standard-ised into series having pitches of 0·35, 0·5, 0·75, 1·0, 1·5, 2·0, 3·0, 4·0 and 6·0 mm., the thread form being as shown in Fig. 2. The basic dimensions are derived from the relations:

Triangular height $(H) = 0{\cdot}8660\ p$; Depth of thread $(h) = 0{\cdot}6495\ p$; Effective Diameter $(E) = D - h$; Minor Diameter $(d) = D - 2h$; and Radius at root $(r) = H/8 = 0{\cdot}1082\ p$.

0·5 MM. PITCH SERIES (METRISCHES FEINGEWINDE D.I.N. 519)

The depth of thread throughout the series is $h = 0{\cdot}325$ mm.; and the radius at root, $r = 0{\cdot}05$ mm. The series comprises 50 tabulated nominal sizes, commencing with 3·5 mm. nominal diameter and progressing in 0·5 mm. steps until 15 mm. diameter. Thence the series progresses in increments of 1·0 mm. A number of preferred sizes are indicated as follows:

1*st* Preference mm.: 5, 6, 8, 10, 12, 16, 20, 24, 30, 36, 42 and 48.
2*nd* ,, mm.: 3·5, 4·5, 7, 9, 11, 14, 18, 22, 27, 33, 39 and 45.
3*rd* ,, mm.: 13, 15, 17, 19, 21, 23, 25, 26, 28, 32, 34, 35, 38, 4[illegible] and 50.

Sizes which it is recommended should be avoided if possible are:
mm. 5·5, 6·5, 7·5, 8·5, 9·5, 10·5, 11·5, 12·5, 13·5, 14·5

The basic dimensions are obtained as follows:
Major Diameter (D) = Nominal Diameter.
Effective Diameter $(E) = D - 0{\cdot}325$ mm.
Minor Diameter $(d) = D - 0{\cdot}650$ mm.

1·5 MM. PITCH SERIES (METRISCHES FEINGEWINDE D.I.N. 516)

The depth of thread is 0·974 mm., and the radius at the root equal to 0·16 mm.

The series comprises tabulated nominal sizes commencing at 11·5 mm and extending up to 300 mm. with an indicated possible extension up to 500 mm. The sizes tabulated are as follows:

(11·5). 12·0. (12·5), 13, (13·5), 14·0, (14·5), 15, 16, 17, 18, 19, 20, 2[illegible] 22, 23, 24, 25, 26, 27, 28, 30, 32, 33, 34, 35, 36, 38, 39, 40, 42, 45, 4[illegible] 50, 52, 55, 56, 58, 60, 62, 64, 65, 68, 70, 72, 75, 76, 78, 80, 82, 85, 8[illegible] thence all the dimensions represented by a whole number of mm. ending in the figures 0, 2, 5 and 8. It is recommended that wherever possible, the bracketed sizes should be avoided.

The basic dimensions are obtained from:
Major Diameter (D) = Nominal Diameter;
Effective Diameter $(E) = D - 0{\cdot}974$ mm.;
Minor Diameter $(d) = D - 1{\cdot}948$ mm.

2 MM. PITCH SERIES (METRISCHES FEINGEWINDE D.I.N. 247)

The depth of thread is 1·299 mm. and the radius at the root, 0·22 mm. The series comprises 128 sizes in fairly regular sequence with an indicated extension of a further 200.

The sizes tabulated are as follows:

From 17 mm. to 28 in steps of 1·0 mm., then the sizes 30, 32, 33, 34, 35, 36, 38, 39, 40, 42, 45, 48, 50, 52, 55, 56, 58, 60, 62, 64, 65, 68, 70, 72, 75, 76, 78, 80, 82, 85, 88, thence in all whole numbers of mm. ending in 0, 2, 5 or 8, up to 300 mm.

Further extension is indicated up to 500 mm.

Selections of preferred sizes are indicated as follows:

irst Preference

20, 24, 30, 36, 42, 48, 56, 64, 72, 80, and all sizes ending in "0" for remainder of the series.

econd Preference

18, 22, 27, 33, 39, 45, 52, 60, 68, 76, 85, and all sizes ending in "5" for the remainder of the series.

The basic dimensions are obtained from:

Major Diameter (D) = Nominal Diameter;
Effective Diameter (E) = $D - 1{\cdot}299$ mm.;
Minor Diameter (d) = $D - 2{\cdot}598$ mm.

MM. PITCH SERIES (METRISCHES FEINGEWINDE D.I.N. 246)

The depth of thread is 1·949 mm. and the radius at the root, 0·32 mm. he series consists of 117 sizes, from 28 to 300 mm. with indicated possible xtension of up to 500 mm. The sizes tabulated are as follows:

28, 30, 32, 33, 34, 35, 36, 38, 39, 40, 42, 45, 48, 50, 52, 55, 56, 58, 60, 62, 64, 65, 68, 70, 72, 75, 76, 78, thence all whole numbers of mm. ending with 0, 2, 5 or 8, up to 300 mm.

Selections of preferred sizes are as follows:

irst Preference

30, 36, 42, 48, 56, 64, 72, 80, and all sizes ending in "0" for the remainder of the series.

econd Preference

33, 39, 45, 52, 60, 68, 76, 85 and all sizes ending in "5" for the remainder of the series.

The basic dimensions are obtained as follows:

Major Diameter (D) = Nominal Diameter;
Effective Diameter (E) = $D - 1{\cdot}949$ mm.;
Minor Diameter (d) = $D - 3{\cdot}898$ mm.

MM. PITCH SERIES (METRISCHES FEINGEWINDE D.I.N. 245)

The depth of thread is 2·598 mm., and the radius at the root is 0·43 mm. he series comprises 108 sizes from 42 to 300 mm., with an indicated ossible extension up to 500 mm. The sizes in mm. nominal diameter e as follows:

40, 42, 45, 48, 50, 52, 55, 56, 58, 60, 62, 64, 65, 68, 70, 72, 75, 76, 78, 80, 82, 85, 88, 90, then all whole numbers of mm. ending in 0, 2, 5 and 8 up to 300 mm.

The basic dimensions are obtained in the following manner:

Major Diameter (D) = Nominal Diameter;
Effective Diameter (E) = $D - 2{\cdot}598$ mm.;
Minor Diameter (d) = $D - 5{\cdot}196$ mm.

6 MM. PITCH SERIES (METRISCHES FEINGEWINDE D.I.N. 244)

The depth of thread is 3·897 mm. and the radius at the root is 0·65 m

The series comprises 67 sizes as follows:

72, 76, 80, 85, 90, then all sizes given in whole numbers of mm. endi in 0 or 5 up to 300 mm., and then all sizes ending in 0 up to 500. T preferred sizes are those whose size in mm. ends in "0".

The basic dimensions are obtained in the following manner:

Major Diameter (D) = Nominal Diameter;
Effective Diameter (E) = D – 3·897 mm.;
Minor Diameter (d) = D – 7·794 mm.

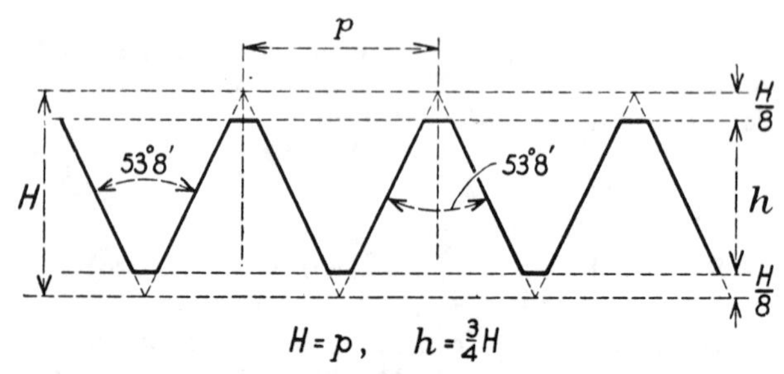

Fig. 3. Löwenherz Thread Profile

LÖWENHERZ THREAD

The Löwenherz thread as first devised by Dr. Leopold Löwenherz w of sharp-V form. It was subsequently truncated at the crest and ro by the amount adopted for the standard metric form. A thread angle 53 deg 8 mins was chosen to give a triangular height equal to the pitc This angle does not exactly fulfil this condition because tan ½ (53 d 8 mins) = 0·5000352 and not 0·5000000. The basic form is shown Fig. 3, and the specified proportions are:

Thread angle (2θ) = 53 deg 8 mins; Flank angle (θ) = 26 deg 34 min Triangular height (H) = 1·0 p; Shortening at crest and root (S) 0·125 p; Depth of thread (h) = 0·75 p.

The dimensions for the usual range of sizes are given in Table G6.

GERMAN BUTTRESS (SÄGENGEWINDE OR SAW TOOTH)

The German Sägengewinde thread is specified in three series, namel "Sägengewinde," "Sägengewinde fein," and "Sägengewinde grob The first has diameters ranging from 22 to 300 mm., with pitches from 5 26 mm., the second has diameters ranging from 10 to 640 mm. with pitch from 2 to 24 mm., and the third from 22 to 400 mm. with pitches from to 48 mm. The basic form is shown in Fig. 4 and has the following pr portions calculated on the construction shown therein:

Thread angle ($\theta_p + \theta_t$) = 33 deg; Pressure flank angle (θ_p) = 3 de Trailing flank angle (θ_t) = 30 deg; H_g = 1·73205 p; h_e = 0·75 S_g = 0·52507 p; a = 0·11777 p; $h = h_e + a$ = 0·86777 p; f 0·26384 p; S_r = 0·45698 p; r = 0·12427 p.

Values of the dimensions for the various pitches are given in Table G

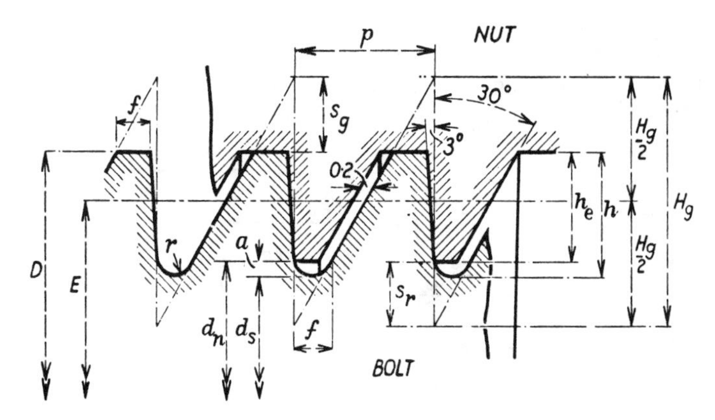

Fig. 4. German Buttress or Sägengewinde Thread Form

he basic diameters are related to the nominal or basic major diameter screw as follows:

Effective Diameter $(E) = D - 0{\cdot}68191\ p$;
Minor Diameter of Screw $(d_s) = D - 1{\cdot}73554\ p$;
Minor Diameter of Nut $(d_r) = D - 1{\cdot}5\ p$.

A minimum longitudinal clearance of 0·2 mm. is provided between the ailing flanks of screw and nut.

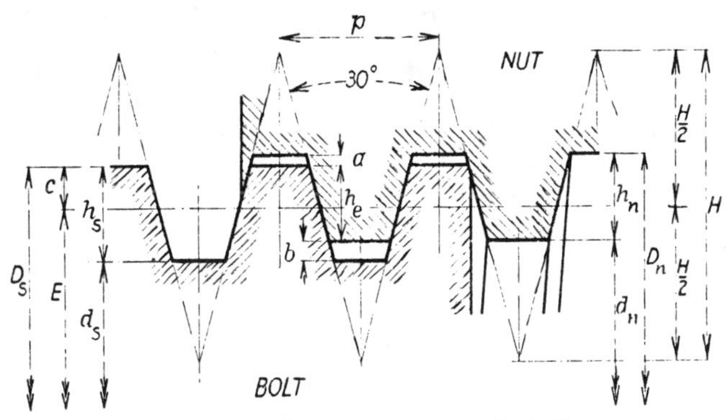

Fig. 5. Trapezoidal Metric Thread—Basic Form

RAPEZOIDAL METRIC THREAD

The form shown in Fig. 5 has the following proportions:

Thread angle $(2\theta) = 30$ deg; Flank angle $(\theta) = 15$ deg; Triangular height $(H) = 1{\cdot}866\ p$; Depth of thread in screw $(h_s) = 0{\cdot}5\ p + a$; Depth of thread in nut $(h_n) = 0{\cdot}5\ p + 2a - b$; Depth of engagement $(h_e) = 0{\cdot}5\ p + a - b$.

Height of thread in screw above pitch line $(c) = 0{\cdot}25\ p$.

The basic major diameter of the screw is equal to the nominal diameter.

The normal series (Swiss and German) contains sizes ranging from 10 to 00 mm. with pitches from 3 to 26 mm. In addition the German specifica-

tions include two further series namely "fein" and "grob" the forme
ranging from 10 to 195 mm. diameter with pitches from 2 to 8 mm., and th
latter from 22 to 400 mm. with pitches from 8 to 48 mm. Basic measure
ments depending on pitch are given in Table G8.

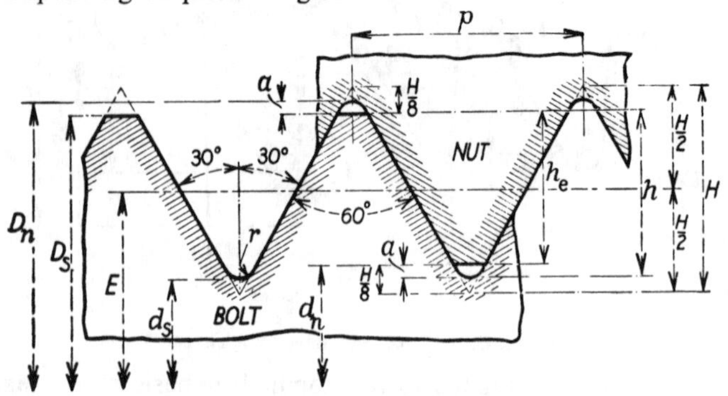

Fig. 6. Swiss Metric Thread—Basic Form

SWISS METRIC THREADS

The standard series of Swiss metric threads are five in numbe
designated the A, B, C, D and E series with the nominal diameters an
pitches as indicated in Table G10 on page 155. The series B, C, D and
constitute the Swiss Fine threads. The basic thread form for these i
shown in Fig. 6, the proportions being derived from the relations:

Thread angle (2θ) = 60 deg.; Flank angle (θ) = 30 deg.; Triangula
height (H) = 0·8660 p; Clearance (a) = 0·050 p; Depth of engage
ment (h_e) = 0·6495 p; Depth of thread (h) = 0·6995 p; Majo
diameter of screw (D_s) = Nominal diameter; Effective diamete
(E) = D_s – 0·6495 p; Minor diameter of screw (d_s) = D_s – 1·3990 p
Major diameter of nut (D_n) = D_s + 0·100 p; Minor diameter o
nut (d_n) = D_s – 1·2990 p; Radius (r) = 0·058 p.

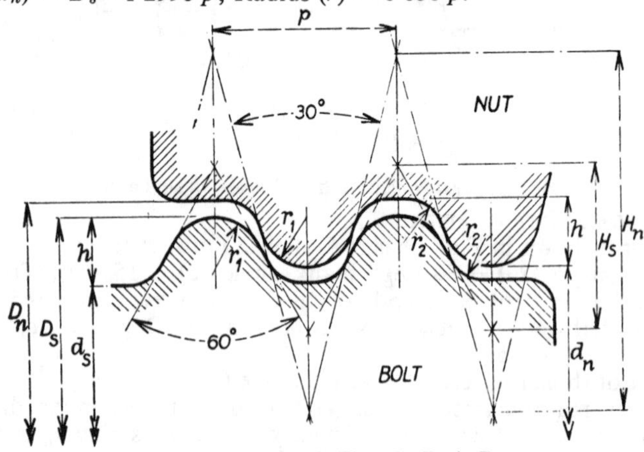

Fig. 7. German Bottle Thread—Basic Form

Values for the various pitches are given in Table G9, and the classification of the series by combination of pitches and nominal diameters is indicated in Table G10.

German Sewing Machine Series

The basic thread form of the series has the particulars:
Thread angle (2θ) = 60 deg; Flank angle (θ) = 30 deg; Triangular height (H) = 0·8660254 p; Shortening at crest and root (S) = 0·1082532 p; Depth of thread (h) = 0·6495191 p.

The crests and roots are flat. The thread is probably derived from the old United States Standard form, since the pitches are specified in inch units. The basic dimensions of the series are given in Table G11.

German Bottle Closure Thread

As indicated in Fig. 7, this thread has a thread angle of 60 deg for the male member, and a corresponding angle of 30 deg for the female member.

The proportions are as follows:
$H_s = 0{\cdot}866\ p$; $H_n = 1{\cdot}866\ p$; $h = 0{\cdot}34\ p$; $r_1 = 0{\cdot}263\ p$; $r_2 = 0{\cdot}2\ p$.

ISO General Purpose Screw Threads

The ISO thread supersedes the ISA metric form, and has been adopted as the standard thread form in the U.K. following the introduction of the metric system. The *basic profile*, shown in Fig. 8, is the theoretical profile and is associated with the basic sizes of the major, pitch and minor diameters of the thread. In practice, deviations are applied to these basic sizes. The proportions are derived from the following relations:

Thread angle (2θ) = 60 deg; Flank angle (θ) = 30 deg; Triangular

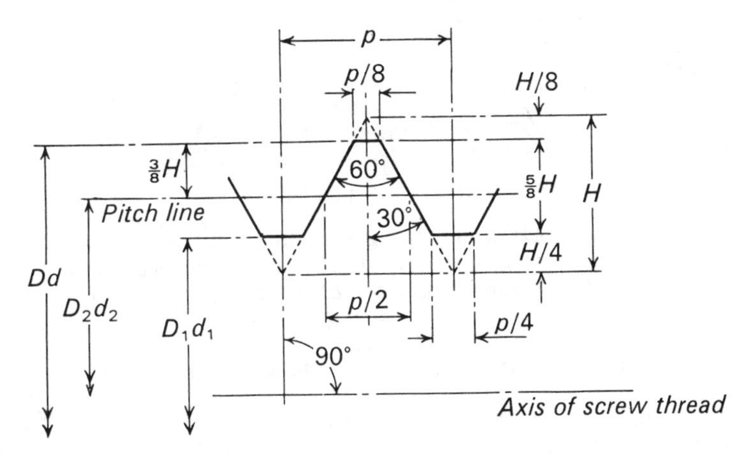

Fig. 8. ISO Metric Thread – Basic Form

height (H) = 0·86603 p; Truncation at crest = $H/8$ = 0·10825 p; Truncation at root = $H/4$ = 0·21651 p; Depth of thread = 5/8 H = 0·54127 p; Basic major diameter = Dd = nominal diameter; Basic effective (pitch) diameter = D_2d_2 = Dd – 0·64952 p; Basic minor diameter = D_1d_1 = Dd – 1·08253 p.

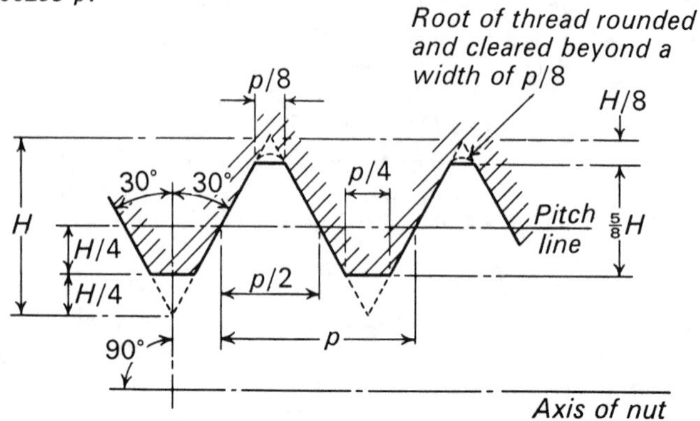

Fig. 9. ISO Metric Nut Thread – Maximum Material Profile

The *maximum material profile of the nut thread is* shown in Fig. 9, and is theoretically the same as the basic profile. In practice, however, in order to avoid sharp corners at the root of the major diameter, the roots of the thread are usually rounded and cleared beyond a width of $p/8$.

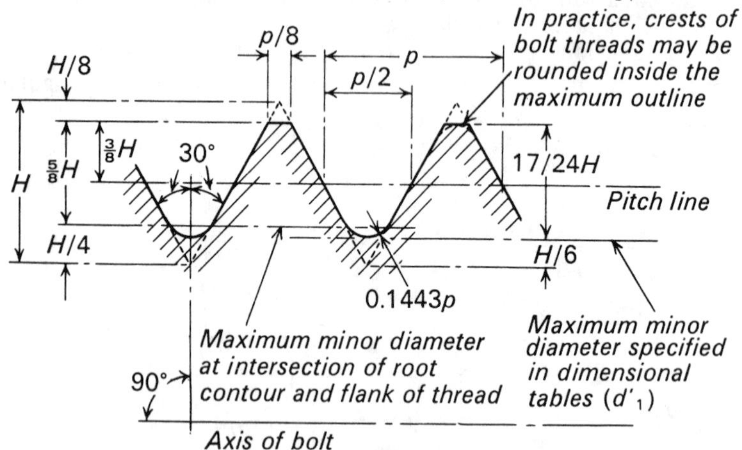

Fig. 10. ISO Metric Bolt Thread – Maximum Material Profile

The *maximum material profile of the bolt thread* is shown in Fig. 10, and differs from the basic profile only in that the root of the thread (at the minor diameter) is rounded to a theoretical radius of 0·1443 p below the flat width of $p/4$.

A *thread series* consists of a series of standard thread diameters, each f which is associated with a particular thread pitch.

A *graded pitch thread series* is a series in which the pitch varies with the hread diameter.

A *constant pitch thread series* is a series in which the pitch is constant respective of the thread diameter.

A graded pitch series known as the *Coarse Series,* together with constant itch threads, are shown in tables G15-17.

At present, ISO has only recognized one thread series with graded itches, namely the Coarse Series, although ISO recommendation R262 ives a directive for the selection of fasteners with both coarse and fine hreads, as indicated in table G18.

olerance grades and positions. The ISO metric screw thread system rovides several positions for tolerance zones, relative to the basic size. hese zones are designated by *G* and *H* for nut threads, and by *e, g* and *h* or bolt threads. The *tolerance position* determines the clearance, or undamental deviation, from the basic size, as indicated in Fig. 11.

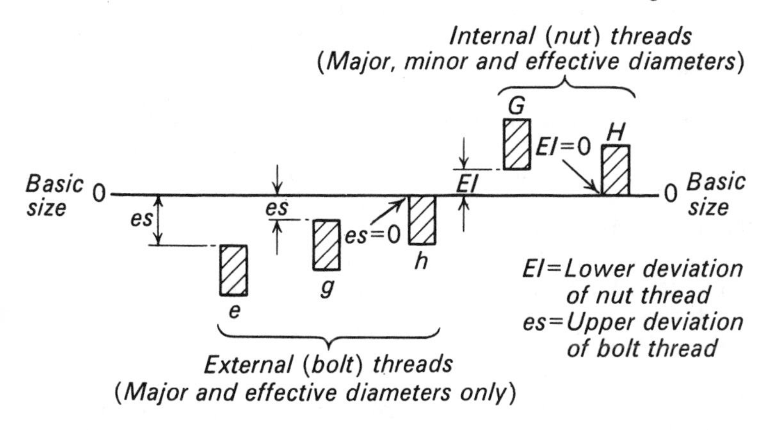

ig. 11. ISO Metric Thread – Relationship between the Basic Size and the Positions of Different Tolerance Zones.

In order to provide tolerance variations, *tolerance grades* have been established and are represented in magnitude by figures. A combination of a tolerance position and a tolerance grade is known as a *tolerance class.*

A *class of fit* is obtained by the combination of a nut of a known olerance class with a bolt of known tolerance class. Three classes of fit have been standardized, as indicated in table G19.

Basic sizes, fundamental deviations and tolerances. The *basic sizes* to which he deviations are applied are given in tables G20 to G27. *Deviations* which are applied to these basic sizes are given in tables G28 to G32. It should be noted that the tolerance on the minor diameter of the external thread (Td_1) presents a special case, since it is a function of the thread pitch (p) and the effective diameter tolerance (Td_2). It can be shown that the tolerance on the minor diameter is given by the expression: $Td_1 = Td_2$ - $0{\cdot}072\,p$, where Td_1, Td_2 and p are in mm.

Example of the calculation of limits of size. It is required to find the limi of size of a bolt of 6 mm. diameter, 1 mm. pitch and tolerance class 6 From table G20, the basic major diameter of the bolt is 6·000 mm., tl basic effective diameter is 5·350 mm., and the basic minor diameter 4·917 mm.

To determine the *limits of size of the major diameter* it is first necessar to find the fundamental deviation *es* for position *g* from table G28. For pitch of 1 mm., the value given is – 26 µm. Therefore the maximum majc diameter = 6·000 – 0·026 = 5·974 mm. From table G30, tolerance *Td* fc quality 6 and a pitch of 1 mm. is 180 µm. Therefore the minimum majc diameter = 5·974 – 0·180 = 5·794 mm.

To determine the *limits of size of the effective diameter* it is agai necessary to find the fundamental deviation *es* for position *g* from tabl G28, and for a pitch of 1mm., this value is again – 26 µm. Therefore th maximum effective diameter = 5·350 – 0·026 = 5·324 mm. From tabl G32, tolerance Td_2 for quality 6, a diameter of 6 mm., and a pitch of 1 mm is 112µm. Therefore the minimum effective diameter = 5·324 – 0·112 = 5·212 mm.

To determine the *limits of size of the minor diameter* it is again necessar to find the fundamental deviation *es* for position *g* from table G28, and fc a pitch of 1 mm. the value is – 170 µm. Therefore the maximum minc diameter = 4·917 – 0·170 = 4·747 mm. From the formula given above $Td_1 = 0{\cdot}112 + (0{\cdot}072 \times 1) = 0{\cdot}184$ mm. Therefore the minimum mino diameter = 4·747 – 0·184 = 4·563 mm.

Designation. A complete designation of a screw thread gives details of th thread system, the size and pitch of the thread, and the tolerances tha are applicable to the thread. Typical designations are given below, anc with the notes, are largely self-explanatory.

Designation for an internal thread (nut):–M 6 × 1–6*H*

Designation for an external thread (bolt):– M 8 × 1.25–6*g*

Thread system symbol for ISO Metric (general purpose)

Nominal size of thread in mm.

Pitch in mm.

Tolerance class designation

Notes. (1) The tolerance classes for the effective and crest diameters onl differ in special cases.

(2) It is the convention in many countries (but not in Great Britain) t omit the pitch from the designation of coarse series threads. Absence o an indication of the pitch implies that a thread of the coarse series i specified.

ength of thread engagement. Tolerances specified for the three standard-ed classes of fit are suitable for lengths of thread engagement from ·24$Pd^{0\cdot2}$ to 6·7$Pd^{0\cdot2}$. These lengths of engagement are given in Table G35. f lengths of engagement are encountered which are significantly greater or ss than those listed, the thread tolerances should be adjusted to obtain a atisfactory connection. Fuller information is given in BS.3643.

TANDARDS IN PREPARATION

At the time of publication of this edition of Machinery's Screw Thread ook a draft specification for ISO miniature screw threads has been pre-ared and approved by the Mechanical Engineering Industry Committee f the BSI. This draft covers thread sizes from 0·3 to 1·4 mm. with pitches rom 0·08 to 0·3 mm. (see also Table H8 on page 196). First choice sizes n the draft are 0·3, 0·4, 0·5, 0·6, 0·8 and 0·9 mm.

It may also be noted that a draft specification for ISO trapezoidal crew threads had been agreed in principle by the ISO/TCI committee nd it was the intention that this draft should serve as basis for a draft ritish Standard.

In addition, a draft British Standard had been prepared for ISO metric up and countersunk head bolts and set screws with hexagon nuts. The bject was to provide a metric series standard corresponding to BS 325; 947 (obsolescent). In the standard, general dimensions and tolerances vill be given for: cup head bolts and screws, cup head bolts with nibs, up head square neck bolts, 90° countersunk head bolts and screws, 90° ountersunk head bolts with nibs, 90° countersunk head square neck olts, and 120° countersunk head square neck bolts. For each group there vill be eight sizes, ranging from 5 to 24 mm.

ISO METRIC THREAD

Basic Sizes in mm.

Nom. Diam. (also Major Diam. of Screw)	Pitch	Effective Diameter	Minor Diam. of Screw	Minor Diam. of Nut
1	0·25	0·838	0·693	0·729
1:1	0·25	0·938	0·793	0·829
1·2	0·25	1·038	0·893	0·929
1·4	0·3	1·205	1·032	1·075
1·6	0·35	1·373	1·171	1·221
1·8	0·35	1·573	1·371	1·421
2	0·4	1·740	1·509	1·567
2·2	0·45	1·908	1·648	1·713
2·5	0·45	2·208	1·948	2·013
3	0·5	2·675	2·387	2·459
3·5	0·6	3·110	2·764	2·850
4	0·7	3·545	3·141	3·242
4·5	0·75	4·013	3·580	3·688
5	0·8	4·480	4·019	4·134
6	1	5·350	4·773	4·917
7	1	6·350	5·773	5·917
8	1·25	7·188	6·466	6·647
10	1·5	9·026	8·160	8·376
12	1·75	10·863	9·853	10·106
14	2	12·701	11·546	11·835
16	2	14·701	13·546	13·835
18	2·5	16·376	14·933	15·294
20	2·5	18·376	16·933	17·294
22	2·5	20·376	18·933	19·294
24	3	22·051	20·319	20·752
27	3	25·051	23·319	23·752
30	3·5	27·727	25·706	26·211
33	3·5	30·727	28·706	29·211
36	4	33·402	31·093	31·670
39	4	36·402	34·093	34·670

Table G1. FRENCH METRIC THREAD

Basic Sizes

Nom. Diam. (mm.)	Pitch (mm.)	Effective Diam. (mm.)	Major Diam. (mm.)	Minor Diam. (mm.)
3	0·6	2·610	3·000	2·160
3·5	0·6	3·110	3·500	2·660
4	0·75	3·513	4·000	2·940
4·5	0·75	4·013	4·500	3·440
5	0·9	4·415	5·000	3·730
5·5	0·9	4·915	5·500	4·230
6	1·0	5·350	6·000	4·590
7	1·0	6·350	7·000	5·590
8	1·25	7·188	8·000	6·240
9	1·25	8·188	9·000	7·240
10	1·5	9·026	10·000	7·890
11	1·5	10·026	11·000	8·890
12	1·75	10·863	12·000	9·540
14	2·0	12·701	14·000	11·180
16	2·0	14·701	16·000	13·180
18	2·5	16·376	18·000	14·480
20	2·5	18·376	20·000	16·480
22	2·5	20·376	22·000	18·480
24	3·0	22·051	24·000	19·780
27	3·0	25·051	27·000	22·780
30	3·5	27·727	30·000	25·070
33	3·5	30·727	33·000	28·070
36	4·0	33·402	36·000	30·370
39	4·0	36·402	39·000	33·370
42	4·5	39·077	42·000	35·670
45	4·5	42·077	45·000	38·670
48	5·0	44·752	48·000	40·960
52	5·0	48·752	52·000	44·960
56	5·5	52·428	56·000	48·260
60	5·5	56·428	60·000	52·260

Table G2. FRENCH AUTOMOBILE THREADS*

Formulae for Basic Dimensions

Pitch	Effective Diam.	Minor Diam. Screw	Major Diam. Nut	Minor Diam. Nut
0·60	D — 0·390	D — 0·779	D + 0·043	D — 0·715
0·75	D — 0·487	D — 0·974	D + 0·054	D — 0·893
0·90	D — 0·585	D — 1·169	D + 0·065	D — 1·072
1·00	D — 0·650	D — 1·299	D + 0·072	D — 1·191
1·25	D — 0·812	D — 1·624	D + 0·090	D — 1·489
1·50	D — 0·974	D — 1·949	D + 0·108	D — 1·786
1·75	D — 1·137	D — 2·273	D + 0·126	D — 2·084
2·00	D — 1·299	D — 2·598	D + 0·144	D — 2·382
2·50	D — 1·624	D — 3·248	D + 0·180	D — 2·977
3·00	D — 1·949	D — 3·897	D + 0·216	D — 3·572
3·50	D — 2·273	D — 4·547	D + 0·253	D — 4·168

*D = Nominal Diameter.

Table G3. FRENCH AUTOMOBILE THREADS

Combination of Pitch and Nominal Diameter

Pas Normal				Pas Fins			
D	*p*	*D*	*p*	*D*	*p*	*D*	*p*
3·0	0·60	11·0	1·50	5·0	0·75	22·0	1·50
3·5	0·60	12·0	1·75	6·0	0·75	24·0	1·50
4·0	0·75	14·0	2·00	7·0	0·75	25·0	1·50
4·5	0·75	16·0	2·00	8·0	1·00	27·0	1·50
5·0	0·90	18·0	2·50	9·0	1·00	28·0	1·50
5·5	0·90	20·0	2·50	10·0	1·00	30·0	1·50
6·0	1·00	22·0	2·50	10·0	1·25	32·0	1·50
7·0	1·00	24·0	3·00	11·0	1·00	33·0	1·50
8·0	1·25	27·0	3·00	11·0	1·25	35·0	1·50
9·0	1·25	30·0	3·50	12·0	1·25	36·0	1·50
10·0	1·50	—	—	12·0	1·50	38·0	1·50
				14·0	1·50	39·0	1·50
				16·0	1·50	40·0	1·50
				17·0	1·50	42·0	1·50
				18·0	1·50		
				20·0	1·50		

able G4. GERMAN METRIC SERIES*

Basic Dimensions in mm.‡

Nominal Diam.† (mm.)	Pitch (mm.)	Effective Diam. (mm.)	Minor Diam. (mm.)	Depth of Thread (mm.)	Radius at Root (mm.)
1	0·250	0·838	0·676	0·162	0·03
1·2	0·250	1·038	0·876	0·162	0·03
1·4	0·300	1·205	1·010	0·195	0·03
1·7	0·350	1·473	1·246	0·227	0·04
2·0	0·400	1·740	1·480	0·260	0·04
2·3	0·400	2·040	1·780	0·260	0·04
2·6	0·450	2·308	2·016	0·292	0·05
3·0	0·500	2·675	2·350	0·325	0·05
3·5	0·600	3·110	2·720	0·390	0·06
4·0	0·700	3·545	3·090	0·455	0·08
5·0	0·800	4·480	3·960	0·520	0·09
6·0	1·000	5·350	4·700	0·650	0·11
7·0	1·000	6·350	5·700	0·650	0·11
8·0	1·250	7·188	6·376	0·812	0·14
9·0	1·250	8·188	7·376	0·812	0·14
10·0	1·500	9·026	8·052	0·974	0·16
11·0	1·500	10·026	9·052	0·974	0·16
12·0	1·750	10·863	9·726	1·137	0·19
14·0	2·000	12·701	11·402	1·299	0·22
16·0	2·000	14·701	13·402	1·299	0·22
18·0	2·500	16·376	14·752	1·624	0·27
20·0	2·500	18·376	16·752	1·624	0·27
22·0	2·500	20·376	18·752	1·624	0·27
24·0	3·000	22·051	20·102	1·949	0·32
27·0	3·000	25·051	23·102	1·949	0·32
30·0	3·500	27·727	25·454	2·273	0·38
33·0	3·500	30·727	28·454	2·273	0·38
36·0	4·000	33·402	30·804	2·598	0·43
39·0	4·000	36·402	33·804	2·598	0·43
42·0	4·500	39·077	36·154	2·923	0·49
45·0	4·500	42·077	39·154	2·923	0·49
48·0	5·000	44·752	41·504	3·248	0·54
52·0	5·000	48·752	45·504	3·248	0·54
56·0	5·500	52·428	48·856	3·572	0·60
60·0	5·500	56·428	52·856	3·572	0·60
64·0	6·000	60·103	56·206	3·897	0·65
68·0	6·000	64·103	60·206	3·897	0·65

*For sizes smaller than 1 mm. see page 159. †Also Basic Major Diameter.
‡Inch dimensions see page 151.

Table G5. **GERMAN METRIC SERIES**

Tolerances in Units of 0·001 mm.

Nom. Diam.	Major Diam. Screw (—)	Effective Diameter Screw (—); Nut (+)			Minor Diameter Nut (+)	
		Close (fein)	Medium (mittel)	Free (grob)	Allowance	Tolerance
1	63	45	56		10	80
1·2	63	45	56		10	80
1·4	71	45	71		10	90
1·7	71	45	71		10	90
2·0	100	50	80		15	125
2·3	100	50	80		15	125
2·6	112	50	80		20	140
3·0	120	50	80		20	150
3·5	140	63	100		20	180
4·0	150	63	100		22	190
5·0	180	63	100	160	26	224
6·0	224	71	112	180	35	280
7·0	224	71	112	180	35	280
8·0	250	71	112	180	40	315
9·0	250	71	112	180	40	315
10·0	280	90	140	224	45	355
11·0	280	90	140	224	45	355
12·0	400	100	160	250	50	450
14·0	475	100	160	250	55	475
16·0	475	100	160	250	55	475
18·0	560	100	160	250	70	560
20·0	560	100	160	250	70	560
22·0	560	100	160	250	70	560
24·0	600	125	200	315	70	600
27·0	600	125	200	315	70	600
30·0	710	125	200	315	90	710
33·0	710	125	200	315	90	710
36·0	800	140	224	355	100	800
39·0	800	140	224	355	100	800
42·0	900	140	224	355	100	900
45·0	900	140	224	355	100	900
48·0	900	140	224	355	100	900
52·0	900	140	224	355	100	900
56·0	950	140	224	355	110	950
60·0	950	140	224	355	110	950
64·0	1000	140	224	355	120	1000
68·0	1000	140	244	355	120	1000

'able G6. LÖWENHERZ THREAD

Basic Dimensons

Nominal Size* (mm.)	Pitch (mm.)	Effective Diam. (mm.)	Minor Diam. (mm.)	Depth of Thread (mm.)	Tapping Drill
1	0·25	0·812	0·625	0·187	No. 71
1·2	0·25	1·012	0·825	0·187	66
1·4	0·30	1·175	0·950	0·225	61
1·7	0·35	1·437	1·175	0·262	56
2·0	0·40	1·700	1·400	0·300	53
2·3	0·40	2·000	1·700	0·300	50
2·6	0·45	2·262	1·925	0·337	47
3·0	0·50	2·625	2·250	0·375	42
3·5	0·60	3·050	2·600	0·450	36
4·0	0·70	3·475	2·950	0·525	31
4·5	0·75	3·937	3·375	0·562	29
5·0	0·80	4·400	3·800	0·600	23
5·5	0·90	4·825	4·150	0·675	18
6·0	1·00	5·250	4·500	0·750	14
7·0	1·10	6·175	5·350	0·825	5·5mm.
8·0	1·20	7·100	6·200	0·900	E
9·0	1·30	8·025	7·050	0·975	7·25mm.
10·0	1·40	8·950	7·900	1·050	P

*Also the Basic Major Diameter.

Table G7. GERMAN BUTTRESS (SÄGENGEWINDE) THREAD

Basic Form Dimensions in mm.

Pitch	Depth of Thread (h)	Depth of Engagement (h_e)	Width of Flat (f)	Clearance (a)	Radius (r)
2	1·736	1·500	0·528	0·236	0·249
3	2·603	2·250	0·792	0·353	0·373
4	3·471	3·000	1·055	0·471	0·497
5	4·339	3·750	1·319	0·589	0·621
6	5·207	4·500	1·583	0·707	0·746
7	6·074	5·250	1·847	0·824	0·870
8	6·942	6·000	2·111	0·942	0·994
9	7·810	6·750	2·375	1·060	1·118
10	8·678	7·500	2·638	1·178	1·243
12	10·413	9·000	3·166	1·413	1·491
14	12·149	10·500	3·694	1·649	1·740
16	13·884	12·000	4·221	1·884	1·988
18	15·620	13·500	4·749	2·120	2·237
20	17·355	15·000	5·277	2·355	2·485
22	19·091	16·500	5·804	2·591	2·734
24	20·826	18·000	6·332	2·826	2·982
26	22·562	19·500	6·860	3·062	3·231
28	24·298	21·000	7·388	3·298	3·480
32	27·769	24·000	8·443	3·769	3·977
36	31·240	27·000	9·498	4·240	4·474
40	34·711	30·000	10·554	4·711	4·971
44	38·182	33·000	11·609	5·182	5·468
48	41·653	36·000	12·664	5·653	5·965

Table G8. TRAPEZOIDAL METRIC THREAD

Basic Form Dimensions in mm.

Pitch	Depth of Thread		Clearance		Depth of Engagement
	Screw	Nut	*(a)*	*(b)*	
2	1·25	1·00	0·25	0·5	0·75
3	1·75	1·50	0·25	0·5	1·25
4	2·25	2·00	0·25	0·5	1·75
5	2·75	2·25	0·25	0·75	2·00
6	3·25	2·75	0·25	0·75	2·50
7	3·75	3·25	0·25	0·75	3·00
8	4·25	3·75	0·25	0·75	3·50
9	4·75	4·25	0·25	0·75	4·00
10	5·25	4·75	0·25	0·75	4·50
12	6·25	5·75	0·25	0·75	5·50
14	7·50	6·5	0·5	1·5	6·00
16	8·50	7·5	0·5	1·5	7·00
18	9·50	8·5	0·5	1·5	8·00
20	10·50	9·5	0·5	1·5	9·00
22	11·50	10·5	0·5	1·5	10·00
24	12·50	11·5	0·5	1·5	11·00
26	13·50	12·5	0·5	1·5	12·00
28	14·50	13·5	0·5	1·5	13·00
32	16·50	15·5	0·5	1·5	15·00
36	18·50	17·5	0·5	1·5	17·00
40	20·50	19·5	0·5	1·5	19·00
44	22·50	21·5	0·5	1·5	21·00
48	24·50	23·5	0·5	1·5	23·00

Table G9.

SWISS METRIC THREAD

Basic Form Dimensions in mm.

Pitch	Depth of Thread h_e+a	Double Depth of Thread $2(h_e+a)$	Clearance a	Depth of Engagement h_e	Radius a Root r
0·2	0·140	0·280	0·010	0·130	0·012
0·25	0·174	0·348	0·012	0·162	0·015
0·30	0·210	0·420	0·015	0·195	0·018
0·35	0·245	0·490	0·018	0·227	0·021
0·40	0·280	0·560	0·020	0·026	0·023
0·45	0·314	0·628	0·022	0·292	0·026
0·5	0·350	0·700	0·025	0·325	0·029
0·6	0·420	0·840	0·030	0·390	0·035
0·7	0·490	0·980	0·035	0·455	0·041
0·75	0·525	1·050	0·038	0·487	0·044
0·8	0·560	1·120	0·040	0·520	0·047
0·9	0·630	1·260	0·045	0·585	0·052
1·0	0·700	1·400	0·050	0·650	0·058
1·25	0·874	1·748	0·062	0·812	0·073
1·5	1·049	2·098	0·075	0·974	0·087
1·75	1·224	2·448	0·087	1·137	0·102
2	1·399	2·798	0·100	1·299	0·116
2·5	1·749	3·498	0·125	1·624	0·145
3·0	2·098	4·196	0·150	1·948	0·174
3·5	2·448	4·896	0·175	2·273	0·203
4·0	2·798	5·596	0·200	2·598	0·232
4·5	3·148	6·296	0·225	2·923	0·261
5·0	3·498	6·996	0·250	3·248	0·290
5·5	3·847	7·694	0·275	3·572	0·319
6	4·197	8·394	0·300	3·897	0·348

able G10. SWISS METRIC THREAD SERIES A, B, C, D & E

Classification of Series

Nom. Diam. mm.	Pitches (mm.)					Nom. Diam. mm.	Pitches (mm.)				
	A	B	C	D	E		A	B	C	D	E
1·0	0·25	0·2				52·0	5	3	2	1·5	1
1·2	0·25	0·2				55·0	–	4	3	2	1·5
1·4	0·30	0·2				56·0	5·5	4	3	2	1·5
1·7	0·35	0·2				58·0	–	4	3	2	1·5
2·0	0·4	0·25				60·0	5·5	4	3	2	1·5
2·3	0·4	0·25				62·0	–	4	3	2	1·5
2·6	0·45	0·35				64·0	6	4	3	2	1·5
3·0	0·5	0·35				65·0	–	4	3	2	1·5
3·5	0·6	0·35				68·0	6	4	3	2	1·5
4·0	0·7	0·5				70·0	–	4	4	3	1·5
(4·5)	0·75	0·5				72·0	6	4	3	2	1·5
5·0	0·8	0·5				75·0	–	4	3	2	1·5
(5·5)	0·9	0·5				76·0	6	4	3	2	1·5
6·0	1	0·75				78·0	–	4	3	2	1·5
(7·0)	1	0·75				80·0	6	4	3	2	1·5
8·0	1·25	1	0·75			82·0	–	4	3	2	1·5
(9·0)	1·25	1	0·75			85·0	6	4	3	2	1·5
10·0	1·5	1	0·75			88·0	–	4	3	2	1·5
(11·0)	1·5	1	0·75			90·0	6	4	3	2	1·5
12·0	1·75	1·5	1			92·0	–	4	3	2	1·5
14·0	2	1·5	1			95·0	6	4	3	2	1·5
15·0	–	1·5	1			98·0	–	4	3	2	1·5
16·0	2	1·5	1			100	6	4	3	2	1·5
17·0	–	1·5	1			102	–	4	3	2	1·5
18·0	2·5	1·5	1			105	6	4	3	2	1·5
20·0	2·5	1·5	1			108	–	4	3	2	1·5
22·0	2·5	1·5	1			110	6	4	3	2	1·5
24·0	3	2	1·5	1		112	–	4	3	2	1·5
25·0	–	2	1·5	1		115	6	4	3	2	1·5
26·0	–	2	1·5	1		118	–	4	3	2	1·5
27·0	3	2	1·5	1		120	6	4	3	2	1·5
28·0	–	2	1·5	1		122	–	4	3	2	1·5
30·0	3·5	2	1·5	1		125	6	4	3	2	1·5
32·0	–	2	1·5	1		128	–	4	3	2	1·5
33·0	3·5	2	1·5	1		130	6	4	3	2	1·5
35·0	–	3	2	1·5	1	132	–	4	3	2	1·5
36·0	4	3	2	1·5	1	135	6	4	3	2	1·5
38·0	–	3	2	1·5	1	138	–	4	3	2	1·5
39·0	4	3	2	1·5	1	140	6	4	3	2	1·5
40·0	–	3	2	1·5	1	142	–	4	3	2	1·5
42·0	4·5	3	2	1·5	1	145	6	5	3	2	1·5
45·0	4·5	3	2	1·5	1	148	–	4	3	2	1·5
48·0	5	3	2	1·5	1	150	6	4	3	2	1·5
50·0	–	3	2	1·5	1						

Series B, C, D and E extend to 300 mm. in dimensions ending in 2, 5, 8 and 0.

Table G11. SEWING MACHINE THREADS

Nom. Diam. (ins.)	Threads per inch	Major Diam. Screw (mm.)	Major Diam. Nut (mm.)	Effective Diam. Screw and Nut (mm.)	Minor Diam. Screw (mm.)	Minor Diam. Nut (mm.)
$\frac{5}{64}$	64	1·984	2·013	1·726	1·468	1·554
$\frac{3}{32}$	100	2·381	2·339	2·216	2·051	2·106
$\frac{3}{32}$	56	2·381	2·414	2·086	1·791	1·889
$\frac{1}{8}$	44	3·175	3·217	2·800	2·425	2·550
$\frac{9}{64}$	40	3·572	3·618	3·160	2·748	2·885
$\frac{11}{64}$	40	4·366	4·412	3·954	3·542	3·679
$\frac{3}{16}$	32	4·763	4·820	4·247	3·731	3·903
$\frac{3}{16}$	28	4·763	4·828	4·174	3·585	3·781
$\frac{15}{64}$	28	5·953	6·018	5·364	4·775	4·971
$\frac{1}{4}$	40	6·350	6·396	5·938	5·526	5·663
$\frac{9}{32}$	20	7·144	7·236	6·319	5·494	5·769
$\frac{9}{32}$	28	7·144	7·209	6·555	5·966	6·162
$\frac{5}{16}$	18	7·938	8·040	7·021	6·104	6·410
$\frac{3}{8}$	28	9·525	9·590	8·936	8·347	8·543
$\frac{7}{16}$	28	11·113	11·178	10·524	9·935	10·131
$\frac{9}{16}$	20	14·288	14·380	13·463	12·638	12·913

able G12. GERMAN BOTTLE CLOSURE THREAD SERIES

Basic Dimensions in mm.

Nominal Diameter	Pitch	Neck Major Diam.	Neck Minor Diam.	Cap Major Diam.	Cap Minor Diam.
8	2	8·00	6·64	8·20	6·84
9	2	9·00	7·64	9·20	7·84
10	2	10·00	8·64	10·20	8·84
11	2	11·00	9·64	11·20	9·84
12	3	12·00	9·96	12·20	10·16
14	3	14·00	11·96	14·20	12·16
16	3	16·00	13·96	16·20	14·16
18	3	18·00	15·96	18·20	16·16
20	3	20·00	17·96	20·20	18·16
22	3	22·00	19·96	22·20	20·16
25	3	25·00	22·96	25·20	23·16
28	3	28·00	25·96	28·20	26·16
30	4	30·00	27·28	30·30	27·58
32	4	32·00	29·28	32·30	29·58
35	4	35·00	32·28	35·30	32·58
40	4	40·00	37·28	40·30	37·58
45	4	45·00	42·28	45·30	42·58
50	4	50·00	47·28	50·30	47·58
55	6	55·00	50·92	55·40	51·32
60	6	60·00	55·92	60·40	56·32
65	6	65·00	60·92	65·40	61·32
68*	6	68·00	63·92	68·40	64·32
70	6	70·00	65·92	70·4	66·32
75	6	75·00	70·92	75·4	71·32
80	6	80·00	75·92	80·4	76·32
82*	6	82·00	77·92	82·4	78·32
85	6	85·00	80·92	85·4	81·32
90	6	90·00	85·92	90·4	86·32
100	8	100·00	94·56	100·5	95·06
110	8	110·00	104·56	110·5	105·06
125	8	125·00	119·56	125·5	120·06
160	12	160·00	151·84	161·0	152·84
200	12	200·00	191·84	201·0	192·84

* For Honeyjars only

Table G13. GERMAN BOTTLE CLOSURE THREADS

Basic Form Dimensions in mm.

Pitch	Depth of Thread	Radius at Crest r_1	Radius at Root r_2
2	0·68	0·53	0·40
3	1·02	0·79	0·60
4	1·36	1·10	0·80
6	2·04	1·60	1·20
8	2·72	2·10	1·60
12	4·08	3·20	2·40

ble G14. **GERMAN METRIC SERIES**

Basic Dimensions in Inches

Nom. Diam. (mm.)	Pitch (mm.)	Pitch (ins.)	Major Diam. (ins.)	Effect. Diam. (ins.)	Minor Diam. (ins.)
1	0·25	0·0098	0·0394	0·0330	0·0266
1·2	0·25	0·0098	0·0472	0·0409	0·0345
1·4	0·30	0·0118	0·0551	0·0474	0·0398
1·7	0·35	0·0138	0·0669	0·0580	0·0491
2·0	0·40	0·0157	0·0787	0·0685	0·0583
2·3	0·40	0·0157	0·0906	0·0803	0·0701
2·6	0·45	0·0177	0·1024	0·0909	0·0794
3·0	0·50	0·0197	0·1181	0·1053	0·0925
3·5	0·60	0·0236	0·1378	0·1224	0·1071
4·0	0·70	0·0276	0·1575	0·1396	0·1217
5·0	0·80	0·0315	0·1969	0·1764	0·1559
6·0	1·00	0·0394	0·2362	0·2106	0·1850
7·0	1·00	0·0394	0·2756	0·2500	0·2244
8·0	1·25	0·0492	0·3150	0·2830	0·2510
9·0	1·25	0·0492	0·3543	0·3224	0·2904
10·0	1·50	0·0591	0·3937	0·3554	0·3170
11·0	1·50	0·0591	0·4331	0·3947	0·3564
12·0	1·75	0·0689	0·4724	0·4277	0·3829
14·0	2·00	0·0787	0·5512	0·5000	0·4489
16·0	2·00	0·0787	0·6299	0·5788	0·5276
18·0	2·50	0·0984	0·7087	0·6447	0·5808
20·0	2·50	0·0984	0·7874	0·7235	0·6595
22·0	2·50	0·0984	0·8661	0·8022	0·7383
24·0	3·00	0·1181	0·9449	0·8681	0·7914
27·0	3·00	0·1181	1·0630	0·9863	0·9095
30·0	3·50	0·1378	1·1811	1·0916	1·0021
33·0	3·50	0·1378	1·2992	1·2097	1·1202
36·0	4·00	0·1575	1·4173	1·3150	1·2128
39·0	4·00	0·1575	1·5354	1·4331	1·3309
42·0	4·50	0·1772	1·6535	1·5385	1·4234
45·0	4·50	0·1772	1·7717	1·6566	1·5415
48·0	5·00	0·1969	1·8898	1·7619	1·6340
52·0	5·00	0·1969	2·0472	1·9194	1·7915
56·0	5·50	0·2165	2·2047	2·0641	1·9235
60·0	5·50	0·2165	2·3622	2·2216	2·0809
64·0	6·00	0·2362	2·5197	2·3663	2·2128
68·0	6·00	0·2362	2·6772	2·5237	2·3703

Table G15.

ISO METRIC THREAD

Standard diameter and pitch combinations (mm.) for nominal diameters from 1 to 30

Nominal (basic major) diameters			Pitches												
First Choice	Second Choice	Third Choice	Coarse Series (Graded Pitches)	Fine Series (Constant Pitches)											
1			0·25												0
	1·1		0·25												0
1·2			0·25												0
	1·4		0·30												0
1·6			0·35												0
	1·8		0·35												0
2			0·40											0·25	
	2·2		0·45											0·25	
2·5			0·45										0·35		
3			0·50										0·35		
	3·5		0·60										0·35		
4			0·70									0·50			
	4·5		0·75									0·50			
5			0·80									0·50			
		5·5										0·50			
6			1·00								0·75				
		7	1·00								0·75				
8			1·25							1	0·75				
		9	1·25							1	0·75				
10			1·50						1·25	1	0·75				
		11	1·50							1	0·75				
12			1·75					1·5	1·25	1					
	14		2·00					1·5	1·25*	1					
		15						1·5		1					
16			2·00					1·5		1					
		17						1·5		1					
	18		2·50				2	1·5		1					
20			2·50				2	1·5		1					
	22		2·50				2	1·5		1					
24			3·00				2	1·5		1					
		25					2	1·5		1					
		26						1·5							
	27		3·00				2	1·5		1					
		28					2	1·5		1					
30			3·50			3‡	2	1·5		1					

Notes

Preference should be given to the diameters in column 1, and if none in this colu
are suitable, a choice should be made next from column 2, and finally from columr

*The pitch of 1·25 mm. for a diameter of 14 mm. is to be used only for sparking plu

‡This pitch should be avoided if possible.

ble G16. **ISO METRIC THREAD**

**Standard diameter and pitch combinations (mm.)
for nominal diameters from 32 to 150**

ominal (basic major) diameters			Pitches					
irst hoice	Second Choice	Third Choice	Coarse Series (Graded Pitches)	Fine Series (Constant Pitches)				
		32					2	1·5
	33		3·5			3‡	2	1.5
		35†						1·5
36			4			3	2	1·5
		38						1·5
	39		4			3	2	1·5
		40				3	2	1·5
42			4·5		4	3	2	1·5
	45		4·5		4	3	2	1·5
48			5		4	3	2	1·5
		50				3	2	1·5
	52		5		4	3	2	1·5
		55			4	3	2	1·5
56			5·5		4	3	2	1·5
		58			4	3	2	1·5
	60		5·5		4	3	2	1·5
		62			4	3	2	1·5
64			6		4	3	2	1·5
		65			4	3	2	1·5
	68		6		4	3	2	1·5
		70		6	4	3	2	1·5
72				6	4	3	2	1·5
		75			4	3	2	1·5
	76			6	4	3	2	1·5
		78					2	
80				6	4	3	2	1·5
		82					2	
	85			6	4	3	2	
90				6	4	3	2	
	95			6	4	3	2	
100				6	4	3	2	
	105			6	4	3	2	
110				6	4	3	2	
	115			6	4	3	2	
	120			6	4	3	2	
125				6	4	3	2	
	130			6	4	3	2	
		135		6	4	3	2	
140				6	4	3	2	
		145		6	4	3	2	
	150			6	4	3	2	

otes

Preference should be given to the diameters in column 1, and if none in this column are suitable, a choice should be made next from column 2, and finally from column 3.

†35 mm. diameter is to be used for locking nuts for ball bearings.

‡This pitch should be avoided if possible.

Table G17. **ISO METRIC THREAD**

Standard diameter and pitch combinations (mm.) for nominal diameters from 155 to 300

Nominal (basic major) diameters			Pitches											
First Choice	Second Choice	Third Choice	Coarse Series (Graded Pitches)	Fine Series (Constant Pitches)										
		155		6	4	3								
160				6	4	3								
		165		6	4	3								
	170			6	4	3								
		175		6	4	3								
180				6	4	3								
		185		6	4	3								
	190			6	4	3								
		195		6	4	3								
200				6	4	3								
		205		6	4	3								
	210			6	4	3								
		215		6	4	3								
220				6	4	3								
		225		6	4	3								
		230		6	4	3								
		235		6	4	3								
	240			6	4	3								
		245		6	4	3								
250				6	4	3								
		255		6	4									
	260			6	4									
		265		6	4									
		270		6	4									
		275		6	4									
280				6	4									
		285		6	4									
		290		6	4									
		295		6	4									
	300			6	4									

Notes

Preference should be given to the diameters in column 1, and if none in this colum are suitable, a choice should be made next from column 2, and finally from column

ble G18. ISO METRIC THREAD

Screws, bolts and nuts with coarse and fine threads (extract from ISO Recommendation R262) (mm.)

Nominal diameters*		Pitch	
Column 1	Column 2	Coarse	Fine
6		1	
	7	1	
8		1·25	1
10		1·5	1·25
12		1·75	1·25
	14	2	1·5
16		2	1·5
	18	2·5	1·5
20		2·5	1·5
	22	2·5	1·5
24		3	2
	27	3	2
30		3·5	2
	33	3·5	2
36		4	3
	39	4	3

'reference should be given to diameters in column 1, and those in column 2 used only f necessary.

ble G19. ISO METRIC THREAD

Types of fit and tolerance classes for nuts and bolts

Type of fit	Nut Tolerance class	Bolt Tolerance class
Close	5H	4h
*Medium	6H	6g
Free	7H	8g

.imits of size for this class of fit for the coarse series are given in Tables G31 and 32.

Table G20. **ISO METRIC THREAD**

Basic Sizes – Coarse Pitch Series (mm.)

Nominal Diameter			Pitch (p)	Basic Major Diameter (Dd)	Basic Effective Diameter (D_2d_2)	Max. Matl. Minor Diameter Ext. Thread (Design) (d_1)	Basic Minor Diamete (Int. an Ext. Thd (D_1d_1)
First Choice	Second Choice	Third Choice					
1			0·25	1·000	0·838	0·693	0·729
	1·1		0·25	1·100	0·938	0·793	0·829
1·2			0·25	1·200	1·038	0·893	0·929
	1·4		0·3	1·400	1·205	1·032	1·075
1·6			0·35	1·600	1·373	1·171	1·221
	1·8		0·35	1·800	1·573	1·371	1·421
2			0·4	2·000	1·740	1·509	1·567
	2·2		0·45	2·200	1·908	1·648	1·713
2·5			0·45	2·500	2·208	1·948	2·013
3			0·5	3·000	2·675	2·387	2·459
	3·5		0·6	3·500	3·110	2·764	2·850
4			0·7	4·000	3·545	3·141	3·242
	4·5		0·75	4·500	4·013	3·580	3·688
5			0·8	5·000	4·480	4·019	4·134
6			1	6·000	5·350	4·773	4·917
		7	1	7·000	6·350	5·773	5·917
8			1·25	8·000	7·188	6·466	6·647
		9	1·25	9·000	8·188	7·466	7·647
10			1·5	10·000	9·026	8·160	8·376
		11	1·5	11·000	10·026	9·160	9·376
12			1·75	12·000	10·863	9·853	10·106
	14		2	14·000	12·701	11·546	11·835
16			2	16·000	14·701	13·546	13·835
	18		2·5	18·000	16·376	14·933	15·294
20			2·5	20·000	18·376	16·933	17·294
	22		2·5	22·000	20·376	18·933	19·294
24			3	24·000	22·051	20·319	20·752
	27		3	27·000	25·051	23·319	23·752
30			3·5	30·000	27·727	25·706	26·211
	33		3·5	33·000	30·727	28·706	29·211
36			4	36·000	33·402	31·092	31·670
	39		4	39·000	36·402	34·092	34·670
42			4·5	42·000	39·077	36·479	37·129
	45		4·5	45·000	42·077	39·479	40·129
48			5	48·000	44·752	41·866	42·587
	52		5	52·000	48·752	45·866	46·587
56			5·5	56·000	52·428	49·252	50·046
	60		5·5	60·000	56·428	53·252	54·046
64			6	64·000	60·103	56·639	57·505
	68		6	68·000	64·103	60·639	61·505

ble G20 A. ISO METRIC THREAD

Fine Pitch Series (mm.)

\t the time of the publication of this edition of Machinery's Screw read Book, the following fine pitch metric threads were listed on in-mation sheet No. G-13A issued by the Society of Motor Manufacturers 1 Traders Ltd. This document had been submitted to the BSI for nsideration as a British Standard.

Nominal Diameter	Pitch
6	0.75
8	1
10	1.25
12	1.25
14	1.5

ble G21. ISO METRIC THREAD

Basic Sizes – Constant Pitch Series (mm.)
Pitches: 0.2, 0.25, 0.35, 0.5 and 0.75

Nominal Diameter			Pitch (p)	Basic Major Diameter (Dd)	Basic Effective Diameter (D_2d_2)	Max. Matl. Minor Diameter Ext. Thread (Design) (d_1)	Basic Minor Diameter (Int. and Ext. Thd.) (D_1d_1)
'irst 10ice	Second Choice	Third Choice					
1			0·2	1·000	0·870	0·755	0·783
	1·1		0·2	1·100	0·970	0·855	0·883
1·2			0·2	1·200	1·070	0·955	0·983
	1·4		0·2	1·400	1·270	1·155	1·183
1·6			0·2	1·600	1·470	1·355	1·383
	1·8		0·2	1·800	1·670	1·555	1·583
2			0·25	2·000	1·838	1·693	1·729
	2·2		0·25	2·200	2·038	1·893	1·929
2·5			0·35	2·500	2·273	2·071	2·121
3			0·35	3·000	2·773	2·571	2·621
	3·5		0·35	3·500	3·273	3·071	3·121
4			0·5	4·000	3·675	3·387	3·459
	4·5		0·5	4·500	4·175	3·887	3·959
5			0·5	5·000	4·675	4·387	4·459
		5·5	0·5	5·500	5·175	4·887	4·959
6			0·75	6·000	5·513	5·080	5·188
		7	0·75	7·000	6·513	6·080	6·188
8			0·75	8·000	7·513	7·080	7·188
		9	0·75	9·000	8·513	8·080	8·188
0			0·75	10·000	9·513	9·080	9·188
		11	0·75	11·000	10·513	10·080	10·188

Table G22. ISO METRIC THREAD

Basic Sizes – Constant Pitch Series (mm.)
Pitches: 1 and 1·25

Nominal Diameter			Pitch (p)	Basic Major Diameter (Dd)	Basic Effective Diameter (D_2d_2)	Max. Matl. Minor Diameter Ext. Thread (Design) (d_1)	Basic Minor Diamete (Int. anc Ext. Thd. (D_1d_1)
First Choice	Second Choice	Third Choice					
8			1	8·000	7·350	6·773	6·917
		9	1	9·000	8·350	7.773	7·917
10			1·25	10·000	9·188	8·466	8·647
10			1	10·000	9·350	8·773	8·917
		11	1	11·000	10·350	9·773	9·917
12			1·25	12·000	11·188	10·466	10·647
12			1	12·000	11·350	10·773	10·917
	14		1·25	14·000	13·188	12·466	12·647
	14		1	14·000	13·350	12·773	12·917
		15	1	15·000	14·350	13·773	13·917
16			1	16·000	15·350	14·773	14·917
		17	1	17·000	16·350	15·773	15·917
	18		1	18·000	17·350	16·773	16·917
20			1	20·000	19·350	18·773	18·917
	22		1	22·000	21·350	20·773	20·917
24			1	24·000	23·350	22·773	22·917
		25	1	25·000	24·350	23·773	23·917
	27		1	27·000	26·350	25·773	25·917
		28	1	28·000	27·350	26·773	26·917
30			1	30·000	29·350	28·773	28·917

Table G23. **ISO METRIC THREAD**

Basic Sizes – Constant Pitch Series (mm.)
Pitch: 1·5

Nominal Diameter			Pitch (p)	Basic Major Diameter (Dd)	Basic Effective Diameter (D_2d_2)	Max. Matl. Minor Diameter Ext. Thread (Design) (d_1)	Basic Minor Diameter (Int. and Ext. Thd.) (D_1d_1)
First Choice	Second Choice	Third Choice					
12			1·5	12·000	11·026	10·160	10·376
	14		1·5	14·000	13·026	12·160	12·376
		15	1·5	15·000	14·026	13·160	13·376
16			1·5	16·000	15·026	14·160	14·376
		17	1·5	17·000	16·026	15·160	15·376
	18		1·5	18·000	17·026	16·160	16·376
20			1·5	20·000	19·026	18·160	18·376
	22		1·5	22·000	21·026	20·160	20·376
24			1·5	24·000	23·026	22·160	22·376
		25	1·5	25·000	24·026	23·160	23·376
		26	1·5	26·000	25·026	24·160	24·376
	27		1·5	27·000	26·026	25·160	25·376
		28	1·5	28·000	27·026	26·160	26·376
30			1·5	30·000	29·026	28·160	28·376
		32	1·5	32·000	31·026	30·160	30·376
	33		1·5	33·000	32·026	31·160	31·376
		35	1·5	35·000	34·026	33·160	33·376
36			1·5	36·000	35·026	34·160	34·276
		38	1·5	38·000	37·026	36·160	36·376
	39		1·5	39·000	38·026	37·160	37·376
		40	1·5	40.000	39·026	38·160	38·376
42			1·5	42·000	41·026	40·160	40·376
	45		1·5	45·000	44·026	43·160	43·376
48			1·5	48·000	47·026	46·160	46·376
		50	1·5	50·000	49·026	48·160	48·376
	52		1·5	52·000	51·026	50·160	50·376
		55	1·5	55·000	54·026	53·160	53·376
56			1·5	56·000	55·026	54·160	54·376
		58	1·5	58·000	57·026	56·160	56·376
	60		1·5	60·000	59·026	58·160	58·376
		62	1·5	62·000	61·026	60·160	60·376
64			1·5	64·000	63·026	62·160	62·376
		65	1·5	65·000	64·026	63·160	63·376
	68		1·5	68·000	67·026	66·160	66·376
		70	1·5	70·000	69·026	68·160	68·376
72			1·5	72·000	71·026	70·160	70·376
		75	1·5	75·000	74·026	73·160	73·376
	76		1·5	76·000	75·026	74·160	74·376
80			1·5	80·000	79·026	78·160	78·376

Table G24. **ISO METRIC THREAD**

Basic Sizes – Constant Pitch Series (mm.)
Pitch: 2

Nominal Diameter			Pitch (p)	Basic Major Diameter (Dd)	Basic Effective Diameter (D_2d_2)	Max. Matl. Minor Diameter Ext. Thread (Design) (d_1)	Basic Minor Diamete (Int. an Ext. Thd (D_1d_1)
First Choice	Second Choice	Third Choice					
	18		2	18·000	16·701	15·546	15·835
20			2	20·000	18·701	17·546	17·835
	22		2	22·000	20·701	19·546	19·835
24			2	24·000	22·701	21·546	21·835
		25	2	25·000	23·701	22·546	22·835
	27		2	27·000	25·701	24·546	24·835
		28	2	28·000	26·701	25·546	25·835
30			2	30·000	28·701	27·546	27·835
		32	2	32·000	30·701	29·546	29·835
	33		2	33·000	31·701	30·546	30·835
36			2	36·000	34·701	33·546	33·835
	39		2	39·000	37·701	36·546	36·835
		40	2	40·000	38·701	37·546	37·835
42			2	42·000	40·701	39·546	39·835
	45		2	45·000	43·701	42·546	42·835
48			2	48·000	46·701	45·546	45·835
		50	2	50·000	48·701	47·546	47·835
	52		2	52·000	50·701	49·546	49·835
		55	2	55·000	53·701	52·546	52·835
56			2	56·000	54·701	53·546	53·835
		58	2	58·000	56·701	55·546	55·835
	60		2	60·000	58·701	57·546	57·835
		62	2	62·000	60·701	59·546	59·835
64			2	64·000	62·701	61·546	61·835
		65	2	65·000	63·701	62·546	62·835
	68		2	68·000	66·701	65·546	65·835
		70	2	70·000	68·701	67·546	67·835
72			2	72·000	70·701	69·546	69·835
		75	2	75·000	73·701	72·546	72·835
	76		2	76·000	74·701	73·546	73·835
		78	2	78·000	76·701	75·546	75·835
80			2	80·000	78·701	77·546	77·835
		82	2	82·000	80·701	79·546	79·835
	85		2	85·000	83·701	82·546	82·835
90			2	90·000	88·701	87·546	87·835
	95		2	95·000	93·701	92·546	92·835
100			2	100·000	98·701	97·546	97·835

[*continued on page* 1

able G24 *ontinued.*

ISO METRIC THREAD

Basic Sizes – Constant Pitch Series (mm.)
Pitch: 2

Nominal Diameter			Pitch (p)	Basic Major Diameter (Dd)	Basic Effective Diameter (D_2d_2)	Max. Matl. Minor Diameter Ext. Thread (Design) (d_1)	Basic Minor Diameter (Int. and Ext. Thd.) (D_1d_1)
First Choice	Second Choice	Third Choice					
	105		2	105·000	103·701	102·546	102·835
110			2	110·000	108·701	107·546	107·835
	115		2	115·000	113·701	112·546	112·835
	120		2	120·000	118·701	117·546	117·835
125			2	125·000	123·701	122·546	122·835
	130		2	130·000	128·701	127·546	127·835
		135	2	135·000	133·701	132·546	132·835
140			2	140·000	138·701	137·546	137·835
		145	2	145·000	143·701	142·546	142·835
	150		2	150·000	148·701	147·546	147·835

Table G25. ISO METRIC THREAD

Basic Sizes – Constant Pitch Series (mm.)

Pitch: 3

Nominal Diameter			Pitch (p)	Basic Major Diameter (Dd)	Basic Effective Diameter (D_2d_2)	Max. Matl. Minor Diameter Ext. Thread (Design) (d_1)	Basic Minor Diamete (Int. an Ext. Thd (D_1d_1)
First Choice	Second Choice	Third Choice					
30			3	30·000	28·051	26·319	26·752
	33		3	33·000	31·051	29·319	29·752
36			3	36·000	34·051	32·319	32·752
	39		3	39·000	37·051	35·319	35·752
		40	3	40·000	38·051	36·319	36·752
42			3	42·000	40·051	38·319	38·752
	45		3	45·000	43·051	41·319	41·752
48			3	48·000	46·051	44·319	44·752
		50	3	50·000	48·051	46·319	46·752
	52		3	52·000	50·051	48·319	48·752
		55	3	55·000	53·051	51·319	51·752
56			3	56·000	54·051	52·319	52·752
		58	3	58·000	56·051	54·319	54·752
	60		3	60·000	58·051	56·319	56·752
		62	3	62·000	60·051	58·319	58·752
64			3	64·000	62·051	60·319	60·752
		65	3	65·000	63·051	61·319	61·752
	68		3	68·000	66·051	64·319	64·752
		70	3	70·000	68·051	66·319	66·752
72			3	72·000	70·051	68·319	68·752
		75	3	75·000	73·051	71·319	71·752
	76		3	76·000	74·051	72·319	72·752
80			3	80·000	78·051	76·319	76·752
	85		3	85·000	83·051	81·319	81·752
90			3	90·000	88·051	86·319	86·752
	95		3	95·000	93·051	91·319	91·752
100			3	100·000	98·051	96·319	96·752
	105		3	105·000	103·051	101·319	101·752
110			3	110·000	108·051	106·319	106·752
	115		3	115·000	113·051	111·319	111·752
	120		3	120·000	118·051	116·319	116·752
125			3	125·000	123·051	121·319	121·752
	130		3	130·000	128·051	126·319	126·752
		135	3	135·000	133·051	131·319	131·752
140			3	140·000	138·051	136·319	136·752
		145	3	145·000	143·051	141·319	141·752

[*continued on page 1*

ꞏable G25
ꞏntinued.

ISO METRIC THREAD

Basic Sizes – Constant Pitch Series (mm.)
Pitch: 3

Nominal Diameter			Pitch (p)	Basic Major Diameter (Dd)	Basic Effective Diameter (D_2d_2)	Max. Matl. Minor Diameter Ext. Thread (Design) (d_1)	Basic Minor Diameter (Int. and Ext. Thd.) (D_1d_1)
First Choice	Second Choice	Third Choice					
	150		3	150·000	148·051	146·319	146·752
		155	3	155·000	153·051	151·319	151·752
160			3	160·000	158·051	156·319	156·752
		165	3	165·000	163·051	161·319	161·752
	170		3	170·000	168·051	166·319	166·752
		175	3	175·000	173·051	171·319	171·752
180			3	180·000	178·051	176·319	176·752
		185	3	185·000	183·051	181·319	181·752
	190		3	190·000	188·051	186·319	186·752
		195	3	195·000	193·051	191·319	191·752
200			3	200·000	198·051	196·319	196·752
		205	3	205·000	203·051	201·319	201·752
	210		3	210·000	208·051	206·319	206·752
		215	3	215·000	213·051	211·319	211·752
220			3	220·000	218·051	216·319	216·752
		225	3	225·000	223·051	221·319	221·752
		230	3	230·000	228·051	226·319	226·752
		235	3	235·000	233·051	231·319	231·752
	240		3	240·000	238·051	236·319	236·752
		245	3	245·000	243·051	241·319	241·752
250			3	250·000	248·051	246·319	246·752

Table G26. **ISO METRIC THREAD**

Basic Sizes – Constant Pitch Series (mm.)
Pitch: 4

Nominal Diameter			Pitch (p)	Basic Major Diameter (Dd)	Basic Effective Diameter (D_2d_2)	Max. Matl. Minor Diameter Ext. Thread (Design) (d_1)	Basic Minor Diamete (Int. and Ext. Thd (D_1d_1)
First Choice	Second Choice	Third Choice					
42			4	42·000	39·402	37·093	37·670
	45		4	45·000	42·402	40·093	40·670
48			4	48·000	45·402	43·093	43·670
	52		4	52·000	49·402	47·093	47·670
		55	4	55·000	52·402	50·093	50·670
56			4	56·000	53·402	51·093	51·670
		58	4	58·000	55·402	53·093	53·670
	60		4	60·000	57·402	55·093	55·670
		62	4	62·000	60·402	58·093	58·670
64			4	64·000	61·402	59·093	59·670
		65	4	65·000	62·402	60·093	60·670
	68		4	68·000	65·402	63·093	63·670
		70	4	70·000	67·402	65·093	65·670
72			4	72·000	69·402	67·093	67·670
		75	4	75·000	72·402	70·093	70·670
	76		4	76·000	73·402	71·093	71·670
80			4	80·000	77·402	75·093	75·670
	85		4	85·000	82·402	80·093	80·670
90			4	90·000	87·402	85·093	85·670
	95		4	95·000	92·402	90·093	90·670
100			4	100·000	97·402	95·093	95·670
	105		4	105·000	102·402	100·093	100·670
110			4	110·000	107·402	105·093	105·670
	115		4	115·000	112·402	110·093	110·670
	120		4	120·000	117·402	115·093	115·670
125			4	125·000	122·402	120·093	120·670
	130		4	130·000	127·402	125·093	125·670
		135	4	135·000	132·402	130·093	130·670
140			4	140·000	137·402	135·093	135·670
		145	4	145·000	142·402	140·093	140·670
	150		4	150·000	147·402	145·093	145·670
		155	4	155·000	152·402	150·093	150·670
160			4	160·000	157·402	155·093	155·670
		165	4	165·000	162·402	160·093	160·670
	170		4	170·000	167·402	165·093	165·670
		175	4	175·000	172·402	170·093	170·670

[*continued on page 1*

able G26 ontinued.

ISO METRIC THREAD

Basic Sizes – Constant Pitch Series (mm.)
Pitch: 4

Nominal Diameter			Pitch (p)	Basic Major Diameter (Dd)	Basic Effective Diameter (D_2d_2)	Max. Matl. Minor Diameter Ext. Thread (Design) (d_1)	Basic Minor Diameter (Int. and Ext. Thd.) (D_1d_1)
First Choice	Second Choice	Third Choice					
180			4	180·000	177·402	175·093	175·670
		185	4	185·000	182 402	180·093	180·670
	190		4	190·000	187·402	185·093	185·670
		195	4	195·000	192·402	190·093	190·670
200			4	200·000	197·402	195·093	195·670
		205	4	205·000	202·402	200·093	200·670
	210		4	210·000	207·402	205·093	205·670
		215	4	215·000	212·402	210·093	210·670
220			4	220·000	217·402	215·093	215·670
		225	4	225·000	222·402	220·093	220·670
		230	4	230·000	227·402	225·093	225·670
		235	4	235·000	232·402	230·093	230·670
	240		4	240·000	237·402	235·093	235·670
		245	4	245·000	242·402	240·093	240·670
250			4	250·000	247·402	245·093	245·670
		255	4	255·000	252·402	250·093	250·670
	260		4	260·000	257·402	255·093	255·670
		265	4	265·000	262·402	260·093	260·670
270			4	270·000	267·402	265·093	265·670
		275	4	275·000	272·402	270·093	270·670
280			4	280·000	277·402	275·093	275·670
		285	4	285·000	282·402	280·093	280·670
		290	4	290·000	287·402	285·093	285·670
		295	4	295·000	292·402	290·093	290·670
	300		4	300·000	297·402	295·093	295·670

Table G27. **ISO METRIC THREAD**

Basic Sizes – Constant Pitch Series (mm.)

Pitch: 6

Nominal Diameter			Pitch (p)	Basic Major Diameter (Dd)	Basic Effective Diameter (D_2d_2)	Max. Matl. Minor Diameter Ext. Thread (Design) (d_1)	Basic Minor Diameter (Int. and Ext. Thd.) (D_1d_1)
First Choice	Second Choice	Third Choice					
		70	6	70·000	60·103	62·639	63·505
72			6	72·000	68·103	64·639	65·505
	76		6	76·000	72·103	68·639	69·505
80			6	80·000	76·103	72·639	73·505
	85		6	85·000	81·103	77·639	78·505
90			6	90·000	86·103	82·639	83·505
	95		6	95·000	91·103	87·639	88·505
100			6	100·000	96·103	92·639	93·505
	105		6	105·000	101·103	97·639	98·505
110			6	110·000	106·103	102·639	103·505
	115		6	115·000	111·103	107·639	108·505
	120		6	120·000	116·103	112·639	113·505
125			6	125·000	121·103	117·639	118·505
	130		6	130·000	126·103	122·639	123·505
		135	6	135·000	131·103	127·639	128·505
140			6	140·000	136·103	132·639	133·505
		145	6	145·000	141·103	137·639	138·505
	150		6	150·000	146·103	142·639	143·505
		155	6	155·000	151·103	147·639	148·505
160			6	160·000	156·103	152·639	153·505
		165	6	165·000	161·103	157·639	158·505
	170		6	170·000	166·103	162·639	163·505
		175	6	175·000	171·103	167·639	168·505
180			6	180·000	176·103	172·639	173·505
		185	6	185·000	181·103	177·639	178·505
	190		6	190·000	186·103	182·639	183·505
		195	6	195·000	191·103	187·639	188·505
200			6	200·000	196·103	192·639	193·505
		205	6	205·000	201·103	197·639	198·505
	210		6	210·000	206·103	202·639	203·505
		215	6	215·000	211·103	207·639	208·505
220			6	220·000	216·103	212·639	213·505
		225	6	225·000	221·103	217·639	218·505
		230	6	230·000	226·103	222·639	223·505
		235	6	235·000	231·103	227·639	228·505
	240		6	240·000	236·103	232·639	233·505

[*continued on page* 175

able G27
ntinued

ISO METRIC THREAD

Basic Sizes – Constant Pitch Series (mm.)

Pitch: 6

Nominal Diameter			Pitch (p)	Basic Major Diameter (Dd)	Basic Effective Diameter (D_2d_2)	Max. Matl. Minor Diameter Ext. Thread (Design) (d_1)	Basic Minor Diameter (Int. and Ext. Thd.) (D_1d_1)
First Choice	Second Choice	Third Choice					
		245	6	245·000	241·103	237·639	238·505
250			6	250·000	246·103	242·639	243·505
		255	6	255·000	251·103	247·639	248·505
	260		6	260·000	256·103	252·639	253·505
		265	6	265·000	261·103	257·639	258·505
		270	6	270·000	266·103	262·639	263·505
		275	6	275·000	271·103	267·639	268·505
280			6	280·000	276·103	272·639	273·505
		285	6	285·000	281·103	277·639	278·505
		290	6	290·000	286·103	282·639	283·505
		295	6	295·000	291·103	287·639	288·505
	300		6	300·000	296·103	292·639	293·505

Table G28. **ISO METRIC THREAD**

Fundamental Deviations for Nut and Bolt Threads

Pitch (p) (mm.)	Fundamental deviations (μm)							
	Nuts (D, D_2 and D_1) (FD)		Bolts (d and d_2) (fd)			Bolts (d_1) ($fd+0.144\ p$)		
	G EI	H EI	e es	g es	h es	e es	g es	h es
0·2	+17	0	−17	−17	0		−46	−29
0·25	+18	0	−18	−18	0		−54	−36
0·3	+18	0		−18	0		−61	−43
0·35	+19	0		−19	0		−70	−51
0·4	+19	0		−19	0		−77	−58
0·45	+20	0		−20	0		−85	−65
0·5	+20	0	−50	−20	0	−122	−92	−72
0·6	+21	0	−53	−21	0	−140	−108	−87
0·7	+22	0	−56	−22	0	−157	−123	−101
0·75	+22	0	−56	−22	0	−164	−130	−108
0·8	+24	0	−60	−24	0	−176	−140	−116
1	+26	0	−60	−26	0	−204	−170	−144
1·25	+28	0	−63	−28	0	−243	−208	−180
1·5	+32	0	−67	−32	0	−284	−249	−217
1·75	+34	0	−71	−34	0	−324	−287	−253
2	+38	0	−71	−38	0	−360	−327	−289
2·5	+42	0	−80	−42	0	−441	−403	−361
3	+48	0	−85	−48	0	−518	−481	−433
3·5	+53	0	−90	−53	0	−595	−558	−505
4	+60	0	−95	−60	0	−672	−637	−577
4·5	+63	0	−100	−63	0	−750	−713	−650
5	+71	0	−106	−71	0	−828	−793	−722
5·5	+75	0	−112	−75	0	−906	−869	−794
6	+80	0	−118	−80	0	−984	−946	−866

Table G29. ISO METRIC THREAD

Tolerances on Minor Diameters of Nuts (T_{D1})

Pitch (p) (mm.)	Tolerance (μm.)				
	Grade 4	Grade 5	Grade 6	Grade 7	Grade 8
0·2	38				
0·25	45	56			
0·3	53	67	85		
0·35	63	80	100		
0·4	71	90	112		
0·45	80	100	125		
0·5	90	112	140	180	
0·6	100	125	160	200	
0·7	112	140	180	224	
0·75	118	150	190	236	
0·8	125	160	200	250	315
1	150	190	236	300	375
1·25	170	212	265	335	425
1·5	190	236	300	375	475
1·75	212	265	335	425	530
2	236	300	375	475	600
2·5	280	355	450	560	710
3	315	400	500	630	800
3·5	355	450	560	710	900
4	375	475	600	750	950
4·5	425	530	670	850	1060
5	450	560	710	900	1120
5·5	475	600	750	950	1180
6	500	630	800	1000	1250

Table G30. ISO METRIC THREAD

Tolerances on Major Diameters of Bolts (T_d)

Pitch (p) (mm.)	Tolerance (µm.)		
	Grade 4	Grade 6	Grade 8
0·2	36	56	
0·25	42	67	
0·3	48	75	
0·35	53	85	
0·4	60	95	
0·45	63	100	
0·5	67	106	
0·6	80	125	
0·7	90	140	
0·75	90	140	
0·8	95	150	236
1	112	180	280
1·25	132	212	335
1·5	150	236	375
1·75	170	265	425
2	180	280	450
2·5	212	335	530
3	236	375	600
3·5	265	425	670
4	300	475	750
4·5	315	500	800
5	335	530	850
5·5	355	560	900
6	375	600	950

Table G31. ISO METRIC THREAD

Tolerances on Effective Diameters of Nuts (T_{D2})

Nominal (Basic Major) Diameter (mm.)		Pitch (p) (mm.)	Tolerance (µm.)				
Above	Up to and including		Grade 4	Grade 5	Grade 6	Grade 7	Grade 8
		0·2	40				
0·99	1·4	0·25	45	56			
		0·3	48	60	75		
		0·2	42				
		0·25	48	60			

[*continued on page* 179

Table G31 *continued*

ISO METRIC THREAD
Tolerances on Effective Diameters of Nuts (T_{D2})

Nominal (Basic Major) Diameter (mm.)		Tolerance (µm.)					
Above	Up to and including	Pitch (p) (mm.)	Grade 4	Grade 5	Grade 6	Grade 7	Grade 8
1·5	2·8	0·35	53	67	85		
		0·4	56	71	90		
		0·45	60	75	95		
2·8	5·6	0·35	56	71	90		
		0·5	63	80	100	125	
		0·6	71	90	112	140	
		0·7	75	95	118	150	
		0·75	75	95	118	150	
		0·8	80	100	125	160	200
5·6	11·2	0·75	85	106	132	170	
		1	95	118	150	190	236
		1·25	100	125	160	200	250
		1·5	112	140	180	224	280
11·2	22·4	1	100	125	160	200	250
		1·25	112	140	180	224	280
		1·5	118	150	190	236	300
		1·75	125	160	200	250	315
		2	132	170	212	265	335
		2·5	140	180	224	280	335
22·4	45	1	106	132	170	212	
		1·5	125	160	200	250	315
		2	140	180	224	280	355
22·4	45	3	170	212	265	335	425
		3·5	180	224	280	355	450
		4	190	236	300	375	475
		4·5	200	250	315	400	500
45	90	1·5	132	170	212	265	335
		2	150	190	236	300	375
		3	180	224	280	355	450
		4	200	250	315	400	500
		5	212	265	335	425	530
		5·5	224	280	355	450	560
		6	236	300	375	475	600
90	180	2	160	200	250	315	400
		3	190	236	300	375	475
		4	212	265	335	425	530
		6	250	315	400	500	630
180	355	3	212	265	335	425	530
		4	236	300	375	475	600
		6	265	335	425	530	670

Table G32. ISO METRIC THREAD

Tolerances on Effective Diameters of Bolts (T_{d2})

Nominal (Basic Major) Diameter		Pitch (mm.)	Tolerance (µm.)						
Above	Up to and including		Grade 3	Grade 4	Grade 5	Grade 6	Grade 7	Grade 8	Grade 9
0·99	1·4	0·2	24	30	38	48			
		0·25	26	34	42	53			
		0·3	28	36	45	56			
1·5	2·8	0·2	25	32	40	50			
		0·25	28	36	45	56			
		0·35	32	40	50	63	80		
		0·4	34	42	53	67	85		
		0·45	36	45	56	71	90		
2·8	5·6	0·35	34	42	53	67	85		
		0·5	38	48	60	75	95		
		0·6	42	53	67	85	106		
		0·7	45	56	71	90	112		
		0·75	45	56	71	90	112		
		0·8	48	60	75	95	118	150	190
5·6	11·2	0·75	50	63	80	100	125		
		1	56	71	90	112	140	180	224
		1·25	60	75	95	118	150	190	236
		1·5	67	85	106	132	170	212	265
11·2	22·4	1	60	75	95	118	150	190	236
		1·25	67	85	106	132	170	212	265
		1.5	71	90	112	140	180	224	280
		1·75	75	95	118	150	190	236	300
		2	80	100	125	160	200	250	315
		2·5	85	106	132	170	212	265	335
22·4	45	1	63	80	100	125	160	200	250
		1·5	75	95	118	150	190	236	300
		2	85	106	132	170	212	265	335
		3	100	125	160	200	250	315	400
		3·5	106	132	170	212	265	335	425
		4	112	140	180	224	280	355	450
		4·5	118	150	190	236	300	375	475
45	90	1·5	80	100	125	160	200	250	315
		2	90	112	140	180	224	280	355
		3	106	132	170	212	265	335	425
		4	118	150	190	236	300	375	475
		5	125	160	200	250	315	400	500
		5·5	132	170	212	265	335	425	530
		6	140	180	224	280	355	450	560
90	180	2	95	118	150	190	236	300	375

[*continued on page* 181

able G32 ISO METRIC THREAD

ontinued **Tolerances on Effective Diameters of Bolts (T_{d2})**

Nominal (Basic Major) Diameter		Pitch (mm.)	Tolerance (µm.)						
Above	Up to and including		Grade 3	Grade 4	Grade 5	Grade 6	Grade 7	Grade 8	Grade 9
		3	112	140	180	224	280	355	450
		4	125	160	200	250	315	400	500
		6	150	190	236	300	375	475	600
180	355	3	125	160	200	250	315	400	500
		4	140	180	224	280	355	450	560
		6	160	200	250	315	400	500	630

Table G33. ISO METRIC THREAD

Limits of Size for Coarse Series – Bolts (mm.)

Tolerance Class 6g

Nominal Diameter			Pitch	Fund. Dev.	Major diameter			Effective diameter			Minor diameter		
First Choice	Second Choice	Third Choice			max.	tol.	min.	max.	tol.	min.	max.	tol.	min.
1			0·25	0·018	0·982	0·067	0·915	0·820	0·053	0·767	0·675	0·071	0·604
	1·1		0·25	0·018	1·082	0·067	1·015	0·920	0·053	0·867	0·775	0·071	0·704
1·2			0·25	0·018	1·182	0·067	1·115	1·020	0·053	0·967	0·875	0·071	0·804
	1·4		0·3	0·018	1·382	0·075	1·307	1·187	0·056	1·131	1·014	0·078	0·936
1·6			0·35	0·019	1·581	0·085	1·496	1·354	0·063	1·291	1·151	0·088	1·063
	1·8		0·35	0·019	1·781	0·085	1·696	1·554	0·063	1·491	1·351	0·088	1·263
2			0·4	0·019	1·981	0·095	1·886	1·721	0·067	1·654	1·490	0·096	1·394
	2·2		0·45	0·020	2·180	0·100	2·080	1·888	0·071	1·817	1·628	0·103	1·525
2·5			0·45	0·020	2·480	0·100	2·380	2·188	0·071	2·117	1·928	0·103	1·825
3			0·5	0·020	2·980	0·106	2·874	2·655	0·075	2·580	2·367	0·111	2·256
	3·5		0·6	0·021	3·479	0·125	3·354	3·089	0·085	3·004	2·743	0·128	2·615
4			0·7	0·022	3·978	0·140	3·838	3·523	0·090	3·433	3·119	0·140	2·979
	4·5		0·75	0·022	4·478	0·140	4·338	3·991	0·090	3·901	3·558	0·144	3·414
5			0·8	0·024	4·976	0·150	4·826	4·456	0·095	4·361	3·995	0·153	3·842
6			1	0·026	5·974	0·180	5·794	5·324	0·112	5·212	4·747	0·184	4·563
		7	1	0·026	6·974	0·180	6·794	6·324	0·112	6·212	5·747	0·184	5·563
8			1·25	0·028	7·972	0·212	7·760	7·160	0·118	7·042	6·438	0·208	6·230
		9	1·25	0·028	8·972	0·212	8·760	8·160	0·118	0·042	7·438	0·208	7·230
10			1·5	0·032	9·968	0·236	9·732	8·994	0·132	8·862	8·128	0·240	7·888
		11	1·5	0·032	10·968	0·236	10·732	9·994	0·132	9·862	9·128	0·240	8·888

NOTE. Preference should be given to diameters in column 1, and if none in this column are suitable, a choice should be made next from column 2, and finally from column 3.

Table G33 *continued.*

ISO METRIC THREAD

Limits of Size for Coarse Series – Bolts (mm.)

Tolerance Class 6g

Nominal Diameter			Pitch	Fund. Dev.	Major diameter			Effective diameter			Minor diameter		
First Choice	Second Choice	Third Choice			max.	tol.	min.	max.	tol.	min.	max.	tol.	min.
12			1·75	0·034	11·966	0·265	11·701	10·829	0·150	10·679	9·819	0·267	9·543
	14		2	0·038	13·962	0·280	13·682	12·663	0·160	12·503	11·508	0·304	11·204
16			2	0·038	15·962	0·280	15·682	14·663	0·160	14·503	13·508	0·304	13·204
	18		2·5	0·042	17·958	0·335	17·623	16·334	0·170	16·164	14·891	0·350	14·541
20			2·5	0·042	19·958	0·335	19·623	18·334	0·170	18·164	16·891	0·350	16·541
	22		2·5	0·042	21·958	0·335	21·623	20·334	0·170	20·164	18·891	0·350	18·541
24			3	0·048	23·952	0·375	23·577	22·003	0·200	21·803	20·271	0·416	19·855
	27		3	0·048	26·952	0·375	26·577	25·003	0·200	24·803	23·271	0·416	22·855
30			3·5	0·053	29·947	0·425	29·522	27·674	0·212	27·462	25·653	0·464	25·189
	33		3·5	0·053	32·947	0·425	32·522	30·674	0·212	30·462	28·653	0·464	28·189
36			4	0·060	35·940	0·475	35·465	33·342	0·224	33·118	31·033	0·512	30·521
	39		4	0·060	38·940	0·475	38·465	36·342	0·224	36·118	34·033	0·512	33·521
42			4·5	0·063	41·937	0·500	41·437	39·014	0·236	38·778	36·416	0·561	35·855
	45		4·5	0·063	44·937	0·500	44·437	42·014	0·236	41·778	39·416	0·561	38·855
48			5	0·071	47·929	0·530	47·399	44·681	0·250	44·431	41·795	0·611	41·184
	52		5	0·071	51·929	0·530	51·399	48·681	0·250	48·431	45·795	0·611	45·184
56			5·5	0·075	55·925	0·560	55·365	52·353	0·265	52·088	49·177	0·662	48·515
	60		5·5	0·075	59·925	0·560	59·365	56·353	0·265	56·088	53·177	0·662	52·515
64			6	0·080	63·920	0·600	63·320	60·023	0·280	59·743	56·559	0·713	55·846
	68		6	0·080	67·920	0·600	67·320	64·023	0·280	63·743	60·559	0·713	59·846

NOTE. Preference should be given to diameters in column 1, and if none in this column are suitable, a choice should be made next from column 2, and finally from column 3.

Table G34. **ISO METRIC THREAD**

Limits of Size for Coarse Series – Nuts (mm.)
Tolerance Class 6H

Nominal Diameter			Pitch	Fund Dev.	Major diam.	Effective Diameter			Minor Diameter		
First choice	Second Choice	Third Choice			min.	max.	tol.	min.	max.	tol.	min.
1			0·25			Tolerances not specified*					
	1·1		0·25								
1·2			0·25								
	1·4		0·3	0	1·400	1·280	0·075	1·025	1·160	0·085	1·075
1·6			0·35	0	1·600	1·458	0·085	1·373	1·321	0·100	1·22
	1·8		0·35	0	1·800	1·650	0·085	1·573	1·521	0·100	1·42
2			0·4	0	2·000	1·830	0·090	1·740	1·679	0·112	1·56
	2·2		0·45	0	2·200	2·003	0·095	1·908	1·838	0·125	1·71
2·5			0·45	0	2·500	2·303	0·095	2·208	2·138	0·125	2·01
3			0·5	0	3·000	2·775	0·100	2·675	2·599	0·140	2·45
	3·5		0·6	0	3·500	3·222	0·112	3·110	3·010	0·160	2·85
4			0·7	0	4·000	3·663	0·118	3·545	3·422	0·180	3·24
	4·5		0·75	0	4·500	4·131	0·118	4·013	3·878	0·190	3·68
5			0·8	0	5·000	4·605	0·125	4·480	4·334	0·200	4·13
6			1	0	6·000	5·500	0·150	5·350	5·153	0·236	4·91
		7	1	0	7·000	6·500	0·150	6·350	6·153	0·236	5·91
8			1·25	0	8·000	7·348	0·160	7·188	6·912	0·265	6·64
		9	1·25	0	9·000	8·348	0·160	8·188	7·912	0·265	7·64
10			1·5	0	10·000	9·206	0·180	9·026	8·676	0·300	8·37
		11	1·5	0	11·000	10·206	0·180	10·026	9·676	0·300	9·37
12			1·75	0	12·000	11·063	0·200	10·863	10·441	0·335	10·10
	14		2	0	14·000	12·913	0·212	12·701	12·210	0·375	11·83
16			2	0	16·000	14·913	0·212	14·701	14·210	0·375	13·83
	18		2·5	0	18·000	16·600	0·224	16·376	15·744	0·450	15·29
20			2·5	0	20·000	18·600	0·224	18·376	17·744	0·450	17·29
	22		2·5	0	22·000	20·600	0·224	20·376	19·744	0·450	19·29
24			3	0	24·000	22·316	0·265	22·051	21·252	0·500	20·75
	27		3	0	27·000	25·316	0·265	25·051	24·252	0·500	23·75
30			3·5	0	30·000	28·007	0·280	27·727	26·771	0·560	26·21
	33		3·5	0	33·000	31·007	0·280	30·727	29·771	0·560	29·21
36			4	0	36·000	33·702	0·300	33·402	32·270	0·600	31·67
	39		4	0	39·000	36·702	0·300	36·402	35·270	0·600	34·67
42			4·5	0	42·000	39·392	0·315	39·077	37·799	0·670	37·12
	45		4·5	0	45·000	42·392	0·315	42·077	40·799	0·670	40·12
48			5	0	48·000	45·087	0·335	44·752	43·297	0·710	42·58
	52		5	0	52·000	49·087	0·335	48·752	47·297	0·710	46·58
56			5·5	0	56·000	52·783	0·355	53·428	50·796	0·750	50·04
	60		5·5	0	60·000	56·783	0·355	56·428	54·796	0·750	54·04
64			6	0	64·000	60·478	0·375	60·103	58·305	0·800	57·50
	68		6	0	68·000	64·478	0·375	64·103	62·305	0·800	61·50

Note. Preference should be given to diameters in column 1, and if none in this colun are suitable, a choice should be made next from column 2, and finally from column

able G35. ISO METRIC THREAD

Normal length of thread engagement (mm.)

Nominal diameter		Pitch	Normal length of Thread Engagement	
Above	Up to and including		Above	Up to and including
0·99	1·50	0·20	0·5	1·4
		0·25	0·6	1·7
		0·30	0·7	2·0
1·50	2·80	0·20	0·5	1·5
		0·25	0·6	1·9
		0·35	0·8	2·6
		0·40	1·0	3·0
		0·45	1·3	3·8
2·80	5·60	0·35	1·0	3·0
		0·50	1·5	4·5
		0·60	1·7	5·0
		0·70	2·0	6·0
		0·75	2·2	6·7
		0·80	2·5	7·5
5·60	11·20	0·75	2·4	7·1
		1·00	3·0	9·0
		1·25	4·0	12·0
		1·50	5·0	15·0
11·20	22·40	1·00	3·8	11·0
		1·25	4·5	13·0
		1·50	5·6	16·0
		1·73	6·0	18·0
		2·00	8·0	24·0
		2·50	10·0	30·0
22·40	45·00	1·00	4·0	12·0
		1·50	6·3	19·0
		2·00	8·5	25·0
		3·00	12·0	36·0
		3·50	15·0	45·0
		4·00	18·0	53·0
		4·50	21·0	63·0
45·00	90·00	1·50	7·5	22·0
		2·00	9·5	28·0
		3·00	15·0	45·0
		4·00	19·0	56·0
		5·00	24·0	71·0
		5·50	28·0	85·0
		6·00	32·0	95·0
90·00	180·00	2·00	12·0	36·0
		3·00	18·0	53·0
		4·00	24·0	71·0
		6·00	36·0	106·0
180·00	355·00	3·00	20·0	60·0
		4·00	26·0	80·0
		6·00	40·0	118·0

Section H

Section H
HOROLOGICAL SCREWS

Swiss horological threads in current use are based on the ISO profil under the designations N.I.H.S.06-02 and N.I.H.S.06-05. For many yea the series of screw threads most generally used in the watch and cloc industry was that specified in Switzerland under the designation N.H.S 56100, now superseded by the threads of ISO form. A similar series specified in France, and a corresponding series, but with a 60-deg threa angle throughout, is specified in Germany. There is an indication that th original Thury thread is used to some extent on the Continent where it sometimes referred to as the Série Normal de Genève, and the B.A. seri (see p. 46) is employed for clock making in this country. A Unifie Miniature Series has also been formulated (see p. 80).

Some little use is made of the Progress System, especially by jobbin watch and clock makers. Such craftsmen also employ screw plates an taps made by various Continental firms having profiles, diameters an pitches of which published information appears to be lacking. In th U.S.A. some watch manufacturers work to their own standards.

Swiss Horological Threads

ISO Series (Fig. 1)

Swiss horological threads based on the ISO profile are specified i N.I.H.S.06-02 for nominal diameters ranging from 0·3 to 1·4 mm. (se Table H1). Alternative forms of crest and root on the screw thread a permitted.

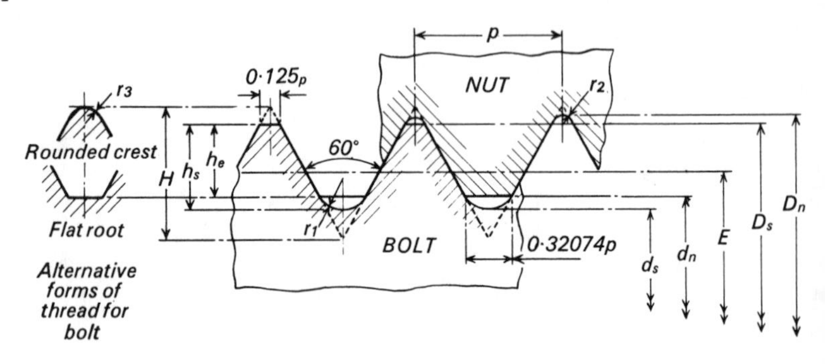

Fig. 1. Swiss Horological Thread Form based on ISO Profile

Thread angle (2θ) = 60 deg; Flank angle (θ) = 30 deg; Triangular heigh (H) = 0·86603p; Major diameter of screw (nominal diameter) = D, Minor diameter of nut (d_n) = D_s-0·96p; Effective diameter (E) = D 0·64952p; Minor diameter of screw (d_s) = D_s-1·12p; Depth of thread o screw (h_s) = 0·56p; Depth of engagement (h_e) = 0·48p; Root radius c ew (r_1) = 0·20074p; Crest radius of screw (r_3) = 0·10825p; Major ımeter (practical) of nut (D_n) = D_s+0·07217p; Root radius of nut (r_2) 0·07217p.

The Swiss standard N.I.H.S.06-05 covers threads of ISO form in ches from 0·08 to 0·3 mm.

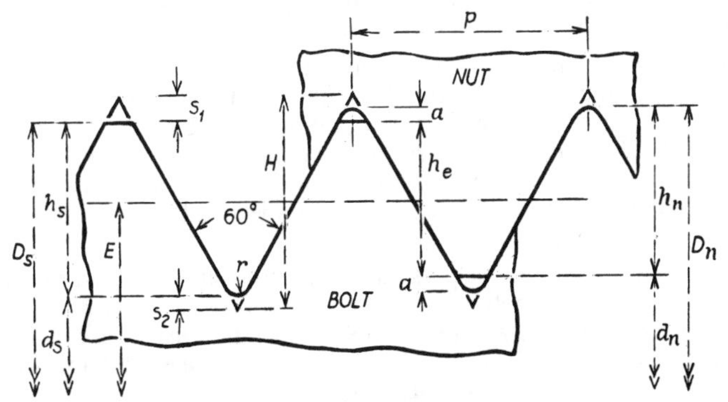

Fig. 2. Swiss 60-deg Horological Thread Form

)-deg Series (Fig. 2)

Thread angle (2θ) = 60 deg; Flank angle (θ) = 30 deg; Triangular height (H) = 0·86603 p; Shortening at crest (S_1) = 0·10825 p; Shortening at root = 0·05778 p; Depth of thread (h) = 0·7 p; Radial clearance at root and crest (a) = 0·05 p, (theoretical value = 0·05047 p); Radius at root (r) = 0·058 p.

'RENCH HOROLOGICAL SCREWS

The French horological screws are in two series as set forth in Tables I2 and H3. These closely correspond to the Swiss, with thread angles f 50 and 60 deg respectively.

0-deg Series

Thread angle (2θ) = 50 deg; Flank angle (θ) = 25 deg; Triangular height (H) = 1·07225 p; Shortening at crest (S_1) = 0·21113 p; Shortening at root (S_2) = 0·15712 p; Depth of thread (h) = 0·70400 p; Depth of engagement (h_e) = 0·650 p. Radial clearance at crest and root (a) = 0·054 p.

0-deg Series

Thread angle (2θ) = 60 deg; Flank angle (θ) = 30 deg; Triangular Height (H) = 0·86603 p; Shortening at crest (S_1) = 0·10802 p; Shortening at root (S_2) = 0·05401 p; Depth of thread (h) = 0·70400 p; Depth of engagement (h_e) = 0·650 p. Radial clearance at crest and root (a) = 0·054 p.

JERMAN THREADS FOR HOROLOGICAL PURPOSES

The threads specified in Germany for horological purposes are those :onstituting the small sizes in the normal metric range. These closely :orrespond to the French and Swiss series. The basic form and proportions ıre given on page 135 under German Metric and the dimensions up to ! mm. are given in Table H4.

THE PROGRESS SYSTEM

The form of the Progress thread is usually defined with the followin proportions and angles.

Thread angle (2θ) = 50 deg; Flank angle (θ) = 25 deg; Shortenin at crest and root (S) = 0·1 p; Depth of thread (h) = 0·8 p.

A little investigation will, however, show that these proportions ar mathematically inconsistent. For θ = 25 deg, the triangular height (H = 1·0722535 p, and a depth of thread h equal to 0·8 p yields a value of equal to 0·1361268 p. The corresponding radius r at crest and root i 0·0996389 p which to the nearest figure in the third place is 0·100 p. I seems probable therefore that the original specification must have give r = 0·1 p and not S = 0·1 p.

In the Progress series of screws each size is given a designating numbe which indicates the diameter of the screw in tenths of a millimetre. Thu a No. 9½ screw has a diameter of 0·95 mm. The list of diameters an pitches usually given is reproduced in Table H5 where each diameter abov 1·5 mm. may have one of two pitches.

THURY THREAD

This thread was originally specified by Professor Thury to have th following proportions:

Thread angle (2θ) = 47 deg 30 mins; Flank angle (θ) = 23 de 45 mins; Triangular height (H) = 1·13635 p; Depth of thread (h) = 0·6 p; Radius at crest (r_1) = $\frac{1}{6}$ p; Radius at root (r_2) = $\frac{1}{5}$ p.

These figures are, however, geometrically inconsistent. For taking θ as 23 deg 45 mins, $H = \frac{1}{2}p \cot \theta$ = 1·1363365 p, and shortening at cres = r_1 (cosec θ – 1) = 0·2471584 p. Also shortening at root = 0·2965901 p

Thus the depth of thread should be:

$$1{\cdot}1363365\,p - 0{\cdot}5437485\,p = 0{\cdot}5925880\,p, \text{ and not } 0{\cdot}6\,p.$$

How these inconsistencies are reconciled in practice is not known, sinc no authorised standard exists for the Thury series.

The Thury screws are numbered exactly as the B.A. screws, witl which they have identical pitches, major diameters and depths of threac with the exception of the following:

Thury No.	Diam.	Pitch	Depth of Thread
22	0·37	0·098	0·059
23	0·33	0·089	0·053
25	0·25	0·072	0·043

able H1.

WISS HOROLOGICAL THREAD – I.S.O. SERIES (N.I.H.S. 06-02)

(Dimensions in mm.)

omi-nal Size	Pitch	Effect-ive Dia-meter E	Screw					Nut		
			Major Dia-meter D_s	Minor Dia-meter d_s	Depth of Thread h_s	Root Radius r_1	Crest Radius r_3	Major Dia-meter D_n	Minor Dia-meter d_n	Root Radius r_2
0·30	0·08	0·248	0·30	0·210	0·045	0·016	0·009	0·306	0·223	0·006
0·35	0·09	0·292	0·35	0·250	0·050	0·018	0·010	0·356	0·264	0·006
0·40	0·10	0·335	0·40	0·288	0·056	0·020	0·011	0·407	0·304	0·007
0·50	0·125	0·419	0·50	0·360	0·070	0·025	0·014	0·509	0·380	0·009
0·60	0·15	0·503	0·60	0·432	0·084	0·030	0·016	0·611	0·456	0·011
0·70	0·175	0·586	0·70	0·504	0·098	0·035	0·019	0·713	0·532	0·013
0·80	0·20	0·670	0·80	0·576	0·112	0·040	0·022	0·814	0·608	0·014
0·90	0·225	0·754	0·90	0·648	0·126	0·045	0·024	0·916	0·684	0·016
1·00	0·250	0·883	1·00	0·720	0·140	0·050	0·027	1·018	0·760	0·018
1·20	0·250	1·038	1·20	0·920	0·140	0·050	0·027	1·218	0·960	0·018
1·40	0·300	1·205	1·40	1·064	0·168	0·060	0·032	1·422	1·112	0·022

Table H2. FRENCH 50-DEG. SERIES (BASIC SIZES)
(Dimensions in mm.)

Nom. Diam.	Pitch	Effective Diam.	Major Diam.	Minor Diam.
0·3	0·075	0·251	0·3	0·194
0·35	0·075	0·301	0·35	0·244
0·4	0·100	0·335	0·40	0·259
0·45	0·100	0·385	0·45	0·309
0·5	0·125	0·419	0·50	0·324
0·55	0·125	0·469	0·55	0·374
0·6	0·150	0·502	0·60	0·389
0·7	0·150	0·602	0·70	0·489
0·8	0·200	0·670	0·80	0·518
0·9	0·200	0·770	0·90	0·618

Table H3. FRENCH 60-DEG. SERIES (BASIC SIZES)
(Dimensions in mm.)

Nom. Diam.	Pitch	Effective Diam.	Major Diam.	Minor Diam.
1	0·250	0·837	1·000	0·650
1·2	0·250	1·037	1·200	0·850
1·4	0·300	1·205	1·400	0·980
1·6	0·300	1·405	1·600	1·180
1·8	0·400	1·540	1·800	1·240
2·0	0·400	1·740	2·000	1·440
2·2	0·450	1·907	2·200	1·570
2·5	0·450	2·207	2·500	1·870

able H4. GERMAN METRIC SERIES (SMALL SIZES)
(Dimensions in mm.)

Nom. Diam.*	Pitch	Effective Diam.	Minor Diam.	Depth of Thread	Radius at Root (approx.)
0·3	0·075	0·251	0·202	0·049	0·01
0·4	0·100	0·335	0·270	0·065	0·01
0·5	0·125	0·419	0·338	0·081	0·01
0·6	0·150	0·503	0·406	0·097	0·02
0·7	0·175	0·586	0·472	0·114	0·02
0·8	0·200	0·670	0·540	0·130	0·02
0·9	0·225	0·754	0·608	0·146	0·02
1·0	0·250	0·838	0·676	0·162	0·03
1·2	0·250	1·038	0·876	0·162	0·03
1·4	0·300	1·205	1·010	0·195	0·03
1·7	0·350	1·473	1·246	0·227	0·04
2·0	0·400	1·740	1·480	0·260	0·04

*Also Basic Major Diameter.

Table H5. **PROGRESS SYSTEM THREADS**

(Dimensions in mm.)

No.	Major Diam.	Pitch	Depth of Thread	Effective Diam.
4	0·40	0·100	0·080	0·320
4½	0·45	0·100	0·080	0·370
5	0·50	0·125	0·100	0·400
5½	0·55	0·125	0·100	0·450
6	0·60	0·150	0·120	0·480
6½	0·65	0·150	0·120	0·530
7	0·70	0·175	0·140	0·560
7½	0·75	0·175	0·140	0·610
8	0·80	0·200	0·160	0·640
8½	0·85	0·200	0·160	0·690
9	0·90	0·225	0·180	0·720
9½	0·95	0·225	0·180	0·770
10	1·0	0·250	0·200	0·800
11	1·1	0·275	0·220	0·880
12	1·2	0·300	0·240	0·960
13	1·3	0·325	0·260	1·040
14	1·4	0·350	0·280	1·120
15	1·5	0·375	0·300	1·200
16	1·6	0·320	0·256	1·344
16	1·6	0·457	0·366	1·234
17	1·7	0·340	0·272	1·428
17	1·7	0·486	0·389	1·311
18	1·8	0·360	0·288	1·512
18	1·8	0·514	0·411	1·389
19	1·9	0·380	0·304	1·596
19	1·9	0·543	0·434	1·466
20	2·0	0·400	0·320	1·680
20	2·0	0·571	0·457	1·543

Table H6. ELGIN WATCH SCREWS

Major Diam. (ins.)	Threads per inch	Major Diam. (ins.)	Threads per inch
0·0132	360	0·0428	120
0·0148	320	0·0448	110
0·0168	260	0·0468	110
0·0208	220	0·0488	140
0·0228	260	0·0488	200
0·0248	220	0·0508	110*
0·0268	180	0·0548	180
0·0288	220	0·0608	110
0·0308	180	0·0608	110*
0·0308	220	0·0708	180*
0·0368	140	0·0768	110*
0·0368	220	0·0772	80*
0·0408	120*	0·0892	80*
0·0408	200		

*Left Handed

Table H7. WALTHAM WATCH SCREWS

No.	Threads per inch	Major Diam. (mm.)	Tap Drill Diam. (mm.)
1	110	1·50	1·32
3	110	1·20	1·02
5	120	1·10	0·95
7	140	1·00	0·85
9	160	0·93	0·71
11	170	1·34	1·22
13	180	1·00	0·85
15	180	0·83	0·71
17	200	0·65	0·54
19	220	0·55	0·45
21	240	0·45	0·34
23	254	0·35	0·27

Table H8. ISO METRIC SERIES (SMALL SIZES)
(Dimensions in mm.)

Nom. Diam.	Pitch	Effective Diam.	Minor Diam. (Screw)	Minor Diam. (Nut)	Depth of Thread
0·25	0·075	0·201	0·158	0·169	0·046
0·3	0·08	0·248	0·202	0·213	0·049
0·35	0·09	0·292	0·240	0·253	0·055
0·4	0·1	0·335	0·277	0·292	0·061
0·45	0·1	0·385	0·327	0·342	0·061
0·5	0·125	0·419	0·347	0·365	0·077
0·55	0·125	0·469	0·397	0·415	0·077
0·6	0·15	0·503	0·416	0·438	0·092
0·7	0·175	0·586	0·485	0·511	0·107
0·8	0·2	0·670	0·555	0·583	0·123
0·9	0·225	0·754	0·624	0·656	0·138

Section J

Section J

WIRE METHODS OF MEASUREMENT; ERRORS DUE TO PITCH AND FLANK ANGLE DEVIATIONS; EFFECT OF EFFECTIVE DIAMETER ON TOOTH THICKNESS

Wire Methods of Measuring Effective Diameter

For precision determination of the effective diameter of screw thread the various modifications of the wire method are widely applied, especiall in the checking of screw-plug gauges. The accuracy of the results obtaine are dependent, not only on reasonable practical care being taken an on the accuracy of equipment used, but upon the validity of the formul used for effecting the necessary calculations. In addition, for very precis measurements, a correction may be required to offset the elastic deform tion of the measuring wires and the screw thread when subjected to th pressure exerted by the anvils of the micrometer or other instrument.

The formulæ available may be classified in order of accuracy as follow

(i) Formulæ in which the effect of rake or helix angle is entire neglected and which can be applied for most purposes with out serious error to screws of small helix angle.

(ii) Formulæ in which the effect of rake or helix angle is approx mately provided for and which can be applied to screws o moderate helix angle.

(iii) Formulæ in which provision is made for the exact effect o helix angle, and which should be used for screws of high hel angle, such as, for example, very coarse-pitch and multi-sta acme threads.

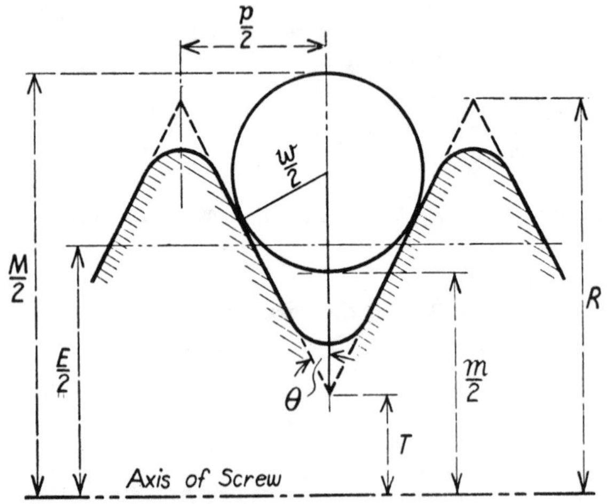

Fig. 1. Gauging the Effective Diameter of Parallel Screw Threads of Symmetrical Prof and Negligible Rake Angle by measuring Over Wires

SCREW OF SMALL HELIX ANGLE

The equations connecting the wire diameter, the effective diameter, and the measurement over the wires for screws of negligible helix angle and having flanks that are symmetrically inclined to the screw axis may be derived as follows:

Referring to Fig 1:

$\frac{1}{2} E = \frac{1}{2} (R + T)$, where E is the effective diameter

$R = T + \frac{1}{2} p \cot \theta$; hence

$$E = 2T + \frac{1}{2} p \cot \theta \qquad \text{(1)}$$

Also, $\frac{1}{2} M = T + \frac{1}{2} w (\operatorname{cosec} \theta + 1)$

i.e. $$2T = M - w (\operatorname{cosec} \theta + 1) \qquad \text{(2)}$$

Substituting (2) in (1) yields

$$E = M - w (\operatorname{cosec} \theta + 1) + \frac{1}{2} p \cot \theta \qquad \text{(3)}$$

or, $$M = E - \frac{1}{2} p \cot \theta + w (\operatorname{cosec} \theta + 1) \qquad \text{(4)}$$

Denoting $\frac{1}{2} \cot \theta$ by X, and $(\operatorname{cosec} \theta + 1)$ by Y, formula (4) becomes

$$M = E - p X + w Y \qquad \text{(5)}$$

For simplifying calculation, values of pX and $w_b Y$ for various screw systems, where w_b is the "best size" wire diameter, may be found in Tables at the end of this section.

Sometimes the measurement is taken under the wires, hence denoting this dimension by m, formulæ (3) and (4) take the following forms:

$$E = m - w (\operatorname{cosec} \theta - 1) + \frac{1}{2} p \cot \theta \qquad \text{(3')}$$

$$m = E - \frac{1}{2} p \cot \theta + w (\operatorname{cosec} \theta - 1) \qquad \text{(4')}$$

or, denoting $(\operatorname{cosec} \theta - 1)$ by Z, and $\frac{1}{2} \cot \theta$ by X

$$m = E - p X + w Z \qquad \text{(5')}$$

Values of $w_b Z$ for the standard screw systems are given in tables at end of this section, and values of X, Y and Z in Table J1.

Table J1. VALUES OF X, Y AND Z FOR DIFFERENT THREAD ANGLES

	θ	$X = \frac{1}{2} \cot \theta$	$Y = \operatorname{cosec} \theta + 1$	$Z = \operatorname{cosec} \theta - 1$
1	30° 0′	0·8660254	3·0000000	1·0000000
2	27° 30′	0·9604911	3·1656806	1·1656806
3	26° 34′	0·9999295	3·2359419	1·2359419
4	25° 0′	1·0722534	3·3662016	1·3662016
5	23° 45′	1·1363365	3·4829503	1·4829503
6	15° 0′	1·8660254	4·8637033	2·8637033
7	14° 30′	1·9333565	4·9939292	2·9939292
8	5° 0′	5·7150260	12·4737133	10·4737133

The above values are applicable to the various thread forms as follows:

(1) Unified; 60-deg Metric; American National; British Standard Cycle; 60-deg Horological; American 60-deg Stub.

(2) All Whitworth forms: B.S.W.; B.S.F.; Brass; B.S.P.; Condu
Copper Tube.
(3) Löwenherz
(4) Progress System; 50-deg Horological
(5) B.A.; Thury
(6) Trapezoidal Metric
(7) Acme; American 29-deg Stub.
(8) American 10-deg Modified Square.

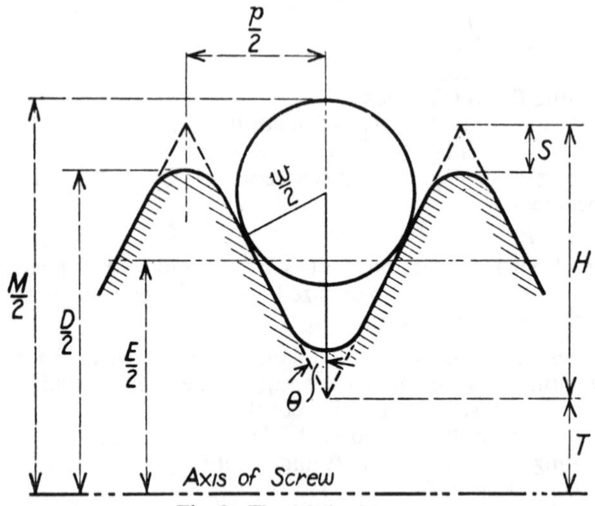

Fig. 2. The 3-Wire Method

THE 3-WIRE METHOD

A method of checking screws that is based on the major diameter often used. This is termed the "3-wire Method" and the necessa basic formulæ may be derived as follows:

Referring to Fig. 2

$$T = \tfrac{1}{2} M - \tfrac{1}{2} w (1 + \operatorname{cosec} \theta) \quad \ldots \quad (6)$$

Also, $T = \tfrac{1}{2} D + s - H$, but

$H = \tfrac{1}{2} p \cot \theta$, hence

$$T = \tfrac{1}{2} D + s - \tfrac{1}{2} p \cot \theta \quad \ldots \quad (7)$$

Equating the right-hand sides of (6) and (7), and re-arrangir yield:

$$M = D + 2 s - p \cot \theta + w (1 + \operatorname{cosec} \theta) \quad \ldots \quad (8)$$

Since s can be expressed in terms of p, equation (8) can be writte $M = D - V p + w Y$.

Useful numerical forms of this equation for various systems ar

Whitworth

$M = D - 1{\cdot}6008183\, p + 3{\cdot}1656806\, w$

Truncated Whitworth

$M = D - 1{\cdot}4529833\, p + 3{\cdot}1656806\, w$

B.A.

$M = D - 1{\cdot}7363364\, p + 3{\cdot}4829503\, w$

Löwenherz

$M = D - 1{\cdot}7498766\, p + 3{\cdot}2359419\, w$

B.S.C.

$M = D - 1{\cdot}3987175\, p + 3w$

American National and Unified

$M = D - 1{\cdot}5155444\, p + 3\, w$

Metric

$M = D - 1{\cdot}5155444\, p + 3\, w$

Acme

$M = D - 2{\cdot}433566\, p + 4{\cdot}9939292\, w$ — Effective Diameter Allowance (see page 108)

Values of Vp for the usual systems are tabulated at the end of this ection.

When in a given thread form, the amount of truncation or shortening rom the full height of the fundamental triangle (triangular height) can e expressed as a fraction $\frac{1}{k}$ of the triangular height H, formula (8) may e written:

$$M = D - \frac{k-1}{k}\, p \cot \theta + w\,(1 + \operatorname{cosec} \theta)$$

e.g. for Whitworth form, $\frac{1}{k} = \frac{1}{6}$

It may happen that the thread angle 2θ is not exactly the standard pecified value. The values of $\frac{1}{2} \cot \theta$ (= X) and $\operatorname{cosec} \theta + 1$ (= Y) will hen have to correspond to the actual value of θ. These may be obtained or a reasonable range from Tables J14 to J16 at the end of this section.

Best Size" Wire

To minimise the effect due to errors in the thread angle 2θ, it is preferble for the wires to make contact at the effective diameter. When the elix angle of the screw is neglected, this requires a fixed diameter of wire or a given combination of angle θ and pitch p, the wire diameter then eing

$$w_b = \tfrac{1}{2}\, p \sec \theta.$$

Calculated values of w_b will be found tabulated at the end of this ection, together with some recommended limits of variation from the heoretical value.

It is of interest to note that the value $\frac{1}{2}p \sec \theta$ can be derived directly rom the assumption that the errors in measurement are zero, notwithtanding a variation in the flank angle θ.

Equation (3) may be put into the form:

$$M - E = w\,(\operatorname{cosec} \theta + 1) - \tfrac{1}{2}\, p \cot \theta.$$

Differentiating with respect to θ,

$$\frac{d\,(M-E)}{d\,\theta} = -\, w \operatorname{cosec} \theta \cot \theta + \tfrac{1}{2}\, p \operatorname{cosec}^2 \theta.$$

If $\frac{d\,(M-E)}{d\,\theta}$ is to be zero, then $w = \frac{1}{2}\, p \sec \theta$.

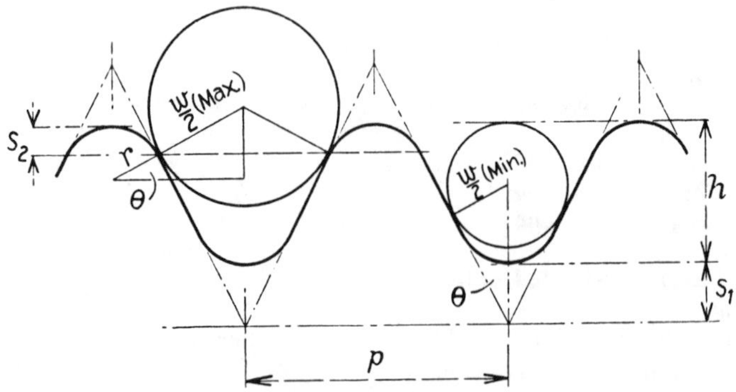

Fig. 3. Practical Maximum and Minimum Diameters of Wires

PRACTICAL LIMITS OF WIRE DIAMETER

The conditions for maximum wire diameter are based on the extrem possible position of the point of contact with the straight portion of th thread flank. Thus, referring to the left-hand of Fig. 3,

$$(\tfrac{1}{2}\, w_{max} + r) \cos \theta = \tfrac{1}{2}p,$$

i.e. $w_{max} = p \sec \theta - 2r$ .. .. (9)

The minimum diameter of the wire is governed by the fact that fc gauging to be practically possible, the top of the wire must not be belo the crests of the adjacent threads. Hence, referring to the right-han side of Fig. 3,

$$\tfrac{1}{2}\, w_{min} \operatorname{cosec} \theta + \tfrac{1}{2}\, w_{min} = h + s_1$$

i.e. $w_{min} = \dfrac{2(h + s_1)}{\operatorname{cosec} \theta + 1}$ .. .. (10)

SCREWS OF MODERATE HELIX ANGLE

The formulæ so far given do not take into account the helix angle of th thread at the points of contact between wire and flanks. The result however, thereby obtained are sufficiently accurate for the normal typ of screw for which the error is in the range 0·00001 to 0·0001-inch. Fc threads of relatively coarser pitch some allowance must be made for th helix angle.

A relation which is useful in such cases is:

$$E = M + \tfrac{1}{2}p \cot \theta - w\,[1 + (\operatorname{cosec}^2 \theta + \cot^2 \theta \tan^2 \lambda)^{\frac{1}{2}}] \qquad (11)$$

where λ is the angle between the axis of the wire and a plane perpendicula to the axis of the thread. For most purposes, λ is taken equal to th helix angle of the thread at the point of contact, which for effective diameter contact is $\lambda = \tan^{-1} \dfrac{l}{\pi E}$, l being the lead.

By expanding the square root as far as the second term, equation (11 takes the form:

$$E = M + \tfrac{1}{2}p \cot \theta - w\,(1 + \operatorname{cosec} \theta) - \tfrac{1}{2}\, w \tan^2 \lambda \cos \theta \cot \theta \quad (12)$$

or, $E = M + \tfrac{1}{2}\, p \cot \theta - w\,(1 + \operatorname{cosec} \theta) - C$ .. .. .. (12′)

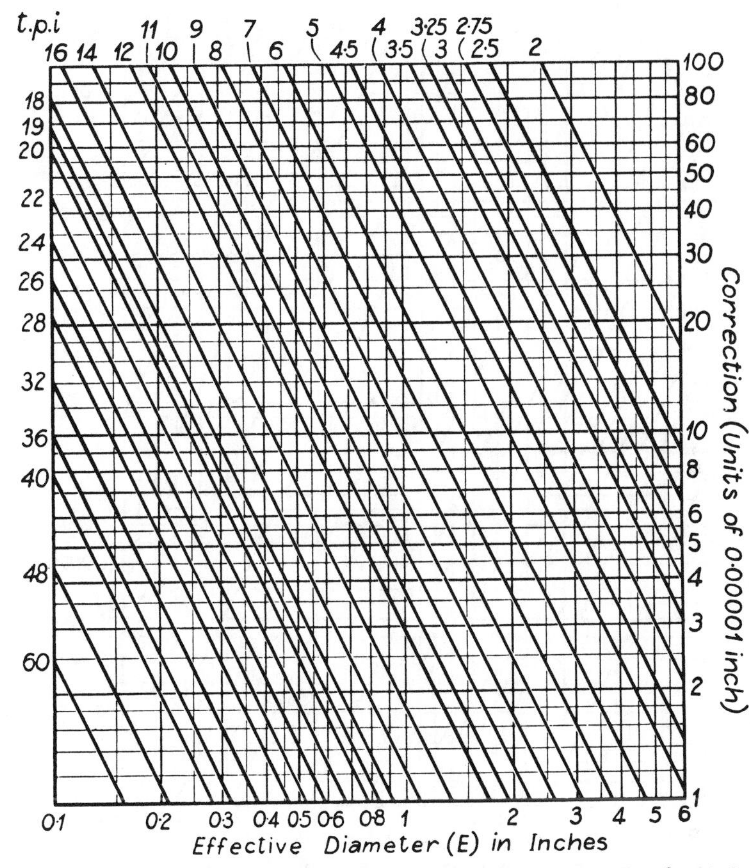

Fig. 4. Rake Correction for Whitworth ($\theta=27\frac{1}{2}$ deg) Threads. For $\theta=30$ deg Threads, multiply the Chart Readings by 0·902

This latter is therefore equivalent to equation (3) page 199, with the dditional negative correction term

$$C = \tfrac{1}{2}\, w \tan^2 \lambda \cos \theta \cot \theta.$$

When contact at the effective diameter is required, and λ is assumed ɔ be equal to the mean helix angle, w may be assumed equal to $= \frac{1}{2} p \sec \theta$, and the correction term becomes:

$$C = \frac{p^3}{4\pi^2 E^2} \cot \theta \qquad \qquad (13)$$

Values of C for different values of p and E for Whitworth form may be ɔund from Fig. 4. Values for $\theta = 30$ deg may be found from Fig. 4, by ıultiplying the readings obtained by 0·902. Corrections for B.A. screws re given in Table J2. Values of C in mm. for metric threads may be read om Fig. 5.

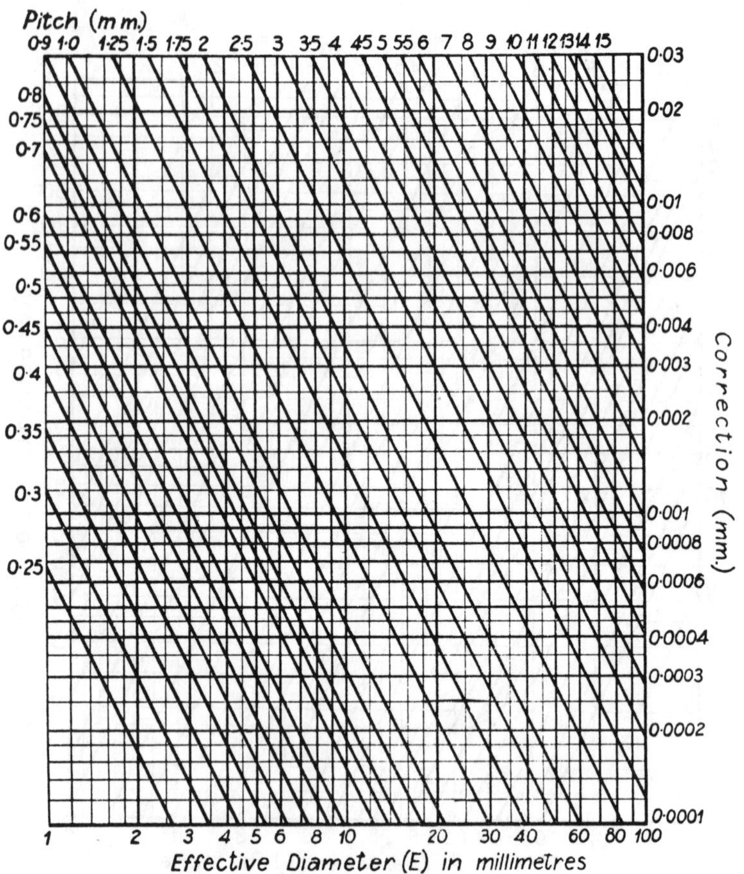

Fig. 5. Rake Correction for Metric Threads

Derivation of the Correction "C" and formula (11)

Assuming the axis of the wires to be inclined at the helix angle of th thread, then the normal thread angle will be θ_n as given by

$$\tan \theta_n = \tan \theta \cos \lambda$$

The term $(1 + \text{cosec}\ \theta)$ in formula (3) then becomes $(1 + \text{cosec}\ \theta_n$ which by the above relation is

$$\begin{aligned} 1 + \text{cosec}\ \theta_n &= 1 + (1 + \cot^2 \theta \sec^2 \lambda)^{\frac{1}{2}} \\ &= 1 + [\text{cosec}\ \theta\ (1 + \cos^2 \theta \tan^2 \lambda)^{\frac{1}{2}}] \\ &= 1 + (\text{cosec}^2\ \theta + \cot^2 \theta \tan^2 \lambda)^{\frac{1}{2}}, \end{aligned}$$

the coefficient of w in the last term of (11). The derivation of "C" the follows as previously given.

Screws of High Helix Angle

When the screw to be measured has such a high helix angle that th correction calculated by equation (13) exceeds about 0·001 inch or mor than the degree of accuracy required, methods of calculation based on th

:act basic geometry must be employed. An approximate solution of the :neral problem was given by H. H. Jeffcott in 1907 upon which the ethods previously dealt with were based.

In 1946, a new set of formulæ were developed in the U.S.A. by r. W. F. Vogel of Wayne University, Detroit, at the request of The Van euren Co. These give a very accurate result by direct calculation. In hat follows, this will be described as method (1). In addition, there has cently been developed in this country a method which is capable of ıy degree of theoretical accuracy; this will be described as method (2).

ethod (1)

The wire diameter is found from

$$w = E \frac{\sin \psi}{\sin \lambda \cos. (\theta' + \psi_2)} \qquad (14)$$

r Whitworth and 60-deg threads, where

$$\psi = \frac{90}{n} \sin^2 \lambda; \ \psi_2 = \frac{90}{n} \sin^3 \lambda; \ \theta' = \tan^{-1} (\tan \theta \cos \lambda); \text{ and}$$

n = number of starts.

· from

$$w = E \frac{\sin \lambda \sec \theta'}{0{\cdot}63662n - \tan \theta' \sin^3 \lambda}, \text{ for Acme threads}$$

$$M = E \frac{\cos \varphi}{\cos \eta} + w \qquad (15)$$

where φ is found from

$$\tan \varphi = \frac{\tan \theta'}{\sin \lambda} \qquad (16)$$

and η is obtained from

$$\eta = \varphi + \beta \qquad (17)$$

The angle β is determined from

$$\tan \beta \frac{\sin \lambda \cos \theta'}{\frac{E}{w} + \sin \theta'} \qquad (18)$$

ethod (2)

This is based on the pair of simultaneous equations:

$$\tfrac{1}{2} w = \left(\frac{p}{4} - \frac{\beta l}{2\pi}\right)\sqrt{1 + \tan^2 \lambda + \tan^2 \theta} \qquad (19a)$$

$$\frac{1}{w} = \frac{1}{E}\left(\frac{\cot \beta \tan \lambda - \tan \theta}{\sqrt{1 + \tan^2 \lambda + \tan^2 \theta}}\right) \qquad (19b)$$

In these, 2β represents the angle between the points of contact ıbtended at the axis and measured in a plane perpendicular to the axis; l = the lead of the screw.

First of all, the angle β is calculated from

$$\tan \beta = f\left(\tfrac{1}{2} + \tfrac{1}{8}g\right) \qquad (20)$$

$$\text{where } f = \frac{p \tan \lambda}{\frac{p}{2} \tan \theta + E \sec^2 \lambda} \qquad (21)$$

$$\text{and } g = \frac{2Ep \tan^2 \lambda \tan \theta}{\left(\frac{p}{2} \tan \theta + E \sec^2 \lambda\right)^2} \quad \ldots \quad (22)$$

The angle β (radians) is then substituted in equation (19a) and th corresponding value of w calculated, which is the diameter of wire t contact at the effective diameter E. A check on the accuracy of result ma be obtained by calculating w from (19b). The value of β from (19a) then inserted in the equation

$$M = 2\left[\left(\frac{p}{4} - \frac{\beta l}{2\pi}\right) \tan \theta + \frac{E}{2}\right] \sec \beta + w \quad \ldots \quad (23)$$

M being the measurement over the wires.

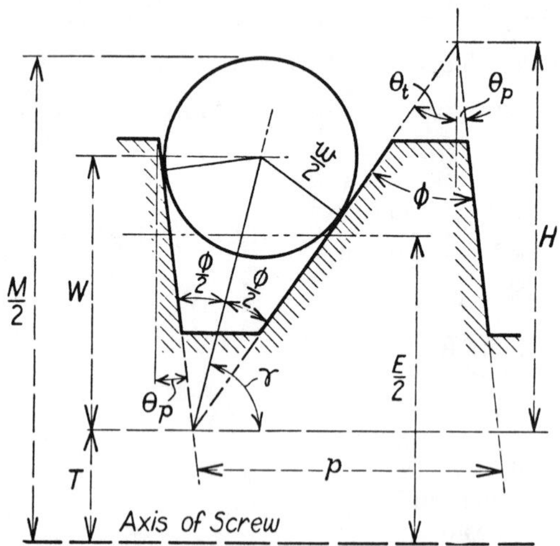

Fig. 6. The Wire Method for Asymmetrical Threads

ASYMMETRICAL THREAD FORMS

When the opposing flanks of a screw thread form are not inclined a an equal angle to the screw axis, a formula for measurement over wire which does not take the helix angle into consideration may be derived a follows.

Referring to Fig. 6.

$\varphi = \theta_t + \theta_p$; and

$\gamma = 90° - \theta_t + \frac{1}{2}(\theta_t + \theta_p) = 90° - \frac{1}{2}(\theta_t - \theta_p)$

Hence, $\sin \gamma = \cos \frac{1}{2}(\theta_t - \theta_p)$

Also, $W = \frac{w}{2} \operatorname{cosec} \frac{\varphi}{2} \sin \gamma$ and

$\frac{1}{2} M = W + T + \frac{1}{2} w$. Therefore

$\frac{1}{2} M - T = \frac{1}{2} w\left(1 + \operatorname{cosec} \frac{\varphi}{2} \sin \gamma\right)$ i.e.

$$M - T = \frac{1}{2} w \left[1 + \operatorname{cosec} \tfrac{1}{2}(\theta_t + \theta_p) \cos \tfrac{1}{2}(\theta_t - \theta_p)\right] \quad \ldots \quad (24)$$

Further

$$\tfrac{1}{2} E - T = \tfrac{1}{2} H = \frac{p}{2\,(\tan\theta_p + \tan\theta_t)} \quad \ldots \quad (25)$$

Subtracting (25) from (24) yields:

$$M - E = w\,[1 + \operatorname{cosec} \tfrac{1}{2}(\theta_t + \theta_p) \cos \tfrac{1}{2}(\theta_t - \theta_p)] - p\,/\,(\tan\theta_p + \tan\theta_t) \quad \ldots \quad (26)$$

Denoting $[1 + \operatorname{cosec} \tfrac{1}{2}(\theta_t + \theta_p) \cos \tfrac{1}{2}(\theta_t - \theta_p)]$ by Y_a, and $1/(\tan\theta_p + \tan\theta_t)$ by X_a, formula (26) may be written:

$$M - E = w\,Y_a - p\,X_a \quad \ldots \quad (27)$$

The size of wire which contacts the θ_p-flank at the effective diameter is given by

$$w_b = p\left(\frac{\tan \tfrac{1}{2}(\theta_p + \theta_t) \sec \theta_p}{\tan\theta_p + \tan\theta_t}\right) \quad \ldots \quad (28)$$

For 7°/45° Buttress Threads

$\theta_p = 7$ deg; $\theta_t = 45$ deg; and $M - E = 3{\cdot}1568906\,w - 0{\cdot}8906428\,p$.

The diameter of wire recommended is $w_b = 0{\cdot}43766\,p$ with a maximum wire of $0{\cdot}47697\,p$.

For 0°/52° Buttress Threads

$\theta_p = 0$ deg; $\theta_t = 52$ deg; and $M - E = 3{\cdot}0503037\,w - 0{\cdot}7812856\,p$.

The diameter of wire recommended is $w_b = 0{\cdot}42008\,p$ and the maximum diameter, $0{\cdot}45910\,p$.

For Telegraph Insulator Thread (Cordeaux)

$\theta_p = 29$ deg; $\theta_t = 35$ deg; and $M - E = 2{\cdot}8844937\,w - 0{\cdot}7971198\,p$.

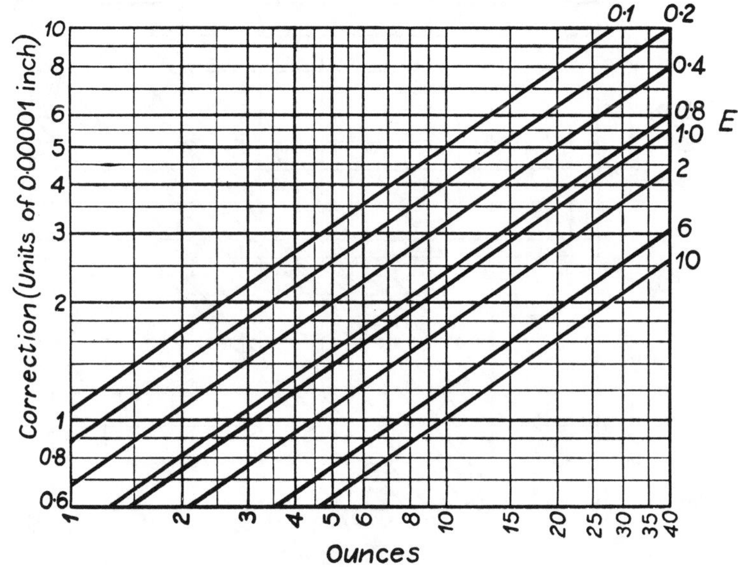

Fig. 7. Compression Correction for θ=30 deg Threads. For θ=27½ deg, multiply by 1⅓

Correction to Compensate for Elastic Deformation

In the precision measurement of screw threads by wire method it ma be advisable to apply a correction to compensate for the elastic deformatic of the wire and that of the screw thread. Beside a flattening action c the wire, there exists a wedging action of the wire in the screw-threa groove. This problem has received the attention of the National Physic Laboratory who have arrived at the empirical relation

$$j = kA^{\frac{2}{3}}E^{-\frac{1}{3}} \quad \ldots \quad (29)$$

where: j = the amount of correction in inches; A = the measurin pressure in lbs., and E = the effective diameter in inches. The factor has the following values:

60-deg threads: $k = 0{\cdot}00003$
55-deg ,, : $k = 0{\cdot}00004$
$47\frac{1}{2}$-deg ,, : $k = 0{\cdot}00006$

Logarithmic plots for 60-deg and B.A. threads are given in Fig. 7 ar 8 respectively. Threads with a 55-deg angle have corrections amountir to 4/3 times those given by Fig. 7.

Taper Threads

Due to the variation in diameter along a taper thread, it is necessa in formulating standards to specify the position of the transverse plane i which the effective diameter shall comply with the specified dimensio It should be noted too that the pitch surface of a taper thread is a con

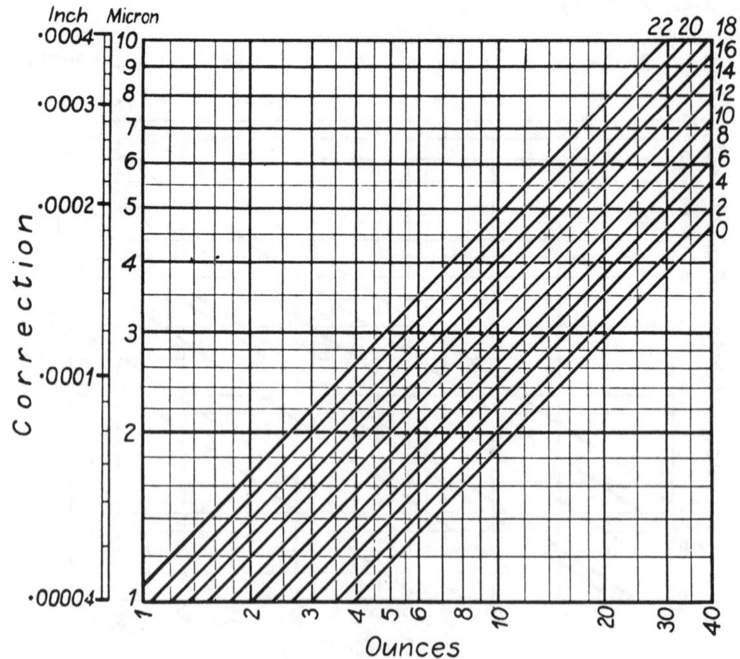

Fig. 8. Elastic Compression Correction for B.A. ($\theta = 23\frac{3}{4}$ deg) Screws

In methods of effective-diameter measurement using wires, one wire ; placed so that the axial section containing the contact points (ignoring ıe rake angle) passes through the root apex of the thread at the specified osition, which is usually given as a dimension from the end of the screw r pipe. Two other wires are placed in the two opposite thread spaces ıat are symmetrically disposed to that containing the first wire.

Two-Wire" Method

With the three wires arranged as explained above, measurements are ıken over the single wire and each of the other wires in turn, keeping ıe anvil planes of the micrometer parallel to the screw axis. If these dimen-ions be denoted by M_1 and M_2, the effective diameter is calculated from

$$E = \tfrac{1}{2}(M_1 + M_2) + \tfrac{1}{2}p(\cot\theta - \tan^2\beta\tan\theta) - w(1 + \operatorname{cosec}\theta) - \tfrac{1}{2}w\tan^2\lambda\cos\theta\cot\theta \qquad (30)$$

The angle β is the semi-conical angle of taper. For threads of smaller elix angle the last term, the correction "C" may be omitted. When equired, this correction may be obtained as for parallel threads.*

For the various systems of taper thread the terms $\frac{1}{2}p(\cot\theta - \tan^2\beta\tan\theta) - w(1 + \operatorname{cosec}\theta)$ are numerically as follows:

British Standard Taper Pipe ($\theta = 27\frac{1}{2}°$; taper = 1 in 16 on diameter)
$0{\cdot}9602368\,p - 3{\cdot}1656806\,w$.

American Taper Pipe ($\theta = 30°$; taper = 1 in 16 on diameter)
$0{\cdot}8657435\,p - 3\,w$.

A.P.I. Threads ($\theta = 30°$)
(Taper 1 in 16) $0{\cdot}8657435\,p - 3\,w$
(,, 1 in 6) $0{\cdot}8640207\,p - 3\,w$
(,, 1 in 4) $0{\cdot}8615149\,p - 3\,w$

Water Well Casing ($\theta = 27\frac{1}{2}°$; taper = 1 in 64 on diameter)
$0{\cdot}9604752\,p - 3{\cdot}1656806\,w$.

DERIVATION OF THE 2-WIRE FORMULA FOR TAPER THREADS

Referring to Fig. 9.

$$\tfrac{1}{2}M = T + \tfrac{1}{2}w + \tfrac{1}{2}w\operatorname{cosec}\theta$$
$$\tfrac{1}{2}E = T + UV + VJ$$

Hence,

$$\tfrac{1}{2}E = \tfrac{1}{2}M + UV + VJ - \tfrac{1}{2}w(1 + \operatorname{cosec}\theta)$$

Now, $NQ = \frac{1}{2}p\tan\beta$; $QS = \frac{1}{2}p\tan\beta\tan\theta$;
$VS = \frac{1}{2}(\frac{1}{2}p - \frac{1}{2}p\tan\beta\tan\theta)$; $UV = VS\cot\theta$;
and $VJ = LV\tan\beta = VS\tan\beta$, so that

$$\begin{aligned} E &= M + 2\,VS(\cot\theta + \tan\beta) - w(1 + \operatorname{cosec}\theta) \\ &= M + \tfrac{1}{2}p(1 - \tan\beta\tan\theta)(\cot\theta + \tan\beta) - w(1 + \operatorname{cosec}\theta) \\ &= M + \tfrac{1}{2}p(\cot\theta - \tan^2\beta\tan\theta) - w(1 + \operatorname{cosec}\theta) \end{aligned}$$

"Three-Wire" Method

If the measurement be taken simultaneously over the three wires, one micrometer anvil contacting two wires, and the other anvil the remaining wire, then the anvil planes will be inclined to the screw axis by angle β. The corresponding formula is therefore:

$$E = (M - w)\sec\beta + \tfrac{1}{2}p(\cot\theta - \tan^2\beta\tan\theta) - w(\operatorname{cosec}\theta + \tfrac{1}{2}\tan^2\lambda\cos\theta\cot\theta) \qquad (31)$$

---

*Cf page 203.

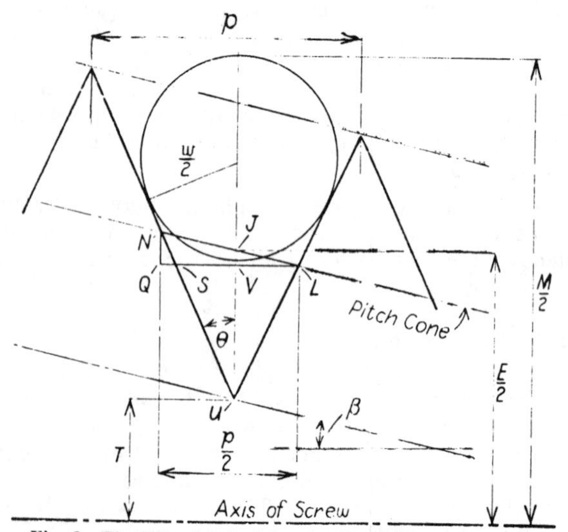

Fig. 9. The Wire Method for the Usual Taper Threads

TAPER THREAD WITH FLANKS EQUALLY INCLINED TO THE PITCH CONE

Where the thread flanks are equally inclined to the pitch cone, som difficulty may be encountered in arranging a set-up to "spot-off" th position of the gauge plane. The conditions of gauging by wires i indicated in Fig. 10.

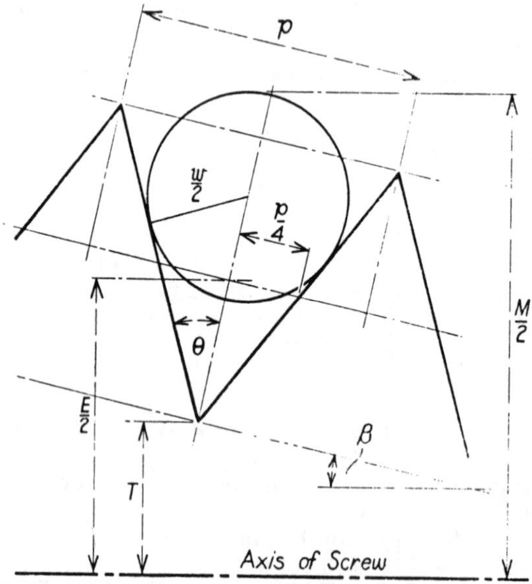

Fig. 10. Taper Threads with Flanks equally inclined to the Pitch Cone

$$\frac{M}{2} = T + \frac{w}{2} \operatorname{cosec} \theta \cos \beta + \frac{w}{2};$$

and $\frac{E}{2} = T + \frac{p}{4} \cot \theta \cos \beta$. Hence

$$E = M + \frac{p}{2} \cot \theta \cos \beta - w (\operatorname{cosec} \theta \cos \beta + 1).$$

·RRORS IN PITCH

Errors in the pitch of the screw thread may be uniform throughout a iven number of pitches, so that the actual pitch is greater or less than the ominal value as required by the standard specification or the stated equirements.

In addition, errors may occur which are cyclic. These may be classified s follows:

(*a*) With a cycle of repetition of like errors equal to an exact revolution of the screw.
(*b*) With a cycle of repetition of like errors equal to an exact sub-multiple of a revolution.
(*c*) With a cycle of repetition which is an exact whole number of revolutions.
(*d*) With a cycle of repetition which is neither an exact whole number of revolutions or sub-multiple of a revolution.

Types (*a*) and (*b*) will not be revealed by checking from pitch to pitch, nd type (*c*) will not be found if, when checking the aggregate length over a umber of pitches, the checking length is equal to the cyclic period of the rror.

It will be appreciated that any number of these types of error are liable o be present at the same time.

The main effect of pitch error is on the maximum possible length of ngagement that can be obtained between a screw having pitch error and nut without pitch error. The apparent effect is that of an increase in the ffective diameter of the screw and a corresponding decrease in the effective iameter of the nut. Engagement between two threads differing slightly n pitch is shown in Fig. 11. To assess the effect quantitatively, if the ccumulative pitch error over the length of engagement be δp, the virtual ifference between the two true effective diameters is $\delta p \cot \theta$. Cal-ulated values of the virtual difference in effective diameter for various alues of accumulative pitch error are given in the Table J3.

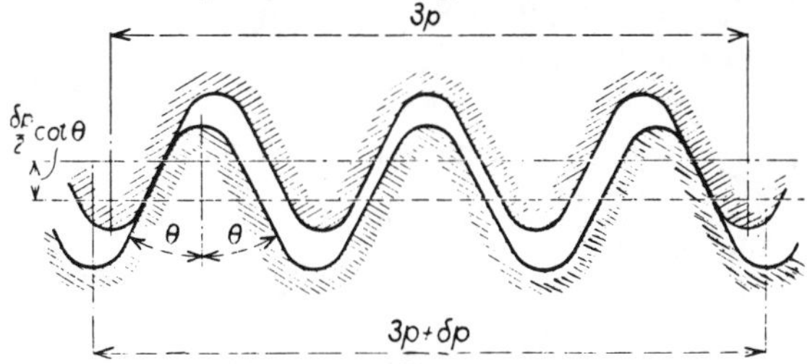

Fig. 11. Virtual Modification of Effective Diameter due to Pitch Error

ERRORS IN FLANK ANGLE

A virtual variation in effective diameter also occurs if the flank angle deviates from its specified value.

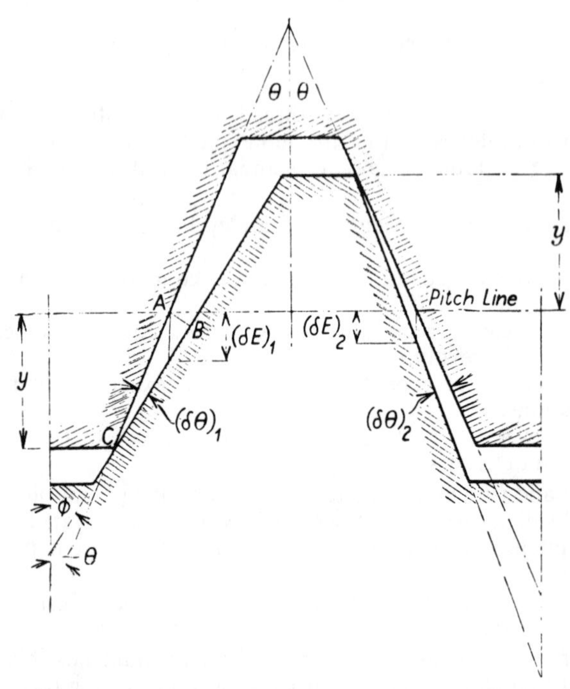

Fig. 12. Virtual Variation of Effective Diameter due to Errors in Flank Angle

Referring to Fig. 12, let the left-hand flank angle have an error $(\delta\theta)$ and the right-hand flank angle have an error $(\delta\theta)_2$, producing corresponding errors in effective diameter of $(\delta E)_1$ and $(\delta E)_2$ respectively. The line $A\ B$ is perpendicular to the flank. Then $(\delta E)_1 = A\ B$ cosec $\varphi = C\ A \sin (\delta\theta)_1$ cosec φ_1 but $C\ A = y \sec \theta$, hence

$(\delta E)_1 = y \sec \theta \operatorname{cosec} \varphi_1 \sin (\delta\theta)_1$.

Similarly $(\delta E)_2 = y \sec \theta \operatorname{cosec} \varphi_2 \sin (\delta\theta)_2$.

Hence total loss in effective diameter

$(\delta E)_1 + (\delta E)_2 = y \sec \theta\ [\operatorname{cosec} \varphi_1 \sin (\delta\theta)_1 + \operatorname{cosec} \varphi_2 \sin (\delta\theta)_2]$

since $\varphi_1 \simeq \varphi_2 \simeq \theta$, this formula may be written:

$\dfrac{y\ [\sin (\delta\theta)_1 + \sin (\delta\theta)_2]}{\sin \theta \cos \theta}$, and since for small values of $\delta\ \theta$,

$\sin \delta\ \theta = \delta\ \theta$

$$(\delta E)_1 + (\delta E)_2 = \frac{2y\ [(\delta\theta)_1 + (\delta\theta)_2]}{\sin 2\theta}$$

The value $\frac{2y}{\sin 2\theta}$ depends on the flank angle and the conditions of engagement of the basic profiles, hence the formula may be written

$$\delta E = l\,p\,[(\delta\theta_t) + (\delta\theta_p)]$$

where: p = pitch; δE = change in effective diameter; and l is a constant depending on the thread form and the maximum depth of engagement on the straight portion of the mating screw threads. As indicated, the changes $\delta\theta_t$ and $\delta\theta_p$ in the flank angles of opposite flanks are to be taken as positive and added. The constant l has the values.

Unified: $l = 0{\cdot}0098$;
Whitworth: $l = 0{\cdot}0105$;
B.A.: $l = 0{\cdot}0091$;
B.S. Metric: $l = 0{\cdot}0115$;
B.S. Cycle: $l = 0{\cdot}0074$.

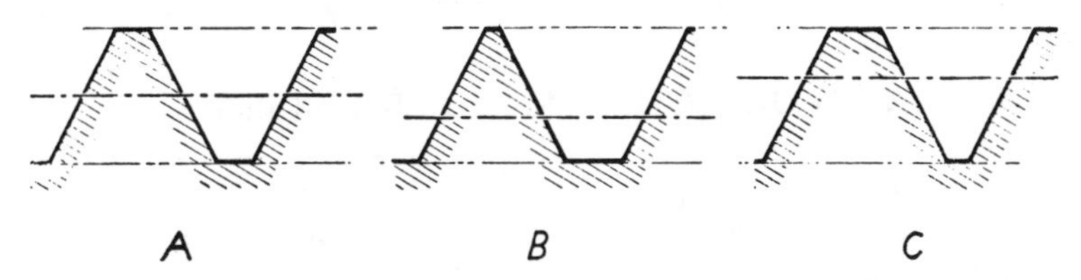

Fig. 13. Changes in Thread Form due to Errors in Effective Diameter, the Major and Minor Diameters Remaining Constant

Errors in Effective Diameter

If the effective diameter is below its basic value while the major and minor diameters are held at their basic values, then the thread thickness will decrease as shown at *B*, in Fig. 13. An increase in effective diameter produces the opposite effect as shown at *C*. It will be noticed that the correct thread profile is not maintained. Such errors can be produced by thread cutting tools having a non-standard amount of truncation and which are fed sufficiently to produce the basic minor diameter.

The result of these changes in internal threads are in the opposite direction—a decrease in effective diameter producing a thick thread, and vice versa.

Table J1. VALUES OF X, Y AND Z FOR DIFFERENT THREAD ANGLES

	θ	$X = \frac{1}{2} \cot \theta$	$Y = \operatorname{cosec} \theta + 1$	$Z = \operatorname{cosec} \theta - 1$
1	30° 0′	0·8660254	3·0000000	1·0000000
2	27° 30′	0·9604911	3·1656806	1·1656806
3	26° 34′	0·9999295	3·2359419	1·2359419
4	25° 0′	1·0722534	3·3662016	1·3662016
5	23° 45′	1·1363365	3·4829503	1·4829503
6	15° 0′	1·8660254	4·8637033	2·8637033
7	14° 30′	1·9333565	4·9939292	2·9939292
8	5° 0′	5·7150260	12·4737133	10·4737133

Table J2. RAKE CORRECTION "C" FOR B.A. SCREWS

B.A. No.	C (mm.)	C (Ins.)	B.A. No.	C (mm.)	C (Ins.)	B.A. No.	C (mm.)	C (Ins.)
0	0·00197	0·000078	9	0·00123	0·000049	18	0·00069	0·000027
1	0·00185	0·000073	10	0·00111	0·000044	19	0·00076	0·000030
2	0·00172	0·000068	11	0·00099	0·000039	20	0·00059	0·000023
3	0·00167	0·000066	12	0·00099	0·000039	21	0·00061	0·000024
4	0·00161	0·000063	13	0·00082	0·000032	22	0·00060	0·000024
5	0·00146	0·000058	14	0·00095	0·000037	23	0·00055	0·000022
6	0·00139	0·000055	15	0·00089	0·000035	24	0·00051	0·000020
7	0·00130	0·000051	16	0·00087	0·000034	25	0·00045	0·000018
8	0·00122	0·000048	17	0·00079	0·000031			

able J3. VIRTUAL EFFECTIVE DIAMETER VARIATION DUE TO ERRORS IN PITCH
(Dimensions in Inches)

Pitch Error	Corresponding Virtual Difference in Effective Diameter		
	Whitworth Thread	60-deg. Threads	B.A. Threads
0·00005	0·00010	0·00009	0·00011
0·0001	0·00019	0·00017	0·00023
0·00015	0·00029	0·00026	0·00034
0·0002	0·00038	0·00035	0·00045
0·00025	0·00048	0·00043	0·00057
0·0003	0·00058	0·00052	0·00068
0·00035	0·00067	0·00061	0·00079
0·0004	0·00077	0·00069	0·00091
0·00045	0·00086	0·00078	0·00102
0·0005	0·00096	0·00086	0·00114
0·00055	0·00106	0·00095	0·00125
0·0006	0·00115	0·00104	0·00136
0·00065	0·00125	0·00112	0·00147
0·0007	0·00134	0·00121	0·00159
0·00075	0·00144	0·00130	0·00170
0·0008	0·00154	0·00138	0·00181
0·00085	0·00163	0·00147	0·00193
0·0009	0·00173	0·00156	0·00205
0·00095	0·00182	0·00164	0·00216
0·0010	0·00192	0·00173	0·00227

Table J4. WHITWORTH—WIRE DIAMETERS
(Inches)

Threads per inch	W_b	$W_{b\,(max)}$	$W_{b\,(min)}$	$W_{(max)}$	$W_{(min)}$
60	0·00940	0·00988	0·00892	0·01422	0·00843
48	0·01174	0·01235	0·01115	0·01777	0·01054
40	0·01409	0·01483	0·01338	0·02132	0·01265
36	0·01566	0·01647	0·01486	0·02369	0·01406
32	0·01762	0·01853	0·01672	0·02665	0·01581
28	0·02013	0·02118	0·01911	0·03046	0·01807
26	0·02168	0·02281	0·02058	0·03280	0·01946
24	0·02349	0·02471	0·02229	0·03554	0·02108
22	0·02562	0·02695	0·02432	0·03877	0·02300
20	0·02818	0·02965	0·02675	0·04265	0·02530
19	0·02967	0·03121	0·02816	0·04489	0·02663
18	0·03132	0·03294	0·02972	0·04738	0·02811
16	0·03523	0·03706	0·03344	0·05331	0·03163
14	0·04026	0·04236	0·03821	0·06092	0·03614
12	0·04697	0·04942	0·04458	0·07108	0·04217
11	0·05124	0·05391	0·04864	0·07754	0·04600
10	0·05637	0·05930	0·05350	0·08530	0·05060
9	0·06263	0·06589	0·05944	0·09477	0·05622
8	0·07046	0·07413	0·06688	0·10662	0·06325
7	0·08053	0·08471	0·07643	0·12185	0·07229
6	0·09395	0·09883	0·08917	0·14216	0·08433
5	0·11274	0·11860	0·10700	0·17060	0·10120
4·5	0·12526	0·13178	0·11889	0·18955	0·11244
4·0	0·14092	0·14825	0·13375	0·21325	0·12650
3·5	0·16105	0·16943	0·15286	0·24371	0·14457
3·25	0·17344	0·18246	0·16462	0·26246	0·15569
3·0	0·18790	0·19767	0·17833	0·28433	0·16867
2·875	0·19607	0·20626	0·18609	0·29669	0·17600
2·75	0·20498	0·21564	0·19455	0·31018	0·18400
2·625	0·21474	0·22590	0·20381	0·32495	0·19276
2·5	0·22548	0·23720	0·21400	0·34120	0·20240

able J5. WHITWORTH—WIRE GAUGING CONSTANTS
(Dimensions in Inches)

Threads per inch	$w_b Y$	$w_b Z$	$p X$	$p V$	Truncated Whit. $p V$
60	0·02976	0·01096	0·01601	0·02668	0·02422
48	0·03717	0·01369	0·02001	0·03335	0·03027
40	0·04460	0·01642	0·02401	0·04002	0·03632
36	0·04957	0·01825	0·02668	0·04447	0·04036
32	0·05578	0·02054	0·03002	0·05003	0·04541
28	0·06373	0·02347	0·03430	0·05717	0·05189
26	0·06863	0·02527	0·03694	0·06157	0·05588
24	0·07436	0·02738	0·04002	0·06670	0·06054
22	0·08110	0·02986	0·04366	0·07277	0·06605
20	0·08921	0·03285	0·04802	0·08004	0·07265
19	0·09393	0·03459	0·05055	0·08425	0·07647
18	0·09915	0·03651	0·05336	0·08893	0·08072
16	0·11153	0·04107	0·06003	0·10005	0·09081
14	0·12745	0·04693	0·06861	0·11434	0·10378
12	0·14869	0·05475	0·08004	0·13340	0·12108
11	0·16221	0·05973	0·08732	0·14553	0·13209
10	0·17845	0·06571	0·09605	0·16008	0·14530
9	0·19827	0·07301	0·10672	0·17787	0·16144
8	0·22305	0·08213	0·12006	0·20010	0·18162
7	0·25493	0·09387	0·13721	0·22869	0·20757
6	0·29742	0·10952	0·16008	0·26680	0·24216
5	0·35690	0·13142	0·19210	0·32016	0·29060
4·5	0·39653	0·14601	0·21344	0·35574	0·32288
4·0	0·44611	0·16427	0·24012	0·40020	0·36325
3·5	0·50983	0·18773	0·27443	0·45738	0·41514
3·25	0·54906	0·20218	0·29554	0·49256	0·44707
3·0	0·59483	0·21903	0·32016	0·53360	0·48433
2·875	0·62069	0·22855	0·33408	0·55681	0·50539
2·75	0·64890	0·23894	0·34927	0·58212	0·52836
2·625	0·67980	0·25032	0·36590	0·60983	0·55352
2·5	0·71380	0·26284	0·38420	0·64033	0·58119

Table J6. **B.A.—WIRE GAUGING CONSTANTS**
(Dimensions in Inches)

B.A. No.	w_b	$w_b Y$	$w_b Z$	$p X$	$p V$
0	0·02151	0·07491	0·03189	0·04474	0·06836
1	0·01935	0·06741	0·02870	0·04026	0·06152
2	0·01742	0·06068	0·02583	0·03624	0·05537
3	0·01570	0·05469	0·02328	0·03266	0·04990
4	0·01419	0·04943	0·02105	0·02953	0·04512
5	0·01269	0·04419	0·01882	0·02640	0·04033
6	0·01140	0·03970	0·01690	0·02371	0·03623
7	0·01032	0·03595	0·01531	0·02147	0·03281
8	0·00925	0·03221	0·01371	0·01924	0·02939
9	0·00839	0·02921	0·01244	0·01745	0·02666
10	0·00753	0·02622	0·01116	0·01566	0·02393
11	0·00667	0·02322	0·00989	0·01387	0·02119
12	0·00602	0·02098	0·00893	0·01253	0·01914
13	0·00538	0·01873	0·00798	0·01118	0·01709
14	0·00495	0·01722	0·00734	0·01029	0·01572
15	0·00452	0·01573	0·00670	0·00939	0·01436
16	0·00409	0·01423	0·00606	0·00850	0·01299
17	0·00366	0·01274	0·00542	0·00761	0·01162
18	0·00322	0·01123	0·00478	0·00671	0·01025
19	0·00301	0·01049	0·00446	0·00626	0·00957
20	0·00258	0·00900	0·00383	0·00537	0·00820
21	0·00237	0·00824	0·00351	0·00492	0·00752
22	0·00215	0·00749	0·00319	0·00447	0·00684

'able J7. B.A.—WIRE GAUGING CONSTANTS
(Dimensions in mm.)

B.A. No.	w_b	$w_b Y$	$w_b Z$	$p X$	$p V$
0	0·5463	1·9027	0·8101	1·1363	1·7363
1	0·4916	1·7122	0·7290	1·0227	1·5627
2	0·4425	1·5412	0·6562	0·9204	1·4064
3	0·3988	1·3890	0·5914	0·8295	1·2675
4	0·3605	1·2556	0·5346	0·7500	1·1460
5	0·3223	1·1225	0·4780	0·6704	1·0244
6	0·2895	1·0083	0·4293	0·6023	0·9203
7	0·2622	0·9132	0·3888	0·5454	0·8334
8	0·2349	0·8181	0·3483	0·4886	0·7466
9	0·2130	0·7419	0·3159	0·4432	0·6772
10	0·1912	0·6659	0·2835	0·3977	0·6077
11	0·1693	0·5897	0·2511	0·3523	0·5383
12	0·1530	0·5329	0·2269	0·3182	0·4862
13	0·1366	0·4758	0·2026	0·2841	0·4341
14	0·1256	0·4375	0·1863	0·2614	0·3994
15	0·1147	0·3995	0·1701	0·2386	0·3646
16	0·1038	0·3615	0·1539	0·2159	0·3299
17	0·0929	0·3236	0·1378	0·1932	0·2952
18	0·0819	0·2853	0·1215	0·1705	0·2605
19	0·0765	0·2664	0·1134	0·1591	0·2431
20	0·0656	0·2285	0·0973	0·1364	0·2084
21	0·0601	0·2093	0·0891	0·1250	0·1910
22	0·0546	0·1902	0·0810	ͻ·1136	0·1736

Table J8. UNIFIED AND AMERICAN NATIONAL WIRE DIAMETERS AND GAUGING CONSTANTS

(Dimensions in Inches)

Threads per inch	w_b	w_{max}*	w_{min}	$w_b\ Y$	$p\ X$	$p\ V$
80	0·00722	0·01263	0·00631	0·02166	0·01083	0·01894
72	0·00802	0·01403	0·00702	0·02406	0·01203	0·02105
64	0·00902	0·01579	0·00789	0·02706	0·01353	0·02368
56	0·01031	0·01804	0·00902	0·03093	0·01546	0·02706
48	0·01203	0·02105	0·01052	0·03609	0·01804	0·03157
44	0·01312	0·02296	0·01148	0·03936	0·01968	0·03444
40	0·01443	0·02526	0·01263	0·04329	0·02165	0·03789
36	0·01604	0·02807	0·01403	0·04812	0·02406	0·04210
32	0·01804	0·03157	0·01579	0·05412	0·02706	0·04736
28	0·02062	0·03608	0·01804	0·06186	0·03093	0·05413
27	0·02138	0·03742	0·01871	0·06414	0·03208	0·05613
24	0·02406	0·04210	0·02105	0·07218	0·03608	0·06315
20	0·02887	0·05052	0·02526	0·08661	0·04330	0·07578
18	0·03208	0·05613	0·02807	0·09624	0·04811	0·08420
16	0·03608	0·06315	0·03157	0·10824	0·05413	0·09472
14	0·04124	0·07217	0·03608	0·12372	0·06186	0·10825
13	0·04441	0·07772	0·03886	0·13323	0·06662	0·11658
12	0·04811	0·08420	0·04210	0·14433	0·07217	0·12629
11½	0·05020	0·08786	0·04393	0·15060	0·07531	0·13179
11	0·05249	0·09185	0·04593	0·15747	0·07873	0·13778
10	0·05773	0·10104	0·05052	0·17319	0·08660	0·15155
9	0·06415	0·11226	0·05613	0·19245	0·09623	0·16839
8	0·07217	0·12630	0·06315	0·21651	0·10825	0·18944
7½	0·07698	0·13472	0·06736	0·23094	0·11547	0·20207
7	0·08248	0·14434	0·07217	0·24744	0·12372	0·21651
6	0·09623	0·16839	0·08420	0·28869	0·14434	0·25259
5	0·11547	0·20207	0·10104	0·34641	0·17321	0·30311
4½	0·12830	0·22453	0·11226	0·38490	0·19245	0·33679
4	0·14434	0·25259	0·12630	0·43302	0·21651	0·37889

* Calculated for flat crest

Table J9. BRITISH STANDARD CYCLE THREAD—WIRE DIAMETERS AND GAUGING CONSTANTS
(Dimensions in Inches)

Threads per inch	w_b	w_{max}	w_{min}	$w_b Y$	$p X$	$p V$
56	0·01031	0·01467	0·00832	0·03093	0·01546	0·02498
44	0·01312	0·01867	0·01059	0·03936	0·01968	0·03179
40	0·01443	0·02053	0·01165	0·04329	0·02165	0·03497
32	0·01804	0·02567	0·01456	0·05412	0·02706	0·04371
30	0·01924	0·02738	0·01553	0·05772	0·02887	0·04662
26	0·02221	0·03159	0·01792	0·06663	0·03331	0·05380
24	0·02406	0·03422	0·01942	0·07218	0·03608	0·05828

Table J10. ACME FORM—WIRE DIAMETERS AND GAUGING CONSTANTS
(Dimensions in Inches)

Threads per inch	w_b	w_{max}	w_{min}	$w_b Y$	$w_b Z$	$p X$
1	0·51645	0·65001	0·48726	2·57912	1·54622	1·93336
$1\frac{1}{3}$	0·38734	0·48751	0·36545	1·93434	1·15966	1·45002
$1\frac{1}{2}$	0·34430	0·43334	0·32484	1·71941	1·03081	1·28891
2	0·25822	0·32501	0·24363	1·28953	0·77311	0·96668
$2\frac{1}{2}$	0·20658	0·26001	0·19491	1·03165	0·61849	0·77334
3	0·17215	0·21667	0·16242	0·85970	0·51541	0·64445
4	0·12911	0·16250	0·12182	0·64478	0·38655	0·48334
5	0·10329	0·13000	0·09745	0·51582	0·30924	0·38667
6	0·08608	0·10834	0·08121	0·42988	0·25770	0·32223
8	0·06456	0·08125	0·06091	0·32239	0·19328	0·24167
10	0·05164	0·06500	0·04873	0·25791	0·15462	0·19334
12	0·04304	0·05417	0·04061	0·21493	0·12885	0·16111
14	0·03689	0·04643	0·03480	0·18422	0·11044	0·13810
16	0·03228	0·04063	0·03045	0·16119	0·09664	0·12084

Table J11. BUTTRESS THREAD (7°/45°)—WIRE DIAMETERS AND GAUGING CONSTANTS (Dimensions in Inches)

Threads per inch	w_b	w_{max}	$w_b Y_a$	$p X_a$
20	0·02188	0·02385	0·06908	0·04453
16	0·02735	0·02981	0·08635	0·05567
12	0·03647	0·03975	0·11514	0·07422
10	0·04377	0·04770	0·13816	0·08906
8	0·05471	0·05962	0·17271	0·11133
6	0·07294	0·07950	0·23027	0·14844
5	0·08753	0·09539	0·27633	0·17813
4	0·10941	0·11924	0·34541	0·22266
3	0·14589	0·15899	0·46055	0·29688
2·5	0·17506	0·19079	0·55266	0·35626
2	0·21883	0·23849	0·69082	0·44532
1·5	0·29177	0·31798	0·92109	0·59376
1·25	0·35013	0·38158	1·10531	0·71251
1	0·43766	0·47697	1·38164	0·89064

Table J12. BUTTRESS THREAD (0°/52°)—WIRE DIAMETERS AND GAUGING CONSTANTS (Dimensions in Inches)

Threads per inch	Pitch	w_b	w_{max}	$w_b Y_a$	$p X_a$
20	0·050000	0·021004	0·022955	0·064069	0·039064
16	0·062500	0·026255	0·028694	0·080086	0·048830
12	0·083333	0·035006	0·038258	0·106779	0·065107
10	0·100000	0·042008	0·045910	0·128137	0·078129
8	0·125000	0·052510	0·057388	0·160171	0·097661
6	0·166667	0·070013	0·076517	0·213561	0·130215
5	0·200000	0·084015	0·091820	0·256271	0·156257
4	0·250000	0·105019	0·114775	0·320340	0·195321
3	0·333333	0·140026	0·153033	0·427122	0·260428
2½	0·400000	0·168031	0·183640	0·512546	0·312514
2	0·500000	0·210039	0·229550	0·640683	0·390643
1½	0·666667	0·280052	0·306067	0·854244	0·520857
1¼	0·800000	0·336062	0·367280	1·025091	0·625028
1	1·000000	0·420077	0·459100	1·281362	0·781286

ble J13. CORDEAUX THREAD—WIRE DIAMETER AND GAUGING CONSTANTS (Dimension in Inches)

Threads per inch	Pitch	w_b	$w_b Y_a$	$p X_a$
6	0·166667	0·094916	0·273785	0·132854
7	0·142857	0·081357	0·234674	0·113874

able J14. WHITWORTH FORM

alues of X ($=\frac{1}{2}\cot\theta$) and Y ($= \operatorname{cosec}\theta + 1$) for Changes in 2θ ($= 55°$)

Mins. (′)	53°		54°		55°		Mins. (′)
	X	*Y*	*X*	*Y*	*X*	*Y*	
30	0·99198	3·22174	0·97081	3·18401	0·95034	3·14770	30
32	0·99126	3·22045	0·97012	3·18277	0·94967	3·14651	32
34	0·99055	3·21918	0·96942	3·18154	0·94900	3·14533	34
36	0·98983	3·21790	0·96873	3·18031	0·94833	3·14414	36
38	0·98912	3·21662	0·96804	3·17909	0·94767	3·14296	38
40	0·98840	3·21535	0·96735	3·17786	0·94700	3·14178	40
42	0·98769	3·21407	0·96666	3·17663	0·94633	3·14060	42
44	0·98698	3·21280	0·96597	3·17541	0·94567	3·13942	44
46	0·98626	3·21153	0·96528	3·17419	0·94500	3·13825	46
48	0·98555	3·21026	0·96460	3·17297	0·94434	3·13707	48
50	0·98484	3·20900	0·96391	3·17175	0·94367	3·13590	50
52	0·98413	3·20773	0·96323	3·17053	0·94301	3·13473	52
54	0·98343	3·20647	0·96254	3·16932	0·94235	3·13356	54
56	0·98272	3·20521	0·96186	3·16810	0·94168	3·13239	56
58	0·98201	3·20395	0·96117	3·16689	0·94102	3·13122	58
	54°		55°		56°		
0	0·98131	3·20269	0·96049	3·16568	0·94036	3·13005	0
2	0·98060	3·20143	0·95981	3·16447	0·93970	3·12889	2
4	0·97990	3·20018	0·95913	3·16326	0·93904	3·12773	4
6	0·97919	3·19892	0·95845	3·16206	0·93839	3·12657	6
8	0·97849	3·19767	0·95777	3·16085	0·93773	3·12540	8
10	0·97779	3·19642	0·95709	3·15965	0·93707	3·12425	10
12	0·97709	3·19517	0·95641	3·15845	0·93642	3·12309	12
14	0·97639	3·19393	0·95573	3·15725	0·93576	3·12193	14
16	0·97569	3·19268	0·95506	3·15605	0·93511	3·12078	16
18	0·97499	3·19144	0·95438	3·15485	0·93445	3·11963	18
20	0·97429	3·19019	0·95371	3·15366	0·93380	3·11847	20
22	0·97359	3·18895	0·95303	3·15246	0·93315	3·11732	22
24	0·97289	3·18772	0·95236	3·15127	0·93250	3·11617	24
26	0·97220	3·18648	0·95169	3·15008	0·93185	3·11503	26
28	0·97150	3·18524	0·95101	3·14889	0·93119	3·11388	28

Table J15. **B.A. FORM**

Values of X ($=\frac{1}{2}\cot\theta$) and Y ($=\operatorname{cosec}\theta + 1$) for Changes in 2θ ($=47\frac{1}{2}°$)

Mins. (′)	46°		47°		48°	
	X	*Y*	*X*	*Y*	*X*	*Y*
0	1·17793	3·55930	1·14992	3·50784	1·12302	3·45859
2	1·17697	3·55755	1·14901	3·50617	1·12214	3·45699
4	1·17602	3·55580	1·14809	3·50449	1·12126	3·45539
6	1·17507	3·55405	1·14718	3·50282	1·12039	3·45378
8	1·17413	3·55231	1·14627	3·50115	1·11951	3·45219
10	1·17318	3·55057	1·14536	3·49948	1·11864	3·45059
12	1·17223	3·54883	1·14445	3·49782	1·11776	3·44900
14	1·17129	3·54709	1·14355	3·49616	1·11689	3·44741
16	1·17035	3·54536	1·14264	3·49450	1·11602	3·44582
18	1·16940	3·54363	1·14174	3·49284	1·11515	3·44423
20	1·16846	3·54190	1·14083	3·49119	1·11428	3·44264
22	1·16753	3·54017	1·13993	3·48953	1·11342	3·44106
24	1·16659	3·53844	1·13903	3·48789	1·11255	3·43948
26	1·16565	3·53672	1·13813	3·48624	1·11168	3·43790
28	1·16472	3·53500	1·13723	3·48459	1·11082	3·43633
30	1·16378	3·53329	1·13634	3·48295	1·10996	3·43476
32	1·16285	3·53157	1·13544	3·48131	1·10910	3·43318
34	1·16192	3·53986	1·13455	3·47967	1·10824	3·43162
36	1·16099	3·52815	1·13365	3·47804	1·10738	3·43005
38	1·16006	3·52645	1·13276	3·47640	1·10652	3·42848
40	1·15913	3·52474	1·13187	3·47477	1·10566	3·42692
42	1·15820	3·52304	1·13098	3·47314	1·10481	3·42536
44	1·15728	3·52134	1·13009	3·47152	1·10395	3·42380
46	1·15635	3·51965	1·12920	3·46989	1·10310	3·42225
48	1·15543	3·51795	1·12831	3·46827	1·10224	3·42069
50	1·15451	3·51626	1·12743	3·46665	1·10139	3·41914
52	1·15359	3·51457	1·12654	3·46504	1·10054	3·41760
54	1·15267	3·51289	1·12566	3·46342	1·09969	3·41605
56	1·15175	3·51120	1·12478	3·46181	1·09884	3·41450
58	1·15084	3·50952	1·12390	3·46020	1·09800	3·41296

able J16. **60-DEG. FORM**

Values of X ($=\frac{1}{2}\cot\theta$) and Y ($=\operatorname{cosec}\theta + 1$) for Changes in 2θ

Mins. (′)	58° X	58° Y	59° X	59° Y	60° X	60° Y	Mins. (′)
30	0·89281	3·04658	0·87482	3·01525	0·85736	2·98502	30
32	0·89221	3·04551	0·87423	3·01422	0·85679	2·98403	32
34	0·89160	3·04445	0·87364	3·01320	0·85622	2·98304	34
36	0·89099	3·04339	0·87305	3·01218	0·85565	2·98205	36
38	0·89038	3·04233	0·87246	3·01116	0·85508	2·98107	38
40	0·88978	3·04128	0·87187	3·01014	0·85451	2·98008	40
42	0·88917	3·04022	0·87129	3·00912	0·85394	2·97910	42
44	0·88857	3·03916	0·87070	3·00810	0·85337	2·97811	44
46	0·88796	3·03811	0·87011	3·00708	0·85280	2·97713	46
48	0·88736	3·03706	0·86953	3·00607	0·85223	2·97615	48
50	0·88675	3·03601	0·86894	3·00505	0·85166	2·97517	50
52	0·88615	3·03496	0·86836	3·00404	0·85109	2·97420	52
54	0·88555	3·03391	0·86777	3·00303	0·85053	2·97322	54
56	0·88495	3·03286	0·86719	3·00202	0·84996	2·97224	56
58	0·88435	3·03182	0·86661	3·00101	0·84940	2·97127	58
	59°		60°		61°		
0	0·88375	3·03077	0·86603	3·00000	0·84888	2·97029	0
2	0·88315	3·02973	0·86544	2·99899	0·84827	2·96932	2
4	0·88255	3·02869	0·86486	2·99799	0·84770	2·96835	4
6	0·88195	3·02765	0·86428	2·99698	0·84714	2·96738	6
8	0·88135	3·02661	0·86370	2·99598	0·84658	2·96641	8
10	0·88076	3·02557	0·86312	2·99498	0·84601	2·96544	10
12	0·88016	3·02453	0·86255	2·99398	0·84545	2·96448	12
14	0·87956	3·02349	0·86197	2·99298	0·84489	2·96351	14
16	0·87897	3·02246	0·86139	2·99198	0·84433	2·96255	16
18	0·87837	3·02143	0·86081	2·99098	0·84377	2·96158	18
20	0·87778	3·02039	0·86024	2·98998	0·84321	2·96062	20
22	0·87719	3·01936	0·85966	2·98899	0·84265	2·95966	22
24	0·87659	3·01833	0·85909	2·98799	0·84210	2·95870	24
26	0·87600	3·01730	0·85851	2·98700	0·84154	2·95774	26
28	0·87541	3·01628	0·85794	2·98601	0·84098	2·95678	28

Table J17. METRIC SERIES—WIRE GAUGING CONSTANTS
(Dimensions in mm.)

Pitch	w_b	$w_b Y$	$w_b Z$	pX	pV*
0·075	0·0433	0·1299	0·0433	0·0650	0·1137
0·100	0·0577	0·1732	0·0577	0·0866	0·1516
0·125	0·0722	0·2165	0·0722	0·1083	0·1894
0·150	0·0866	0·2598	0·0866	0·1299	0·2273
0·175	0·1010	0·3031	0·1010	0·1516	0·2652
0·200	0·1155	0·3464	0·1155	0·1732	0·3031
0·225	0·1299	0·3897	0·1299	0·1949	0·3410
0·250	0·1443	0·4330	0·1443	0·2165	0·3789
0·300	0·1732	0·5196	0·1732	0·2598	0·4547
0·350	0·2021	0·6062	0·2021	0·3031	0·5304
0·400	0·2309	0·6928	0·2309	0·3464	0·6062
0·450	0·2598	0·7794	0·2598	0·3897	0·6820
0·500	0·2887	0·8660	0·2887	0·4330	0·7578
0·600	0·3464	1·0392	0·3464	0·5196	0·9093
0·700	0·4041	1·2124	0·4041	0·6062	1·0609
0·750	0·4330	1·2990	0·4330	0·6495	1·1367
0·800	0·4619	1·3856	0·4619	0·6928	1·2124
0·900	0·5196	1·5588	0·5196	0·7794	1·3640
1·000	0·5774	1·7321	0·5774	0·8660	1·5155
1·250	0·7217	2·1651	0·7217	1·0825	1·8944
1·500	0·8660	2·5981	0·8660	1·2990	2·2733
1·750	1·0104	3·0311	1·0104	1·5155	2·6522
2·000	1·1547	3·4641	1·1547	1·7321	3·0311
2·500	1·4434	4·3301	1·4434	2·1651	3·7889
3·000	1·7321	5·1962	1·7321	2·5981	4·5466
3·500	2·0207	6·0622	2·0207	3·0311	5·3044
4·000	2·3094	6·9282	2·3094	3·4641	6·0622
4·500	2·5981	7·7942	2·5981	3·8971	6·8200
5·000	2·8868	8·6603	2·8868	4·3301	7·5777
5·500	3·1754	9·5263	3·1754	4·7631	8·3355
6·000	3·4641	10·3923	3·4641	5·1962	9·0933

*based on $S_1 = \frac{1}{8}H$.

able J18. METRIC SERIES—WIRE GAUGING CONSTANTS
(Dimensions in Inches)

Pitch (mm.)	w_b	$w_b Y$	$w_b Z$	pX	pV*
0·075	0·00170	0·00511	0·00170	0·00256	0·00448
0·100	0·00227	0·00682	0·00227	0·00341	0·00597
0·125	0·00284	0·00852	0·00284	0·00426	0·00746
0·150	0·00341	0·01023	0·00341	0·00511	0·00895
0·175	0·00398	0·01193	0·00398	0·00597	0·01044
0·200	0·00455	0·01364	0·00455	0·00682	0·01193
0·225	0·00511	0·01534	0·00511	0·00767	0·01343
0·250	0·00568	0·01705	0·00568	0·00852	0·01492
0·300	0·00682	0·02046	0·00682	0·01023	0·01790
0·350	0·00796	0·02387	0·00796	0·01193	0·02088
0·400	0·00909	0·02728	0·00909	0·01364	0·02387
0·450	0·01023	0·03069	0·01023	0·01534	0·02685
0·500	0·01137	0·03410	0·01137	0·01705	0·02983
0·600	0·01364	0·04091	0·01364	0·02046	0·03580
0·700	0·01591	0·04773	0·01591	0·02387	0·04177
0·750	0·01705	0·05114	0·01705	0·02557	0·04475
0·800	0·01818	0·05455	0·01818	0·02728	0·04773
0·900	0·02046	0·06137	0·02046	0·03069	0·05370
1·000	0·02273	0·06819	0·02273	0·03410	0·05967
1·250	0·02841	0·08524	0·02841	0·04262	0·07458
1·500	0·03410	0·10229	0·03410	0·05114	0·08950
1·750	0·03978	0·11933	0·03978	0·05967	0·10442
2·000	0·04546	0·13638	0·04546	0·06819	0·11933
2·500	0·05683	0·17048	0·05683	0·08524	0·14917
3·000	0·06819	0·20457	0·06819	0·10229	0·17900
3·500	0·07956	0·23867	0·07956	0·11933	0·20883
4·000	0·09092	0·27276	0·09092	0·13638	0·23867
4·500	0·10229	0·30686	0·10229	0·15343	0·26850
5·000	0·11365	0·34095	0·11365	0·17048	0·29834
5·500	0·12502	0·37505	0·12502	0·18753	0·32817
6·000	0·13638	0·40915	0·13638	0·20457	0·35800

*based on $S_1 = \frac{1}{8}H$.

Section K

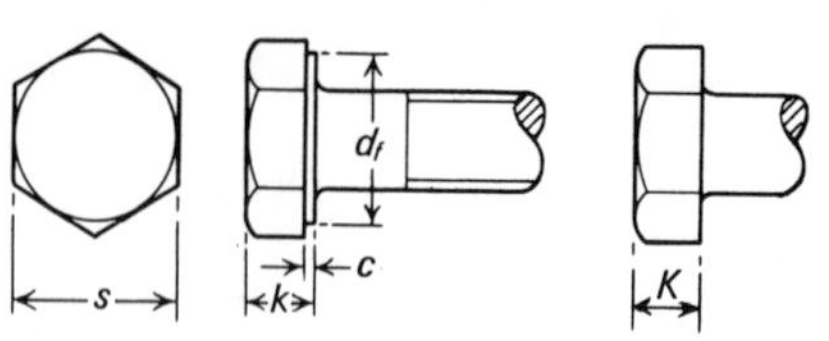

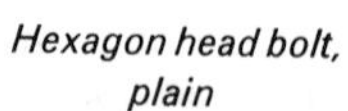

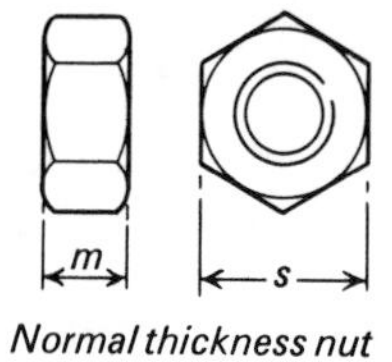

Hexagon head bolt, washer faced *Hexagon head bolt, plain* *Normal thickness nut* *Thin nut*

Table K1.

ISO METRIC PRECISION HEXAGON BOLTS AND NUTS (B.S. 3692)

Nominal Size	Width across flats (mm.)	Bolt Head			Nut	
		Overall depth – Plain and washer-faced types (mm.)	Washer-faced type		Depth of Normal Nut (mm.)	Depth of Thin Nut (mm.)
			Diameter of washer (mm.)	Depth of washer (mm.)		
M1·6	3·2/3·08	1·225/0·975	—	—	1·3/1·05	—
M2	4·0/3·88	1·525/1·275	—	—	1·6/1·35	—
M2·5	5·0/4·88	1·825/1·575	—	—	2·0/1·75	—
M3	5·5/5·38	2·125/1·875	5·08/4·83	0·1	2·4/2·15	—
M4	7·0/6·85	2·925/2·675	6·55/6·30	0·1	3·2/2·90	—
M5	8·0/7·85	3·65/3·35	7·55/7·30	0·2	4·0/3·7	—
M6	10·0/9·78	4·15/3·85	9·48/9·23	0·3	5·0/4·7	—
M8	13·0/12·73	5·65/5·35	12·43/12·18	0·4	6·5/6·14	5·0/4·7
M10	17·0/16·73	7·18/6·82	16·43/16·18	0·4	8·0/76·4	6·0/5·7
M12	19·0/18·67	8·18/7·82	18·37/18·12	0·4	10·0/96·4	7·0/6·64
(M14)	22·0/21·67	9·18/8·82	21·37/21·12	0·4	11·0/10·57	8·0/7·64
M16	24·0/23·67	10·18/9·82	23·37/23·02	0·4	13·0/12·57	8·0/7·64
(M18)	27·0/26·67	12·215/11·785	26·27/26·02	0·4	15·0/14·57	9·0/8·64
M20	30·0/29·67	13·215/12·785	29·27/28·8	0·4	16·0/15·57	9·0/8·64
(M22)	32·0/31·61	14·215/13·785	31·21/30·74	0·4	18·0/17·57	10·0/9·64
M24	36·0/35·38	15·215/14·785	34·98/34·51	0·5	19·0/18·48	10·0/9·64
(M27)	41·0/40·38	17·215/16·785	39·98/39·36	0·5	22·0/21·48	12·0/11·57
M30	46·0/45·38	19·26/18·74	44·98/44·36	0·5	24·0/23·48	12·0/11·57
(M33)	50·0/49·38	21·26/20·74	48·98/48·36	0·5	26·0/25·48	14·0/13·57
M36	55·0/54·26	23·26/22·74	53·86/53·24	0·5	29·0/28·48	14·0/13·57
(M39)	60·0/59·26	25·26/24·74	58·86/58·24	0·6	31·0/30·38	16·0/15·57
M42	65·0/64·26	26·26/25·74	63·76/63·04	0·6	34·0/33·38	16·0/15·57
(M45)	70·0/69·26	28·26/27·74	68·76/68·04	0·6	36·0/35·38	18·0/17·57
M48	75·0/74·26	30·26/29·74	73·76/73·04	0·6	38·0/37·38	18·0/17·57
(M52)	80·0/79·26	33·31/32·69	—	—	42·0/41·38	20·0/19·48
M56	85·0/84·13	35·31/34·69	—	—	45·0/44·38	—
(M60)	90·0/89·13	38·31/37·69	—	—	48·0/47·38	—
M64	95·0/94·13	40·31/39·69	—	—	51·0/50·26	—
(M68)	100·0/99·13	43·31/42·69	—	—	54·0/53·26	—

Note: Sizes in brackets are non-preferred.

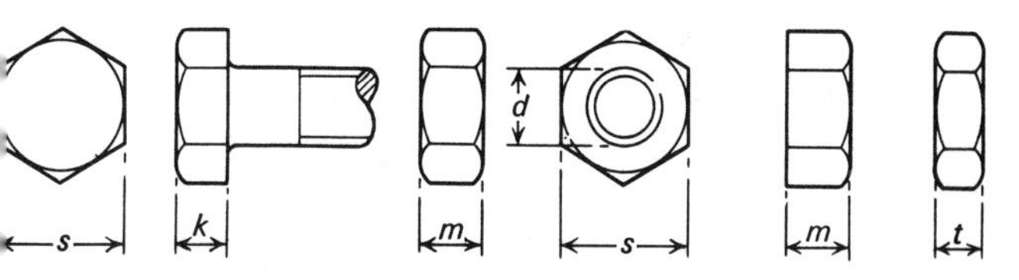

able K2.

ISO METRIC BLACK HEXAGON BOLTS AND NUTS (B.S. 4190)

Nominal Size	Width across flats (mm.)	Depth			
		Bolt head (mm.)	Normal nut – plain (mm.)	Normal nut – faced one side (mm.)	Thin nut – faced both sides (mm.)
M5	8·0/7·64	3·875/3·125	4·375/3·625	4·0/3·52	—
M6	10·0/9·64	4·375/3·625	5·375/4·625	5·0/4·52	—
M8	13·0/12·57	5·875/5·125	6·875/6·125	6·5/5·92	5·0/4·52
M10	17·0/16·57	7·45/6·55	8·45/7·55	8·0/7·42	6·0/5·52
M12	19·0/18·48	8·45/7·55	10·45/9·55	10·0/9·42	7·0/6·42
M16	24·0/23·16	10·45/9·55	13·55/12·45	13·0/1·23	9·0/8·42
M20	30·0/29·16	13·9/12·1	16·55/15·45	16·0/15·3	9·0/8·42
(M22)	32·0/31·0	14·9/13·1	18·55/17·45	18·0/17·3	10·0/9·42
M24	36·0/35·0	15·9/14·1	16·55/18·35	19·0/18·16	10·0/9·42
(M27)	41·0/40·0	17·9/16·1	22·65/21·35	22·0/21·15	12·0/11·30
M30	46·0/45·0	20·05/17·95	24·65/23·35	24·0/23·16	12·0/11·30
(M33)	50·0/49·0	22·05/19·95	26·65/25·35	26·0/25·16	14·0/13·30
M36	55·0/53·8	24·05/21·95	29·65/28·35	29·0/28·16	14·0/13·30
(M39)	60·0/58·8	26·05/23·95	31·8/30·2	31·0/30·0	16·0/15·3
M42	65·0/63·8	27·05/24·95	34·8/33·2	34·0/33·0	16·0/15·3
(M45)	70·0/68·8	29·05/26·95	36·8/35·2	36·0/35·0	18·0/17·3
M48	75·0/73·8	31·05/28·95	38·8/37·2	38·0/37·0	18·0/17·3
(M52)	80·0/78·8	34·25/31·75	42·8/41·2	42·0/41·0	20·0/19·16
M56	85·0/83·6	36·25/33·75	45·8/44·2	45·0/44·0	—
(M60)	90·0/88·6	39·25/36·75	48·8/47·2	48·0/47·0	—
M64	95·0/93·6	41·25/38·75	51·95/5,005	51·0/49·8	—
(M68)	100·0/98·6	44·25/41·75	54·95/5·305	54·0/52·8	—

Note: Sizes in brackets are non-preferred.

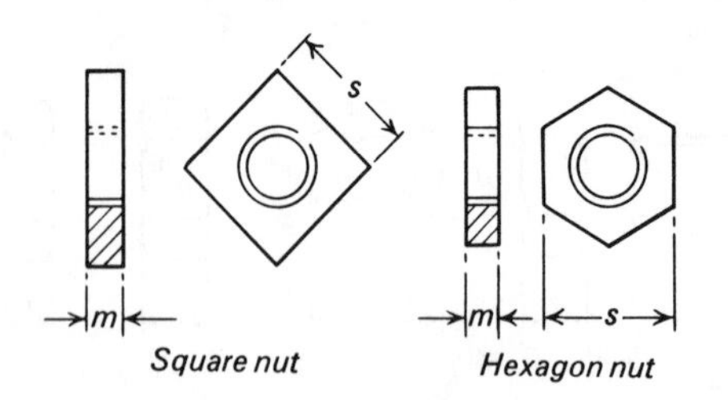

Table K3.

ISO METRIC PRESSED NUTS – HEXAGON AND SQUARE (B.S. 4183)

Nominal Size	Width across flats (mm.)	Thickness (mm.)
M1·6	3·2/3·02	1·0/0·75
M2	4·0/3·82	1·2/0·95
(M2·2)	4·5/4·32	1·2/0·95
M2·5	5·0/4·82	1·6/1·35
M3	5·5/5·32	1·6/1·35
(M3·5)	6·0/5·82	2·0/1·75
M4	7·0/6·78	2·0/1·75
M5	8·0/7·78	2·5/2·25
M6	10·0/9·78	3·0/2·75
M8	13·0/12·73	4·0/3·7
M10	17·0/16·73	5·0/4·7

Note: Sizes in brackets are non-preferred.

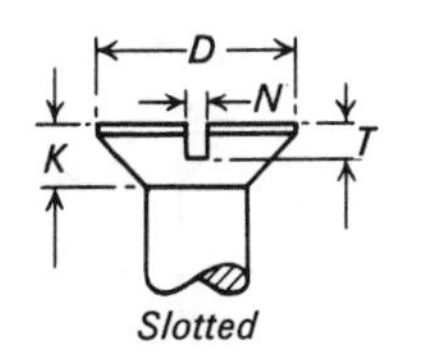

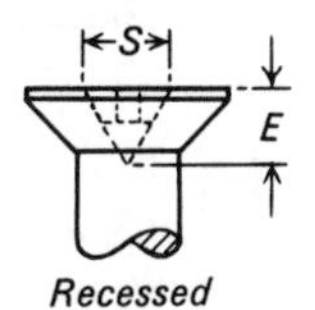

Countersunk head

able K4.

ꟷSO METRIC COUNTERSUNK HEAD SCREWS – SLOTTED AND RECESSED (B.S. 4183)

Nominal Size	Head		Slot		Recess	
	Diameter (mm.) D	Height (mm.) K	Width (mm.) N	Depth (mm.) T	Depth (mm.) E	Nominal max. width across cruciform (mm.) S
M1	2·0/1·75	0·5/0·45	0·45/0·31	0·3/0·2	—	—
M1·2	2·4/2·1	0·6/0·54	0·5/0·36	0·36/0·24	—	—
(M1·4)	2·8/2·45	0·7/0·63	0·5/0·36	0·42/0·28	—	—
M1·6	3·2/2·8	0·8/0·72	0·6/0·46	0·48/0·32	—	—
M2·0	4·0/3·5	1·0/0·9	0·7/0·56	0·6/0·4	—	—
(M2·2)	4·4/3·85	1·1/0·99	0·8/0·66	0·66/0·44	—	—
M2·5	5·0/4·38	1·25/1·12	0·8/0·66	0·75/0·5	1·6/1·19	2·39
M3	6·0/5·25	1·5/1·35	1·0/0·86	0·9/0·6	1·73/1·32	2·52
(M3·5)	7·0/6·1	1·75/1·57	1·0/0·86	1·05/0·7	1·98/1·52	3·51
M4	8·0/7·0	2·0/1·8	1·2/1·06	1·2/0·8	2·18/1·72	3·71
(M4·5)	9·0/7·85	2·25/2·03	1·2/1·06	1·35/0·9	2·69/2·23	4·22
M5	10·0/8·75	2·5/2·25	1·51/1·26	1·5/1·0	2·9/2·44	4·42
M6	12·0/10·5	3·0/2·7	1·91/1·66	1·8/1·2	3·45/2·99	6·1
M8	16·0/14·0	4·0/3·6	2·31/2·06	2·4/1·6	4·28/3·82	7·85
M10	20·0/17·5	5·0/4·5	2·81/2·56	3·0/2·0	5·84/5·38	9·42
M12	24·0/21·0	6·0/5·4	3·31/3·06	3·6/2·4	6·63/6·17	10·18
(M14)	28·0/24·5	7·0/6·3	3·31/3·06	4·2/2·8	—	—
M16	32·0/28·0	8·0/7·2	4·37/4·07	4·8/3·2	—	—
(M18)	36·0/31·5	9·0/8·1	4·37/4·07	5·4/3·6	—	—
M20	40·0/35·0	10·0/5·37	5·37/5·07	6·0/4·0	—	—

Note: Included angle of head = 90°/92°.
Sizes in brackets are non-preferred.

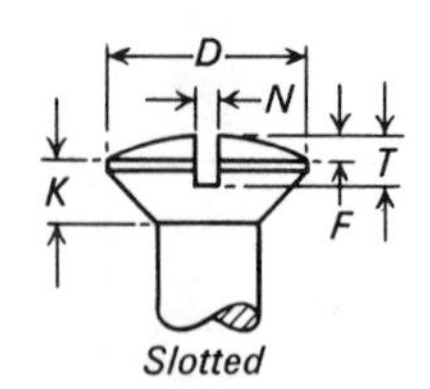

Slotted

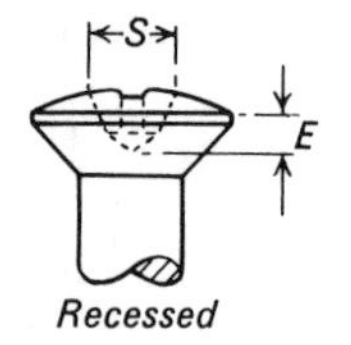

Recessed

Countersunk/raised head

Table K5.

ISO METRIC RAISED COUNTERSUNK HEAD SCREWS – SLOTTED AND RECESSED (B.S. 4183)

Nominal Size	Head			Slot		Recess	
	Diameter (mm.) D*	Height of Conical Portion (mm.) K	Nominal Height of Dome (mm.) F	Width (mm.) N	Depth (mm.) T	Depth (mm.) E	Nomin max. width across crucifor (mm.) S
M1	2·0/1·75	0·5/0·45	0·25	0·45/0·31	0·5/0·4	—	—
M1·2	2·4/2·1	0·6/0·54	0·3	0·5/0·36	0·6/0·48	—	—
(M1·4)	2·8/2·45	0·7/0·63	0·35	0·5/0·36	0·7/0·56	—	—
M1·6	3·2/2·8	0·8/0·72	0·4	0·6/0·46	0·8/0·64	—	—
M2	4·0/3·5	1·0/0·9	0·5	0·7/0·56	1·0/0·8	—	—
(M2·2)	4·4/3·85	1·1/0·99	0·55	0·8/0·66	1·1/0·88	—	—
M2·5	5·0/4·38	1·25/1·12	0·6	0·8/0·66	1·25/1·0	1·98/1·57	2·77
M3	6·0/5·25	1·5/1·35	0·75	1·0/0·86	1·5/1·2	2·18/1·78	2·97
(M3·5)	7·0/6·1	1·75/1·57	0·9	1·0/0·86	1·75/1·4	2·51/2·05	4·06
M4	8·0/7·0	2·0/1·8	1·0	1·2/1·06	2·0/1·6	2·77/2·31	4·32
(M4·5)	9·0/7·85	2·25/2·03	1·1	1·2/1·06	2·25/1·8	3·56/3·10	5·08
M5	10·0/8·75	2·5/2·25	1·25	1·51/1·26	2·5/2·0	3·81/3·35	5·28
M6	12·0/10·5	3·0/2·7	1·5	1·91/1·66	3·0/2·4	4·47/4·01	7·11
M8	16·0/14·0	4·0/3·6	2·0	2·31/2·06	4·0/3·2	5·21/4·75	8·79
M10	20·0/17·5	5·0/4·5	2·5	2·81/2·56	5·0/4·0	7·37/6·91	10·92
M12	24·0/21·0	6·0/5·4	3·0	3·31/3·06	6·0/4·8	8·23/7·77	11·76
(M14)	28·0/24·5	7·0/6·3	3·5	3·31/3·06	7·0/5·6	—	—
M16	32·0/28·0	8·0/7·2	4·0	4·37/4·07	8·0/6·4	—	—
(M18)	36·0/31·5	9·0/8·1	4·5	4·37/4·07	9·0/7·2	—	—
M20	40·0/35·0	10·0/9·0	5·0	5·37/5·07	10·0/8·0	—	—

Notes: Included angle of head = 90°/92°.
Sizes in brackets are non-preferred.
*See diagrams above.

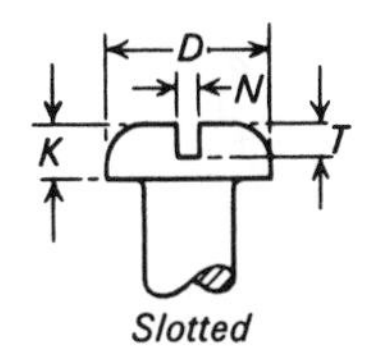

Slotted

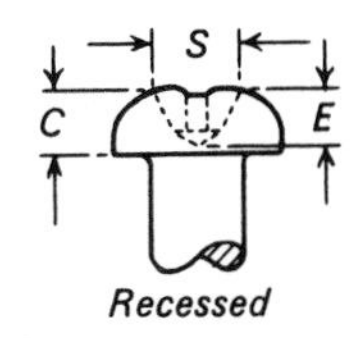

Recessed

Pan head

able K6.

ʖO METRIC PAN HEAD SCREWS – SLOTTED AND RECESSED (B.S. 4183)

Iominal Size	Head	Slotted Head			Recessed Head		
	Diameter (mm) D*	Width Height (mm) K	Slot Width (mm) N	Slot Depth (mm) T	Height (mm) C	Recess Depth (mm) E	Nominal max. width across Cruciform (mm) S
12·5	5·0/4·7	1·5/1·36	0·8/0·66	0·9/0·6	1·75/1·61	1·85/1·44	2·64
13	6·0/5·7	1·8/1·66	1·0/0·86	1·08/0·72	2·1/1·96	2·11/1·7	2·89
M3·5)	7·0/6·64	2·1/1·96	1·0/0·86	1·26/0·84	2·45/·231	2·34/1·88	3·91
14	8·0/7·64	2·4/2·26	1·2/1·06	1·44/0·96	2·8/2·66	2·72/2·26	4·27
M4·5)	9·0/8·64	2·7/2·56	1·2/1·06	1·62/1·08	3·15/2·97	2·97/2·51	4·52
15	10·0/9·64	3·0/2·86	1·51/1·26	1·8/1·2	3·5/3·32	3·10/2·64	4·67
16	12·0/11·57	3·6/3·42	1·91/1·66	2·16/1·44	4·2/3·98	4·06/3·61	6·76
18	16·0/15·57	4·8/4·62	2·31/2·06	2·88/1·92	5·6/5·42	4·85/4·39	8·46
110	20·0/19·48	6·0/5·82	2·81/2·56	3·6/2·4	7·0/6·78	6·40/5·94	9·96

Note: Sizes in brackets are non-preferred.
*See diagrams above.

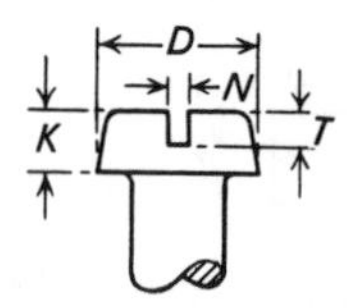

Table K7. *Cheese head*

ISO METRIC CHEESE HEAD SCREWS – SLOTTED (B.S. 4183

Nominal Size	Head Diameter (mm) D*	Head Thickness (mm) K	Slot Width (mm) N	Slot Depth (mm) T
M1	2·0/1·75	0·7/0·56	0·45/0·31	0·44/0·3
M1·2	2·3/2·05	0·8/0·66	0·50/0·36	0·49/0·35
(M1·4)	2·6/2·35	0·9/0·76	0·50/0·36	0·6/0·4
M1·6	3·0/2·75	1·0/0·86	0·60/0·46	0·65/0·45
M2	3·8/3·5	1·3/1·16	0.70/0·56	0·85/0·6
(M2·2)	4·0/3·7	1·5/1·36	0·80/0·66	1·0/0·7
M2·5	4·5/4·2	1·6/1·46	0·80/0·66	1·0/0·7
M3	5·5/5·2	2·0/1·86	1·0/0·86	1·3/0·9
(M3·5)	6·0/5·7	2·4/2·26	1·0/0·86	1·4/1·0
M4	7·0/6·64	2·6/2·46	1·2/1·06	1·6/1·2
(M4·5)	8·0/7·64	3·1/2·92	1·2/1·06	1·8/1·4
M5	8·5/8·14	3·3/3·12	1·51/1·26	2·0/1·5
M6	10·0/9·64	3·9/3·72	1·91/1·66	2·3/1·8
M8	13·0/12·57	5·0/4.82	2·31/2·06	2·8/2·3
M10	16·0/15·57	6·0/5·82	2·81/2·56	3·2/2·7
M12	18·0/17·57	7·0/6·78	3·31/3·06	3·8/3·2
(M14)	21·0/20·48	8·0/7·78	3·31/3·06	4·2/3·6
M16	24·0/23·48	9·0/8·78	4·37/4·07	4·6/4·0
(M18)	27·0/26·48	10·0/9·78	4·37/4·07	5·1/4·5
M20	30·0/29·48	11·0/10·73	5·27/5·07	5·6/5·0

Note: Sizes in brackets are non-preferred.
*See diagram above.

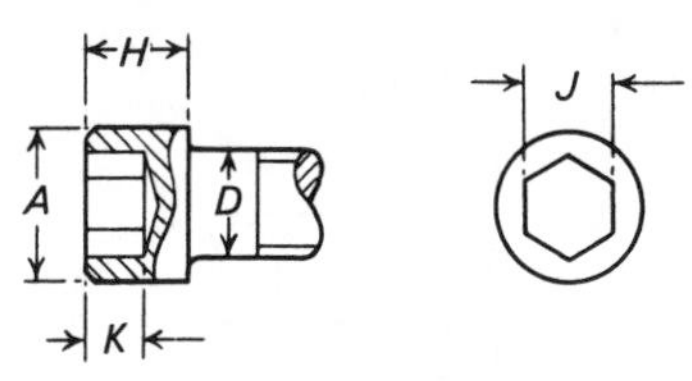

Cap screw

Table K8.

ISO METRIC HEXAGON SOCKET CAP SCREWS (B.S. 4168)

Nominal Size	Shank Diameter (mm) D	Head: Diameter (mm) A	Head: Height (mm) H	Socket: Nominal width across Flats (mm) J	Socket: Depth (mm) K
M3	3·0/2·86	5·5/5·2	3·0/2·86	2·5	1·7/1·3
M4	4·0/3·82	7·0/6·64	4·0/3·82	3·0	2·4/2·0
M5	5·0/4·82	8·5/8·14	5·0/4·82	4·0	3·1/2·7
M6	6·0/5·82	10·0/9·64	6·0/5·82	5·0	3·78/3·3
M8	8·0/7·78	13·0/12·57	8·0/7·78	6·0	4·78/4·3
M10	10·0/9·78	16·0/15·57	10·0/9·78	8·0	6·25/5·5
M12	12·0/11·73	18·0/17·57	12·0/11·73	10·0	7·5/6·6
(M14)	14·0/13·73	21·0/20·48	14·0/13·73	12·0	8·7/7·8
M16	16·0/15·73	24·0/23·48	16·0/15·73	14·0	9·7/8·8
(M18)	18·0/17·73	27·0/26·48	18·0/17·73	14·0	10·7/9·8
M20	20·0/19·67	30·0/29·48	20·0/19·67	17·0	11·8/10·7
(M22)	22·0/21·67	33·0/32·38	22·0/21,67	17·0	12·4/11·3
M24	24·0/23·67	36·0/35·38	24·0/23·67	19·0	14·0/12·9

Note: Sizes in brackets are non-preferred.

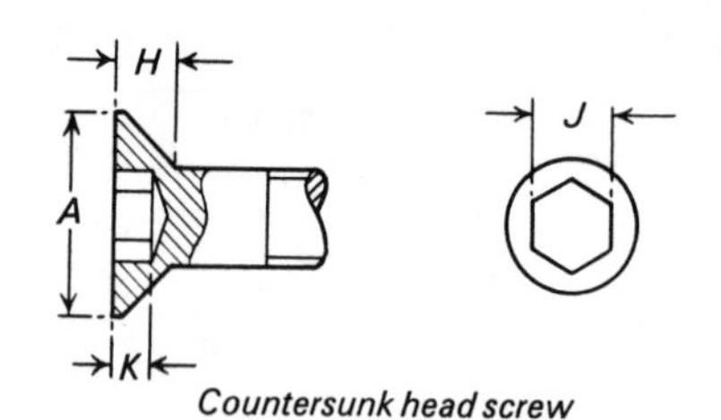

Countersunk head screw

Table K9.

ISO METRIC HEXAGON SOCKET COUNTERSUNK HEAD SCREWS (B.S. 4168)

Nominal Size	Head		Socket	
	Diameter (mm) A*	Height (mm) H	Nominal width across Flats (mm) J	Minimum Depth (mm) K
M3	6·72/5·82	1·86	2·0	1·05
M4	8·96/7·78	2·48	2·5	1·49
M5	11·2/9·78	3·1	3·0	1·86
M6	13·44/11·73	3·72	4·0	2·16
M8	17·92/15·73	4·96	5·0	2·85
M10	22·4/19·67	6·2	6·0	3·6
M12	26·88/23·67	7·44	8·0	4·35
(M14)	30·24/26·67	8·12	10·0	4·65
M16	33·6/29·67	8·8	10·0	4·89
(M18)	36·96/32·61	9·48	12·0	5·25
M20	40·32/35·61	10·16	12·0	5·45

Included angle of conical head = 90°/92°.
*See diagram above.

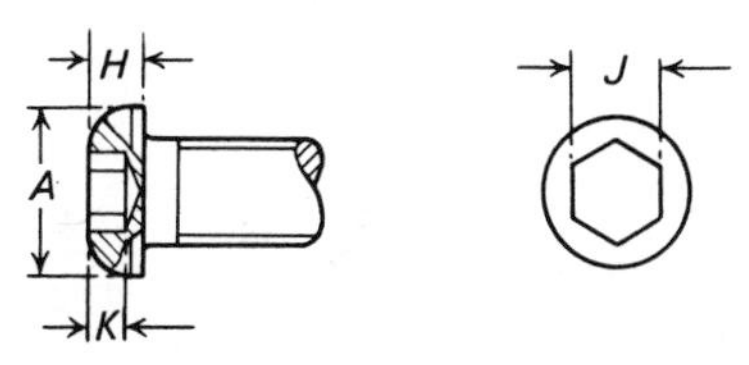

Button head screw

able K10.

ISO METRIC HEXAGON SOCKET BUTTON HEAD SCREWS (B.S. 4168)

	Head		Socket	
Nominal Size	Diameter (mm) A*	Height (mm) H	Nominal width across Flats (mm) J	Minimum Depth (mm) K
M3	5·5/5·32	1·6/1·4	2·0	1·04
M4	7·5/7·28	2·1/1·85	2·5	1·3
M5	9·5/9·28	2·7/2·45	3·0	1·56
M6	10·5/10·23	3·2/2·95	4·0	2·08
M8	14·0/13·73	4·3/3·95	5·0	2·6
M10	18·0/17·73	5·3/4·95	6·0	3·12
M12	21·0/20·67	6·4/5·9	8·0	4·16

*See diagrams above.

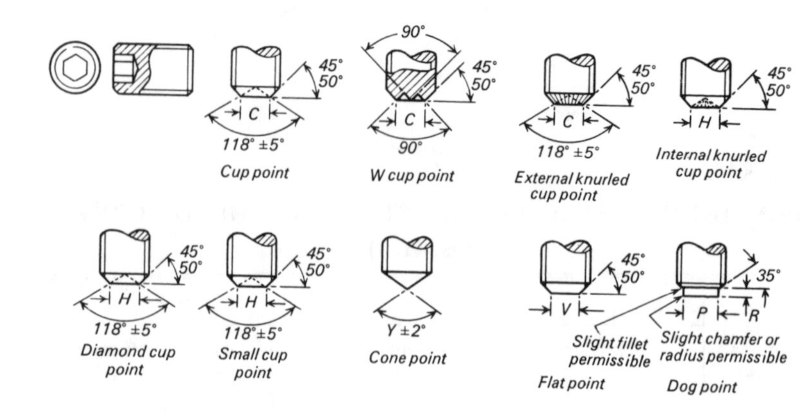

Table K11.

ISO METRIC HEXAGON SOCKET SET SCREWS (B.S. 4168)

	Socket		Point					
Nominal Size	Nominal width across Flats (mm)	Min. Depth (mm)	Small Diameter of Cone-*C* (mm)	Small Diameter of Cone-*H* (mm)	Angle *Y* 90° for these lengths and greater; 118° for shorter lengths (mm)	Small Diameter of Cone-*V* (mm)	Diameter of dog-*P* (mm)	Length of dog-*R* (mm)
M3	1·5	1·2	1·4/1·0	1·5/1·1	5·0	2·0/1·6	2·0/1·86	2·4/2·25
M4	2·0	1·6	2·0/1·6	2·1/1·6	6·0	2·6/2·2	2·5/2·36	2·8/2·6
M5	2·5	2·0	2·5/2·1	2·5/2·1	8·0	3·4/2·92	3·5/3·32	2·8/2·6
M6	3·0	2·4	3·0/2·6	3·0/2·6	9·0	4·0/3·52	4·5/4·32	3·35/3·1
M8	4·0	3·2	5·0/4·52	4·0/3·52	12·0	5·5/5·02	6·0/5·82	4·7/4·5
M10	5·0	4·0	7·0/6·42	5·0/4·52	16·0	7·0/6·42	7·0/6·78	4·9/4·7
M12	6·0	4·8	8·0/7·42	6·0/5·52	18·0	8·5/7·92	9·0/8·78	6·4/6·2
(M14)	6·0	4·8	9·0/8·42	7·0/6·42	22·0	10·0/9·42	10·0/9·78	6·15/5·9
M16	8·0	6·4	10·0/9·42	8·0/7·42	25·0	12·0/11·3	12·0/11·73	8·15/7·9
(M18)	10·0	8·0	12·0/11·3	9·0/8·42	28·0	13·5/12·8	13·0/12·73	7·85/7·6
M20	10·0	8·0	14·0/13·3	10·0/9·42	30·0	15·0/14·3	15·0/14·73	7·85/7·6
(M22)	12·0	9·6	15·0/14·3	11·0/10·3	35·0	16·5/15·8	17·0/16·73	9·9/9·65
M24	12·0	9·6	16·0/15·3	12·0/11·3	38·0	18·0/17·3	18·0/17·73	9·65/9·4

Note: Sizes in brackets are non-preferred.

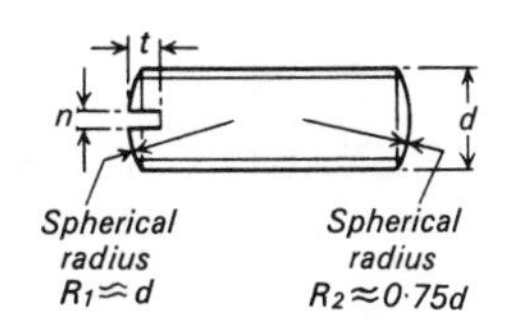

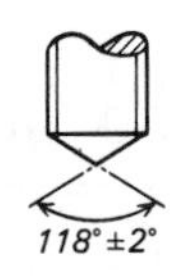

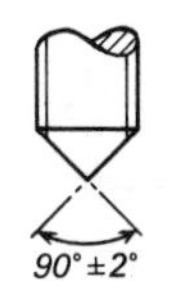

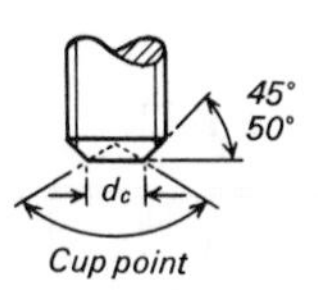

Cone points

Length of screw ≤ Nominal diameter   Length of screw > Nominal diameter

able K12.

ISO METRIC SLOTTED GRUB SCREWS (B.S. 4219)

Nominal Size	Slot		Cup point
	Width-n (mm)	Depth-t (mm)	Cone diameter-d_c (mm)
M1·6	0·5/0·36	1·0/0·8	—
M2	0·5/0·36	1·0/0·8	—
M2·5	0·6/0·46	1·2/1·0	—
M3	0·7/0·56	1·4/1·0	1·4/1·0
M4	0·8/0·66	1·6/1·2	2·0/1·6
M5	1·0/0·86	2·0/1·6	2·5/2·1
M6	1·2/1·06	2·2/1·8	3·0/2·6
M8	1·51/1·26	2·7/2·3	5·0/4·52
M10	1·91/1·66	3·2/2·8	6·0/5·52
M12	2·31/2·06	4·24/3·76	8·0/7·42

Table K13. FRENCH ISO METRIC HEXAGON NUTS

Nom. Diam. (mm.)	Across Flats (mm.)	Depth of Nut			Nom. Diam. (mm.)	Across Flats (mm.)	Depth of Nut		
		Normal Nut (mm.)	Lock Nut (mm.)	Deep Nut (mm.)			Normal Nut (mm.)	Lock Nut (mm.)	Deep Nut (mm.
1·6	3·2	1·3	—	1·6	24·0	36·0	19·0	12·0	24·0
2·0	4·0	1·6	—	2·0	27·0	41·0	22·0	13·5	27·0
2·5	5·0	2·0	—	2·5	30·0	46·0	24·0	15·0	30·0
3·0	5·5	2·4	—	3·0	33·0	50·0	26·0	16·5	33·0
3·5	6·0	2·8	—	3·5	36·0	54·0	29·0	18·0	36·0
4·0	7·0	3·2	—	4·0	39·0	58·0	31·0	19·5	39·0
5·0	8·0	4·0	—	5·0	42·0	63·0	34·0	21·0	42·0
6·0	10·0	5·0	3·0	6·0	45·0	67·0	36·0	22·5	45·0
7·0	11·0	5·5	3·5	7·0	48·0	71·0	38·0	24·0	48·0
8·0	13·0	6·5	4·0	8·0	52·0	77·0	42·0	26·0	52·0
10·0	17·0	8·0	5·0	10·0	56·0	82·0	45·0	28·0	56·0
12·0	19·0	10·0	6·0	12·0	60·0	88·0	48·0	30·0	60·0
14·0	22·0	11·0	7·0	14·0	64·0	94·0	51·0	32·0	64·0
16·0	24·0	13·0	8·0	16·0	68·0	100·0	54·0	34·0	68·0
18·0	27·0	15·0	9·0	18·0	72·0	105·0	58·0	36·0	72·0
20·0	30·0	16·0	10·0	20·0	76·0	110·0	61·0	38·0	76·0
22·0	32·0	18·0	11·0	22·0	80·0	116·0	64·0	40·0	80·0

able K14. UNIFIED (U.N.C. AND U.N.F.) PRECISION HEXAGON BOLTS, SCREWS AND NUTS (B.S. 1768: 1963)

(Dimensions in Inches)

Nom. Size	Width across Flats		Thickness					
			Head		Ord. Nut		Lock Nut	
	Max.	Min.	Max.	Min.	Max.	Min.	Max.	Min.
$\frac{1}{4}$	0·4375	0·4305	0·163	0·153	0·224	0·214	0·161	0·151
$\frac{5}{16}$	0·5000	0·4930	0·211	0·201	0·271	0·261	0·192	0·182
$\frac{3}{8}$	0·5625	0·5545	0·243	0·233	0·333	0·323	0·224	0·214
$\frac{7}{16}$	(see below)		0·291	0·281	0·380	0·370	0·255	0·245
$\frac{1}{2}$	0·7500	0·7420	0·323	0·313	0·442	0·432	0·317	0·307
$\frac{9}{16}$	(see below)		0·371	0·361	0·489	0·479	0·349	0·339
$\frac{5}{8}$	0·9375	0·9295	0·403	0·393	0·552	0·542	0·380	0·370
$\frac{3}{4}$	1·1250	1·1150	0·483	0·463	0·651	0·631	0·432	0·412
$\frac{7}{8}$	1·3125	1·3005	0·563	0·543	0·760	0·740	0·494	0·474
1	1·5000	1·4880	0·627	0·597	0·874	0·844	0·562	0·532

Width across Flats for $\frac{7}{16}$ and $\frac{9}{16}$ sizes.

$\frac{7}{16}$: bolt=0·6250/0·6170; nut=0·6875/0·6795.
$\frac{9}{16}$: bolt=0·8125/0·8045; nut=0·8750/0·8670.

Length of Thread (Minimum).

Up to 6 ins. length of bolt=(2 × diameter)+$\frac{1}{4}$ in.
Over 6 ins. length=(2 × diameter)+$\frac{1}{2}$ in.
Bolts too short for minimum length to be full threaded to 2 pitches from underside of head.

Table K15. UNIFIED (U.N.C. AND U.N.F.)

HEAVY HEXAGON BOLTS, SCREWS AND NUTS

(B.S. 1769: 1951)

(Dimensions in Inches)

Nom. Size	Width across Flats		Thickness					
			Head† (Faced)		Ord. Nut (Faced one side)		Lock Nut (Faced both sides)	
	Max.	Min.	Max.	Min.	Max.	Min.	Max.	Min.
½	0·8750	0·8550	0·426	0·386	0·504	0·464	0·317	0·277
⅝	1·0625	1·0375	0·522	0·478	0·631	0·587	0·381	0·337
¾	1·2500	1·2250	0·618	0·570	0·758	0·710	0·446	0·398
⅞	1·4375	1·4075	0·714	0·662	0·885	0·833	0·510	0·458
1	1·6250	1·5950	0·778	0·722	1·012	0·956	0·575	0·519
1⅛	1·8125	1·7825	0·874	0·814	1·139	1·079	0·639	0·579
1¼	2·0000	1·9550	0·970	0·906	1·251	1·187	0·751	0·687
1⅜	2·1875	2·1425	1·065	0·997	1·378	1·310	0·815	0·747
1½	2·3750	2·3300	1·161	1·089	1·505	1·433	0·880	0·808
1¾	2·7500	2·6900	1·352	1·272	1·759	1·679	1·009	0·929
2	3·1250	3·0650	1·482	1·394	2·013	1·925	1·138	1·050

†The bolt head in the as-forged condition has the following thickness (max./min.)—
½ in. (0·458/0·418); ⅝ in. (0·553/0·509); ¾ in. (0·649/0·601); ⅞ in. (0·745/0·693);
1 in. (0·840/0·784); 1⅛ in. (0·936/0·876); 1¼ in. (1·032/0·968); 1⅜ in. (1·128/1·060);
1½ in. (1·224/1·152); 1¾ in. (1·415/1·335); 2 in. (1·606/1·518).

Length of Thread.

Up to 6 ins. length of bolt = (2 × diameter) + ¼ in.
Over 6 in. length of bolt = (2 × diameter) + ½ in.

Table K16. **UNIFIED MACHINE SCREWS SLOTTED-SCREW HEADS (B.S. 1981: 1953) (Dimensions in Inches)**

Nominal Size	Diameter of head: C'sunk and Raised C'sunk (max.)	(min.)	Pan Heads (max.)	(min.)	Raised Cheese Heads (max.)	(min.)	Width of slot: All (max.)	(min.)
1/4	0·473	0·450	0·492	0·473	0·414	0·389	0·075	0·064
5/16	0·593	0·565	0·615	0·594	0·518	0·490	0·084	0·072
3/8	0·712	0·681	0·740	0·716	0·622	0·590	0·094	0·081
*7/16	0·753	0·719	0·863	0·838	0·625	0·589	0·094	0·081
*1/2	0·808	0·770	0·987	0·958	0·750	0·710	0·106	0·091
*5/8	1·041	0·996	1·125	1·090	0·875	0·827	0·133	0·116
*3/4	1·275	1·223	1·250	1·209	1·000	0·945	0·149	0·131

Nom. Size	Depth of head: C'sunk, Raised C'sunk (max.)	Pan Heads (max.)	Pan Heads (min.)	Raised Cheese (max.)	Raised Cheese (min.)	Depth of slot: C'sunk	Raised C'sunk	Pan Heads	Raised Cheese
1/4	0·153	0·144	0·130	0·237	0·207	0·058	0·113	0·079	0·098
5/16	0·191	0·178	0·162	0·295	0·262	0·073	0·143	0·101	0·124
3/8	0·230	0·212	0·195	0·355	0·315	0·086	0·172	0·122	0·149
*7/16	0·223	0·247	0·227	0·368	0·321	0·086	—	0·133	0·153
*1/2	0·223	0·281	0·260	0·412	0·362	0·086	—	0·152	0·171
*5/8	0·298	0·350	0·325	0·521	0·461	0·113	—	0·189	0·217
*3/4	0·372	0·419	0·390	0·612	0·542	0·141	—	0·226	0·254

*No dimensions for these sizes are specified for raised countersunk heads.

Table K17. GRUB-SCREW SLOTS (B.S. 768: 1958) (B.S.W., B.S.F., U.N.C., and U.N.F.)

Nominal Size	Width of Slot		Depth of Slot
	(max.)	(min.)	
†$\frac{1}{8}$	0·029	0·023	0·043
*$\frac{3}{16}$	0·040	0·032	0·055
$\frac{1}{4}$	0·056	0·045	0·063
$\frac{5}{16}$	0·063	0·051	0·078
$\frac{3}{8}$	0·077	0·064	0·094
$\frac{7}{16}$	0·085	0·072	0·109
$\frac{1}{2}$	0·096	0·081	0·125
$\frac{9}{16}$	0·107	0·091	0·141
$\frac{5}{8}$	0·119	0·102	0·156
$\frac{3}{4}$	0·147	0·129	0·188

*This size is specified for B.S.W. and B.S.F. only; not for Unified.
†This size is specified for B.S.W. only.

Table K18. AMERICAN REGULAR HEXAGON NUTS
(Dimensions in Inches)

Nom. Size	Width across Flats		Thickness of Nut							
			Unfinished				Semi-finished			
			Ord. Nut		Lock Nut		Ord. Nut		Lock Nut	
	Max.	Min.	Max.	Min.	Max.	Min.	Max.	Min.	Max.	Min.
$\frac{1}{4}$	0·4375	0·425	0·235	0·203	0·172	0·140	0·219	0·187	0·157	0·125
$\frac{5}{16}$	0·5625	0·547	0·283	0·249	0·204	0·170	0·267	0·233	0·189	0·155
$\frac{3}{8}$	0·6250	0·606	0·346	0·310	0·237	0·201	0·330	0·294	0·221	0·185
$\frac{7}{16}$	0·7500	0·728	0·394	0·356	0·269	0·231	0·378	0·340	0·253	0·215
$\frac{1}{2}$	0·8125	0·788	0·458	0·418	0·332	0·292	0·442	0·402	0·317	0·277
$\frac{9}{16}$	0·8750	0·847	0·521	0·479	0·365	0·323	0·505	0·463	0·349	0·307
$\frac{5}{8}$	1·0000	0·969	0·569	0·525	0·397	0·353	0·553	0·509	0·381	0·337
$\frac{3}{4}$	1·1250	1·088	0·680	0·632	0·462	0·414	0·665	0·617	0·446	0·398
$\frac{7}{8}$	1·3125	1·269	0·792	0·740	0·526	0·474	0·776	0·724	0·510	0·458
1	1·5000	1·450	0·903	0·847	0·590	0·534	0·887	0·831	0·575	0·519
$1\frac{1}{8}$	1·6875	1·631	1·030	0·970	0·655	0·595	0·999	0·939	0·639	0·579
$1\frac{1}{4}$	1·8750	1·812	1·126	1·062	0·782	0·718	1·094	1·030	0·751	0·687
$1\frac{3}{8}$	2·0625	1·994	1·237	1·169	0·846	0·778	1·206	1·138	0·815	0·747
$1\frac{1}{2}$	2·2500	2·175	1·348	1·276	0·911	0·839	1·317	1·245	0·880	0·808
$1\frac{5}{8}$	2·4375	2·356	—	—	—	—	1·429	1·353	0·944	0·868
$1\frac{3}{4}$	2·6250	2·538	—	—	—	—	1·540	1·460	1·009	0·929
$1\frac{7}{8}$	2·8125	2·719	—	—	—	—	1·651	1·567	1·073	0·989
2	3·0000	2·900	—	—	—	—	1·763	1·675	1·138	1·050
$2\frac{1}{4}$	3·3750	3·262	—	—	—	—	1·970	1·874	1·251	1·155
$2\frac{1}{2}$	3·7500	3·625	—	—	—	—	2·193	2·089	1·505	1·401
$2\frac{3}{4}$	4·1250	3·988	—	—	—	—	2·415	2·303	1·634	1·522
3	4·5000	4·350	—	—	—	—	2·638	2·518	1·763	1·643

Table K19. AMERICAN REGULAR HEXAGON BOLTS
(Dimensions in Inches)

Nom. Size	Width across Flats		Thickness of Head			
			Unfinished		Semi-finished	
	Max.	Min.	Max.	Min.	Max.	Min.
$\frac{1}{4}$	0·4375	0·425	0·188	0·150	0·163	0·150
$\frac{5}{16}$	0·5000	0·484	0·235	0·195	0·211	0·195
$\frac{3}{8}$	0·5625	0·544	0·268	0·226	0·243	0·226
$\frac{7}{16}$	0·6250	0·603	0·316	0·272	0·291	0·272
$\frac{1}{2}$	0·7500	0·725	0·364	0·302	0·323	0·302
$\frac{5}{8}$	0·9375	0·906	0·444	0·378	0·403	0·378
$\frac{3}{4}$	1·1250	1·088	0·524	0·455	0·483	0·455
$\frac{7}{8}$	1·3125	1·269	0·604	0·531	0·563	0·531
1	1·5000	1·450	0·700	0·591	0·627	0·591
$1\frac{1}{8}$	1·6875	1·631	0·780	0·658	0·718	0·658
$1\frac{1}{4}$	1·8750	1·812	0·876	0·749	0·813	0·749
$1\frac{3}{8}$	2·0625	1·994	0·940	0·810	0·878	0·810
$1\frac{1}{2}$	2·2500	2·175	1·036	0·902	0·974	0·902
$1\frac{5}{8}$	2·4375	2·356	1·100	0·962	1·038	0·962
$1\frac{3}{4}$	2·6250	2·538	1·196	1·054	1·134	1·054
$1\frac{7}{8}$	2·8125	2·719	1·260	1·114	1·198	1·114
2	3·0000	2·900	1·388	1·175	1·263	1·175
$2\frac{1}{4}$	3·3750	3·262	1·548	1·327	1·423	1·327
$2\frac{1}{2}$	3·7500	3·625	1·708	1·479	1·583	1·479
$2\frac{3}{4}$	4·1250	3·988	1·869	1·632	1·744	1·632
3	4·5000	4·350	2·060	1·815	1·935	1·815
$3\frac{1}{4}$	4·8750	4·712	2·251	1·936	2·064	1·936
$3\frac{1}{2}$	5·2500	5·075	2·380	2·057	2·193	2·057
$3\frac{3}{4}$	5·6250	5·437	2·572	2·241	2·385	2·241
4	6·0000	5·800	2·764	2·424	2·576	2·424

able K20. AMERICAN HEAVY NUTS

(Dimensions in Inches)

Nom. Size	Width Across Flats		Thickness of Nut							
			Unfinished				Semi-finished			
			Ord. Nut		Lock Nut		Ord. Nut		Lock Nut	
	Max.	Min.	Max.	Min.	Max.	Min.	Max.	Min.	Max.	Min.
$\frac{1}{4}$	0·5000	0·488	0·266	0·218	0·204	0·156	0·250	0·218	0·188	0·156
$\frac{5}{16}$	0·5625	0·546	0·330	0·280	0·236	0·186	0·314	0·280	0·220	0·186
$\frac{3}{8}$	0·6875	0·669	0·393	0·341	0·268	0·216	0·377	0·341	0·252	0·216
$\frac{7}{16}$	0·7500	0·728	0·456	0·403	0·300	0·247	0·441	0·403	0·285	0·247
$\frac{1}{2}$	0·8750	0·850	0·520	0·464	0·332	0·277	0·504	0·464	0·317	0·277
$\frac{9}{16}$	0·9375	0·909	—	—	—	—	0·568	0·526	0·349	0·307
$\frac{5}{8}$	1·0625	1·031	0·647	0·587	0·397	0·337	0·631	0·587	0·381	0·337
$\frac{3}{4}$	1·2500	1·212	0·774	0·710	0·462	0·398	0·758	0·710	0·446	0·398
$\frac{7}{8}$	1·4375	1·394	0·901	0·833	0·526	0·458	0·885	0·833	0·510	0·458
1	1·6250	1·575	1·028	0·956	0·590	0·519	1·012	0·956	0·575	0·519
1$\frac{1}{8}$	1·8125	1·756	1·155	1·079	0·655	0·579	1·139	1·079	0·639	0·579
1$\frac{1}{4}$	2·0000	1·938	1·282	1·187	0·782	0·687	1·251	1·187	0·751	0·687
1$\frac{3}{8}$	2·1875	2·119	1·409	1·310	0·846	0·747	1·378	1·310	0·815	0·747
1$\frac{1}{2}$	2·3750	2·300	1·536	1·433	0·911	0·808	1·505	1·433	0·880	0·808
1$\frac{5}{8}$	2·5625	2·481	1·663	1·556	0·976	0·868	1·632	1·556	0·944	0·868
1$\frac{3}{4}$	2·7500	2·662	1·790	1·679	1·040	0·929	1·759	1·679	1·009	0·929
1$\frac{7}{8}$	2·9375	2·844	1·917	1·802	1·104	0·989	1·886	1·802	1·073	0·989
2	3·1250	3·025	2·044	1·925	1·169	1·050	2·013	1·925	1·138	1·050
2$\frac{1}{4}$	3·5000	3·388	2·298	2·155	1·298	1·155	2·251	2·155	1·251	1·155
2$\frac{1}{2}$	3·8750	3·750	2·552	2·401	1·552	1·401	2·505	2·401	1·505	1·401
2$\frac{3}{4}$	4·2500	4·112	2·806	2·647	1·681	1·522	2·759	2·647	1·634	1·522
3	4·6250	4·475	3·060	2·893	1·810	1·643	3·013	2·893	1·763	1·643
3$\frac{1}{4}$	5·0000	4·838	3·314	3·124	1·939	1·748	3·252	3·124	1·876	1·748
3$\frac{1}{2}$	5·3750	5·200	3·568	3·370	2·068	1·870	3·506	3·370	2·006	1·870
3$\frac{3}{4}$	5·7500	5·562	3·822	3·616	2·197	1·990	3·760	3·616	2·134	1·990
4	6·1250	5·925	4·076	3·862	2·326	2·112	4·014	3·862	2·264	2·112

Table K21. **AMERICAN HEAVY BOLTS**
(Dimensions in Inches)

Nom. Size	Width across Flats		Thickness of Head			
			Unfinished		Semi-finished	
	Max.	Min.	Max.	Min.	Max.	Min.
½	0·8750	0·850	0·364	0·302	0·323	0·302
⅝	1·0625	1·031	0·444	0·378	0·403	0·378
¾	1·2500	1·212	0·524	0·455	0·483	0·455
⅞	1·4375	1·394	0·604	0·531	0·563	0·531
1	1·6250	1·575	0·700	0·591	0·627	0·591
1⅛	1·8125	1·756	0·780	0·658	0·718	0·658
1¼	2·0000	1·938	0·876	0·749	0·813	0·749
1⅜	2·1875	2·119	0·940	0·810	0·878	0·810
1½	2·3750	2·300	1·036	0·902	0·974	0·902
1⅝	2·5625	2·481	1·100	0·962	—	—
1¾	2·7500	2·662	1·196	1·054	1·134	1·054
1⅞	2·9375	2·844	1·260	1·114	—	—
2	3·1250	3·025	1·388	1·175	1·263	1·175
2¼	3·5000	3·388	1·548	1·327	1·423	1·327
2½	3·8750	3·750	1·708	1·479	1·583	1·479
2¾	4·2500	4·112	1·869	1·632	1·744	1·632
3	4·6250	4·475	2·060	1·815	1·935	1·815

'able K22. AMERICAN FINISHED HEXAGON BOLTS AND NUTS
(Dimensions in Inches)

Nom. Size	Width Across Flats		Depth							
			Bolt Head		Ord. Nut		Locknut		Thick Nut	
	Max.	Min.	Max.	Min.	Max.	Min.	Max.	Min.	Max.	Min.
$\frac{1}{4}$	0·4375	0·428	0·163	0·150	0·226	0·212	0·163	0·150	0·288	0·274
$\frac{5}{16}$	0·5000	0·489	0·211	0·195	0·273	0·258	0·195	0·180	0·336	0·320
$\frac{3}{8}$	0·5625	0·551	0·243	0·226	0·337	0·320	0·227	0·210	0·415	0·398
$\frac{7}{16}$	0·6250*	0·612*	0·291	0·272	0·385	0·365	0·260	0·240	0·463	0·444
$\frac{1}{2}$	0·7500	0·736	0·323	0·302	0·448	0·427	0·323	0·302	0·573	0·552
$\frac{9}{16}$	0·8125*	0·798*	0·371	0·348	0·496	0·473	0·324	0·301	0·621	0·598
$\frac{5}{8}$	0·9375	0·922	0·403	0·378	0·559	0·535	0·387	0·363	0·731	0·706
$\frac{3}{4}$	1·1250	1·100*	0·483	0·455	0·665	0·617	0·446	0·398	0·827	0·798
$\frac{7}{8}$	1·3125	1·285*	0·563	0·531	0·776	0·724	0·510	0·458	0·922	0·890
1	1·5000	1·469*	0·627	0·591	0·887	0·831	0·575	0·519	1·018	1 982
$1\frac{1}{8}$	1·6875	1·631	0·718	0·658	0·999	0·939	0·639	0·579	1·176	1·136
$1\frac{1}{4}$	1·8750	1·812	0·813	0·749	1·094	1·030	0·751	0·687	1·272	1·228
$1\frac{3}{8}$	2·0625	1·994	0·878	0·810	1·206	1·138	0·815	0·747	1·399	1·351
$1\frac{1}{2}$	2·2500	2·175	0·974	0·902	1·317	1·245	0·880	0·808	1·526	1·474
$1\frac{5}{8}$	2·4375	2·356	1·038	0·962	1·429	1·353	0·944	0·868	—	—
$1\frac{3}{4}$	2·6250	2·538	1·134	1·054	1·540	1·460	1·009	0·929	—	—
$1\frac{7}{8}$	2·8125	2·719	1·198	1·114	1·651	1·567	1·073	0·989	—	—
2	3·0000	2·900	1·263	1·175	1·763	1·675	1·138	1·050	—	—
$2\frac{1}{4}$	3·3750	3·262	1·423	1·327	1·970	1·874	1·251	1·155	—	—
$2\frac{1}{2}$	3·7500	3·625	1·583	1·479	2·193	2·089	1·505	1·401	—	—
$2\frac{3}{4}$	4·1250	3·988	1·744	1·632	2·415	2·303	1·634	1·522	—	—
3	4·5000	4·350	1·935	1·815	2·638	2·518	1·763	1·643	—	—

*These dimensions are true only for the bolts; the corresponding dimensions for the nuts are: $\frac{7}{16}$", 0·6875/0·675; $\frac{9}{16}$", 0·8750/0·861; $\frac{3}{4}$", 1·088; $\frac{7}{8}$", 1·269; 1", 1·450.

Table K23. AMERICAN HEXAGON HEAD MACHINE SCREWS (Dimensions in Inches)

Nominal Size	Width across Flats		Height of Head	
	Max.	Min.	Max.	Min.
2	0·125	0·120	0·050	0·040
3	0·187	0·181	0·055	0·044
4	0·187	0·181	0·060	0·049
5	0·187	0·181	0·070	0·058
6	0·250	0·244	0·093	0·080
8	0·250	0·244	0·110	0·096
10	0·312	0·305	0·120	0·105
12	0·312	0·305	0·155	0·139

Table K24. AMERICAN HEXAGON AND SQUARE MACHINE SCREW NUTS (Dimensions in Inches)

Nominal Size	Width across Flats		Thickness	
	Max.	Min.	Max.	Min.
2	0·1875	0·180	0·066	0·057
3	0·1875	0·180	0·066	0·057
4	0·2500	0·241	0·098	0·087
5	0·3125	0·302	0·114	0·102
6	0·3125	0·302	0·114	0·102
8	0·3438	0·332	0·130	0·117
10	0·3750	0·362	0·130	0·117
12	0·4375	0·423	0·161	0·148

Table K25. WHITWORTH (B.S.W. AND B.S.F.) PRECISION HEXAGON BOLTS, SCREWS AND NUTS (B.S. 1083; 1951)

(Dimensions in Inches)

Nom. Size	Width across Flats		Thickness						Nominal Length* of Thread
			Head		Ord. Nut		Lock Nut		
	Max.	Min.	Max.	Min.	Max.	Min.	Max.	Min.	
$\frac{1}{4}$	0·445	0·438	0·19	0·18	0·200	0·190	0·133	0·123	$\frac{1}{2}$
$\frac{5}{16}$	0·525	0·518	0·22	0·21	0·250	0·240	0·166	0·156	$\frac{5}{8}$
$\frac{3}{8}$	0·600	0·592	0·27	0·26	0·312	0·302	0·208	0·198	$\frac{3}{4}$
$\frac{7}{16}$	0·710	0·702	0·33	0·32	0·375	0·365	0·250	0·240	$\frac{7}{8}$
$\frac{1}{2}$	0·820	0·812	0·38	0·37	0·437	0·427	0·291	0·281	1
$\frac{9}{16}$	0·920	0·912	0·44	0·43	0·500	0·490	0·333	0·323	$1\frac{1}{8}$
$\frac{5}{8}$	1·010	1·000	0·49	0·48	0·562	0·552	0·375	0·365	$1\frac{1}{4}$
$\frac{3}{4}$	1·200	1·190	0·60	0·59	0·687	0·677	0·458	0·448	$1\frac{1}{2}$
$\frac{7}{8}$	1·300	1·288	0·66	0·65	0·750	0·740	0·500	0·490	$1\frac{3}{4}$
1	1·480	1·468	0·77	0·76	0·875	0·865	0·583	0·573	2
$1\frac{1}{8}$	1·670	1·658	0·88	0·87	1·000	0·990	0·666	0·656	$2\frac{1}{4}$
$1\frac{1}{4}$	1·860	1·845	0·98	0·96	1·125	1·105	0·750	0·730	$2\frac{1}{2}$
†$1\frac{3}{8}$	2·050	2·035	1·09	1·07	1·250	1·230	0·833	0·813	$2\frac{3}{4}$
$1\frac{1}{2}$	2·220	2·200	1·20	1·18	1·375	1·355	0·916	0·896	3
$1\frac{3}{4}$	2·580	2·555	1·42	1·40	1·625	1·605	1·083	1·063	$3\frac{1}{2}$
2	2·760	2·735	1·53	1·51	1·750	1·730	1·166	1·146	4

*B.S.W. Bolts and B.S.F. Long Bolts. Tolerance of $+\frac{3}{32}-0$ for sizes $\frac{1}{4}$ to $\frac{9}{16}$ incl.; $+\frac{1}{8}-0$ for sizes $\frac{5}{8}$ to 1 in. incl.; $+\frac{1}{4}-0$ for sizes $1\frac{1}{8}$ to 2 in. incl.

†Not standard with B.S.W. Thread.

Table K26.* **WHITWORTH B.S.W.**

BRIGHT HEXAGON BOLTS, SET-SCREWS AND NUTS
(Old Standard—B.S. 190: 1924)
(Dimensions in Inches)

Nom. Size	Width across Flats		Thickness						Length† of Thread
			Head		Ord. Nut		Lock Nut		
	Max.	Min.	Max.	Min.	Max.	Min.	Max.	Min.	Min.
$\frac{1}{4}$	0·525	0·520	0·22	0·21	0·250	0·240	0·166	0·156	0·500
$\frac{5}{16}$	0·600	0·595	0·27	0·26	0·312	0·302	0·208	0·198	0·625
$\frac{3}{8}$	0·710	0·705	0·33	0·32	0·375	0·365	0·250	0·240	0·750
$\frac{7}{16}$	0·820	0·815	0·38	0·37	0·437	0·427	0·291	0·281	0·875
$\frac{1}{2}$	0·920	0·915	0·44	0·43	0·500	0·490	0·333	0·323	1·000
$\frac{9}{16}$	1·010	1·002	0·49	0·48	0·562	0·552	0·375	0·365	1·125
$\frac{5}{8}$	1·100	1·092	0·55	0·54	0·625	0·615	0·416	0·406	1·250
$\frac{3}{4}$	1·300	1·292	0·66	0·65	0·750	0·740	0·500	0·490	1·500
$\frac{7}{8}$	1·480	1·468	0·77	0·76	0·875	0·865	0·583	0·573	1·750
1	1·670	1·658	0·88	0·87	1·000	0·990	0·666	0·656	2·000
$1\frac{1}{8}$	1·860	1·845	0·98	0·96	1·125	1·105	0·750	0·730	2·250
$1\frac{1}{4}$	2·050	2·035	1·09	1·07	1·250	1·230	0·833	0·813	2·500
$1\frac{1}{2}$	2·410	2·390	1·31	1·29	1·500	1·480	1·000	0·980	3·000
$1\frac{3}{4}$	2·760	2·735	1·53	1·51	1·750	1·730	1·166	1·146	3·500
2	3·150	3·125	1·75	1·73	2·000	1·980	1·333	1·313	4·000

*This standard has been officially superseded by B.S. 1083, but applications still exist where it is found preferable to use the larger hexagon sizes specified by B.S. 190.
†These lengths apply only to B.S.W. Bolts.

able K27. WHITWORTH (B.S.W. AND B.S.F.)
BLACK BOLTS AND NUTS—HEXAGON AND SQUARE (B.S. 916: 1953*)
(Dimensions in Inches)

Nom. Size	Width across Flats		Thickness					
			Head (faced under)		Black Ord. Nut		Black Lock Nut (Hex.)	
	Max.	Min.	Max.	Min.	Max.	Min.	Max.	Min.
¼	0·445	0·435	0·176	0·156	0·220	0·200	0·185	0·180
5/16	0·525	0·515	0·218	0·198	0·270	0·250	0·210	0·200
⅜	0·600	0·585	0·260	0·240	0·332	0·312	0·260	0·250
7/16	0·710	0·695	0·302	0·282	0·395	0·375	0·275	0·265
½	0·820	0·800	0·343	0·323	0·467	0·437	0·300	0·290
9/16	0·920	0·900	0·375	0·345	0·530	0·500	0·333	0·323
⅝	1·010	0·985	0·417	0·387	0·602	0·562	0·410	0·375
¾	1·200	1·175	0·500	0·470	0·728	0·687	0·490	0·458
⅞	1·300	1·270	0·583	0·553	0·810	0·750	0·550	0·500
1	1·480	1·450	0·666	0·636	0·935	0·875	0·630	0·583
1⅛	1·670	1·640	0·75	0·71	1·060	1·000	0·720	0·666
1¼	1·860	1·815	0·83	0·79	1·205	1·125	0·810	0·750
1⅜	2·050	2·005	0·92	0·88	1·330	1·250	0·890	0·833
1½	2·220	2·175	1·00	0·96	1·455	1·375	0·980	0·916
1⅝	2·410	2·365	1·08	1·02	1·580	1·500	1·060	1·000
1¾	2·580	2·520	1·17	1·11	1·725	1·625	1·160	1·083
2	2·760	2·700	1·33	1·27	1·850	1·750	1·250	1·166
2¼	3·150	3·090	1·50	1·42	1·975	1·875	1·430	1·250
2½	3·550	3·490	1·67	1·59	2·225	2·125	1·600	1·416
2¾	3·890	3·830	1·83	1·75	2·475	2·375	1·770	1·580
3	4·180	4·080	2·00	1·90	2·775	2·625	1·980	1·750

LENGTH OF THREAD OF BLACK BOLTS†

Length of Bolt	Length of Thread	
	over ½ in. dia.	up to ½ in. dia.
Up to and including 4 in.	1½D	2 D
Above 4 in. and up to 8 in.	2 D	2 D
Above 8 in.	2½D	2½D

*This standard also gives thickness dimensions for nuts machined on one or two faces.
†For bolts faced under the head, the thread length is two diameters for all sizes.

Table K28.* WHITWORTH BLACK BOLTS AND NUTS

(Old Standard B.S. 28)

(Dimensions in Inches)

Nom. Size	Width across Flats		Thickness					
			Head		Ord. Nut		Lock Nut	
	Max.	Min.	Max.	Min.	Max.	Min.	Max.	Min.
1/4	0·525	0·505	0·24	0·22	0·27	0·25	0·19	0·17
5/16	0·600	0·580	0·29	0·27	0·33	0·31	0·23	0·21
3/8	0·710	0·690	0·35	0·33	0·40	0·38	0·27	0·25
7/16	0·820	0·800	0·40	0·38	0·46	0·44	0·31	0·29
1/2	0·920	0·900	0·46	0·44	0·52	0·50	0·35	0·33
9/16	1·010	0·990	0·51	0·49	0·58	0·56	0·40	0·38
5/8	1·100	1·080	0·57	0·55	0·65	0·63	0·44	0·42
11/16	1·200	1·180	0·62	0·60	0·71	0·69	0·48	0·46
3/4	1·300	1·280	0·68	0·66	0·77	0·75	0·52	0·50
13/16	1·390	1·370	0·73	0·71	0·83	0·81	0·56	0·54
7/8	1·480	1·460	0·79	0·77	0·90	0·88	0·60	0·58
15/16	1·580	1·560	0·84	0·82	0·96	0·94	0·65	0·63
1	1·670	1·650	0·90	0·88	1·02	1·00	0·69	0·67
1 1/8	1·860	1·830	1·01	0·98	1·16	1·13	0·78	0·75
1 1/4	2·050	2·020	1·12	1·09	1·28	1·25	0·86	0·83
1 3/8	2·220	2·190	1·23	1·20	1·41	1·38	0·95	0·92
1 1/2	2·410	2·380	1·34	1·31	1·53	1·50	1·03	1·00
1 5/8	2·580	2·550	1·45	1·42	1·66	1·63	1·11	1·08
1 3/4	2·760	2·730	1·56	1·53	1·78	1·75	1·20	1·17
1 7/8	3·020	2·990	1·67	1·64	1·91	1·88	1·28	1·25
2	3·150	3·120	1·78	1·75	2·03	2·00	1·36	1·33
2 1/8	3·340	3·300	1·89	1·86	2·16	2·13	1·45	1·42
2 1/4	3·550	3·510	2·00	1·97	2·28	2·25	1·53	1·50
2 3/8	3·750	3·710	2·11	2·08	2·41	2·38	1·61	1·58
2 1/2	3·890	3·850	2·22	2·19	2·53	2·50	1·70	1·67
2 5/8	4·050	4·010	2·33	2·30	2·66	2·63	1·78	1·75
2 3/4	4·180	4·140	2·44	2·41	2·78	2·75	1·86	1·83
2 7/8	4·340	4·300	2·55	2·52	2·91	2·88	1·95	1·92
3	4·530	4·490	2·66	2·63	3·03	3·00	2·03	2·00
3 1/8	4·690	4·650	2·76	2·73	3·16	3·13	2·11	2·08
3 1/4	4·850	4·810	2·87	2·84	3·28	3·25	2·20	2·17
3 3/8	5·010	4·970	2·98	2·95	3·41	3·38	2·28	2·25
3 1/2	5·180	5·140	3·09	3·06	3·53	3·50	2·36	2·33
3 5/8	5·360	5·320	3·20	3·17	3·66	3·63	2·45	2·42
3 3/4	5·550	5·510	3·31	3·28	3·78	3·75	2·53	2·50
3 7/8	5·750	5·710	3·42	3·39	3·91	3·88	2·61	2·58
4	5·950	5·910	3·53	3·50	4·03	4·00	2·70	2·67

*These dimensions have been officially superseded by B.S. 916, but in certain applications the use of the dimensions given above has been found preferable.

Table K29. **WHITWORTH (B.S.W. and B.S.F.)**

SLOTTED—SCREW HEADS (B.S. 450: 1958)

(Dimensions in Inches)

Nominal Size	Diameter of head						Width of slot	
	Countersunk, Raised C'sunk, and (Round†)		Pan head		Slotted Cheese Head		All	
	(max.)	(min.)†	(max.)	(min.)	(max.)	(min.)	(max.)	(min.)
$\frac{1}{8}$	0·219	0·201	0·245	0·231	0·188	0·180	0·039	0·032
$\frac{3}{16}$	0·328	0·307	0·373	0·357	0·281	0·270	0·050	0·042
$\frac{7}{32}$	0·383	0·360	0·425	0·407	0·328	0·315	0·055	0·046
$\frac{1}{4}$	0·438	0·412	0·492	0·473	0·375	0·360	0·061	0·051
$\frac{5}{16}$	0·547	0·518	0·615	0·594	0·469	0·450	0·071	0·061
$\frac{3}{8}$	0·656	0·624	0·740	0·716	0·562	0·540	0·082	0·072
*$\frac{7}{16}$	0·766	0·729	0·863	0·838	0·656	0·630	0·093	0·082
*$\frac{1}{2}$	0·875	0·835	0·987	0·958	0·750	0·720	0·104	0·092
*$\frac{9}{16}$	0·984	0·941	1·031	0·999	0·844	0·810	0·115	0·103
*$\frac{5}{8}$	1·094	1·046	1·125	1·090	0·938	0·900	0·126	0·113
*$\frac{3}{4}$	1·312	1·257	1·250	1·209	1·125	1·080	0·148	0·134

Nominal Size	Depth of head					Depth of slot				
	C'sunk, Raised c'sunk	Round and Cheese Hds.		Pan Heads		C'sunk	Raised C'sunk	Round	Pan	Cheese
	(max.)	(max.)	(min.)	(max.)	(min.)					
$\frac{1}{8}$	0·056	0·087	0·082	0·075	0·065	0·027	0·047	0·048	0·040	0·039
$\frac{3}{16}$	0·084	0·131	0·124	0·110	0·099	0·041	0·071	0·072	0·061	0·059
$\frac{7}{32}$	0·098	0·153	0·145	0·125	0·112	0·048	0·083	0·084	0·069	0·069
$\frac{1}{4}$	0·113	0·175	0·165	0·144	0·130	0·055	0·095	0·096	0·078	0·079
$\frac{5}{16}$	0·141	0·219	0·207	0·178	0·162	0·069	0·120	0·120	0·095	0·098
$\frac{3}{8}$	0·169	0·262	0·249	0·212	0·195	0·083	0·144	0·144	0·112	0·118
*$\frac{7}{16}$	0·197	0·306	0·291	0·247	0·227	0·097	—	0·168	0·129	0·138
*$\frac{1}{2}$	0·225	0·350	0·333	0·281	0·260	0·111	—	0·192	0·145	0·157
*$\frac{9}{16}$	0·253	0·394	0·375	0·315	0·293	0·125	—	0·217	0·162	0·177
*$\frac{5}{8}$	0·281	0·437	0·417	0·350	0·325	0·138	—	0·240	0·179	0·197
*$\frac{3}{4}$	0·338	0·525	0·500	0·419	0·390	0·166	—	0·288	0·213	0·236

*No dimensions for these sizes are specified for raised countersunk heads.

†Minimum diameter for round heads are: $\frac{1}{8}$=0·206; $\frac{3}{16}$=0·312; $\frac{7}{32}$=0·365; $\frac{1}{4}$=0·417; $\frac{5}{16}$=0·524; $\frac{3}{8}$=0·629; $\frac{7}{16}$=0·735; $\frac{1}{2}$=0·840; $\frac{9}{16}$=0·946; $\frac{5}{8}$=1·051; $\frac{3}{4}$=1·262.

SET SCREWS—BRIGHT SQUARE HEAD (B.S.W. AND B.S.F.

(WITH FLAT CHAMFERED ENDS)

B.S.W. and B.S.F. bright square headed set screws are specified i three sizes of head, viz. No. 1 (Small), No. 2 (Medium) and No. 3 (Large and are threaded with a full form thread to within a distance from th underside of the head not exceeding two thread pitches. The end of th screw is chamfered to an included conical angle of 45 deg. to provide diameter of the circular end equal to the minor diameter of the screw les $\frac{1}{32}$ in. The sizes of head are as follows:

Table K30. SET SCREWS—BRIGHT SQUARE HEAD (B.S.W. AND B.S.F.)

(B.S. 451: 1932)

(Dimensions in Inches)

Nom. Size	Width across Flats			Depth of Head (All)
	No. 1	No. 2	No. 3	
$\frac{1}{4}$	0·250	0·313	0·375	0·250
$\frac{5}{16}$	0·313	0·375	0·438	0·313
$\frac{3}{8}$	0·375	0·438	0·500	0·375
$\frac{7}{16}$	0·438	0·500	0·625	0·438
$\frac{1}{2}$	0·500	0·563	0·750	0·500
$\frac{5}{8}$	0·625	0·750	0·875	0·625
$\frac{3}{4}$	0·750	0·875	1·000	0·750
$\frac{7}{8}$	0·875	1·000	1·125	0·875
1	1·000	1·125	1·250	1·000

Table K31. B.A. HEXAGON BOLTS, SCREWS AND NUTS
(B.S. 57: 1951)
(Dimensions in Inches)

B.A. No.	Width across Flats		Thickness						Length of Thread of Bolts
			Head		Ord. Nut		Lock Nut		
	Max.	Min.	Max.	Min.	Max.	Min.	Max.	Min.	Min.
0	0·413	0·408	0·177	0·169	0·213	0·203	0·157	0·147	0·71
1	0·365	0·360	0·156	0·149	0·188	0·178	0·139	0·129	0·63
2	0·324	0·319	0·139	0·132	0·167	0·157	0·123	0·113	0·55
3	0·282	0·277	0·121	0·115	0·153	0·143	0·108	0·098	0·48
4	0·248	0·243	0·106	0·100	0·135	0·125	0·094	0·084	0·43
5	0·220	0·216	0·094	0·089	0·120	0·110	0·084	0·074	0·38
6	0·193	0·189	0·083	0·078	0·105	0·095	0·073	0·063	0·33
7	0·172	0·169	0·074	0·070	0·094	0·087	—	—	0·29
8	0·152	0·149	0·065	0·061	0·082	0·075	0·058	0·051	0·26
9	0·131	0·128	0·056	0·052	0·071	0·064	—	—	0·22
10	0·117	0·114	0·050	0·046	0·064	0·057	—	—	0·20
11	0·103	0·101	—	—	0·056	0·051	—	—	—
12	0·090	0·088	—	—	0·049	0·044	—	—	—
13	0·083	0·081	—	—	0·045	0·040	—	—	—
14	0·069	0·067	—	—	0·037	0·032	—	—	—
15	0·062	0·060	—	—	0·034	0·029	—	—	—
16	0·056	0·054	—	—	0·029	0·024	—	—	—

The preferred sizes given by B.S. 57 are B.A. 2 and 4 in hexagon screws and bolts, B.A. 2, 4, 6, 8, 10, 12 in nuts, with second choice for nuts in sizes B.A. 0, 1, 3 and 5.

Hexagon bolts and screws in sizes B.A. 0, 1, 3, 5, 6, 7, 8, 9 and 10 and nuts in sizes B.A. 7, 9, 11, 13, 14, 15 and 16 are not normally stocked.

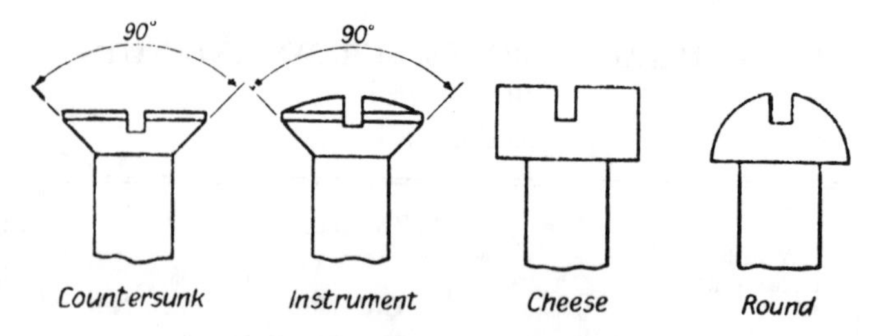

British Association (B.A.) Screw Heads

Table K32. **B.A. SCREW HEADS**
(Dimensions in Inches)

B.A. No.	Diam. of Head	Depth of Head, Total			Width of slot	
		Cheese & Round	C'sunk*	Instrument		
	Max.	Max.	Max.	Max.	Max.	Min.
0	0·413	0·167	0·099	0·147	0·064	0·056
1	0·366	0·148	0·089	0·131	0·058	0·050
2	0·319	0·130	0·077	0·113	0·052	0·044
3	0·283	0·113	0·071	0·106	0·047	0·039
4	0·252	0·101	0·065	0·094	0·040	0·034
5	0·221	0·088	0·058	0·084	0·040	0·034
6	0·194	0·078	0·051	0·074	0·033	0·027
7	0·173	0·069	0·047	0·069	0·033	0·027
8	0·157	0·063	0·043	0·062	0·030	0·024
9	0·128	0·052	0·035	0·049	0·030	0·024
10	0·112	0·045	0·030	0·043	0·024	0·019
11	0·110	0·045	0·033	0·046	0·024	0·019
12	0·095	0·038	0·028	0·040	0·020	0·015
13	0·081	0·032	0·023	0·032	0·020	0·015
14	0·064	0·026	0·019	0·027	0·015	0·011
15	0·064	0·026	0·021	0·029	0·015	0·011
16	0·058	0·023	0·019	0·025	0·013	0·009

*Also nominal depth of countersunk portion, including land, of instrument head.

USA Standard Sheet Metal, Self-Tapping and Metallic Drive Screws. — Table 1 shows the various types of "self-tapping" screw threads covered by the USA Standard USAS B18.6.4-1966. Designations of the USASI, and the corresponding manufacturers' and federal designations are also shown as well as references to tables where recommended hole sizes for these threads are given. Types A, AB, B, BP and C when turned into a hole of proper size form a thread by a displacing action. Types D, F, G, T, BF, and BT when turned into a hole of proper size form a thread by a cutting action. Type U when driven into a hole of proper size forms a series of multiple threads by a displacing action. These screws have the following descriptions and applications:

Type A: Spaced-thread screw with gimlet point primarily for use in light sheet metal, resin-impregnated plywood, and asbestos compositions. This type is no longer recommended. Use Type AB.

Type AB: Spaced-thread screw with same pitches as Type B but with gimlet point, primarily for similar uses as for Type A.

Type B: Spaced-thread screw with a blunt point with pitches generally somewhat finer than Type A. Used for light and heavy sheet metal, non-ferrous castings, plastics, resin-impregnated plywood, and asbestos compositions.

Type BP: Spaced-thread screw, the same as Type B but having a cone point. Used primarily in assemblies where holes are misaligned.

Type C: Screws having machine screw diameter-pitch combinations with threads approximately American National form and with blunt tapered points. Used where the use of a machine screw thread is preferable to the use of the spaced-thread types of thread forming screws. Also useful when chips from machine screw thread-cutting screws are objectionable. It should be recognized that in specific applications, this type of screw may require extreme driving torques due to long thread engagement or use in hard metals.

Types F, G, D and T: Thread-cutting screws with threads approximating machine screw threads, with blunt point, and with tapered entering threads having one or more cutting edges and chip cavities. The tapered threads of the Type F may be complete or incomplete at the producer's option; all other types have incomplete tapered threads. These screws can be used in materials such as aluminum, zinc, and lead die-castings; steel sheets and shapes; cast iron; brass, and plastics.

Types BF and BT: Thread-cutting screws with spaced threads as in Type B, with blunt points, and one or more cutting grooves. Used in plastics, die-castings, metal-clad and resin-impregnated plywoods, and asbestos.

Type U: Multiple-threaded drive screw with large helix angle, having a pilot, for use in metal and plastics. This screw is forced into the work by pressure and is intended for making permanent fastenings.

USA Standard Head Types for Tapping and Metallic Drive Screws. — Many of the head types used with "self-tapping" screw threads are similar to the head types of American Standard machine screws and include the following:

Round Head: The round head has a semi-elliptical top and flat bearing surface. The pan head is now considered preferable.

Flat Head: The flat head has a flat top surface and a conical bearing surface with an included angle of approximately 82 degrees or, for another style, 100 degrees.

Oval Head: The oval head has a rounded top surface and a conical bearing surface with an included angle of approximately 82 degrees.

Flat and Oval Trim Heads: Flat and oval trim heads are similar to the 82-degree flat and oval heads except that the size of head for a given size screw is one or two sizes smaller than the regular flat and oval sizes, and for oval trim heads there is a controlled radius where the top surface meets the conical bearing surface.

Acknowledgement, Data and information from American National Standards Institute (ANSI) Standards B18.6.4, Slotted and Recessed Head Tapping Screws and Metallic Drive Screws, B27.1, Lock Washers, and B27.2 Plain Washers, appear in this book on pages 261 through 283, and have been extracted with the permission of the publisher, the American Society of Mechanical Engineers, United Engineering Center, 345 East 47 Street, New York, New York 10017.

Undercut Flat and Oval Heads: For short lengths of flat and oval head tapping screws, the heads are undercut to 70 per cent of normal side height to afford greater length of thread on the screws.

Fillister Head: The fillister head has a rounded top surface, cylindrical sides and a flat bearing surface.

Truss Head: The truss head has a low rounded top surface with a flat bearing surface. For a given screw size, the diameter of the truss head is larger than the diameter of the corresponding round head.

Pan Head: The slotted pan head has a flat top surface rounded into cylindrical sides and a flat bearing surface. The recessed pan head has a rounded top and flat bearing surface. This head type is now preferred to the round head.

Hexagon Head: The hexagon head has a flat or indented top surface with hexagonal sides and a flat bearing surface.

Hexagon Washer Head: The hexagon washer head has an indented top surface and hexagonal sides and a round flat washer bearing surface which projects beyond the hexagon and is formed integrally with the head.

All of the heads are provided with either a slot or one of three types of cross recesses with the exception of the round head which does not have the Type IA cross recess, the oval trim and flat trim heads which are provided with cross recesses only and the hexagon head and hexagon washer heads which may be slotted only if specified.

Table K33. Type Designations of Sheet Metal Screws, Self-Tapping Screws and Metallic Drive Screws

TYPE	DESIGNATION				TYPE	DESIGNATION			
	USA STD.	MFG.	FED'R'L.	TABLE NO.*		USA STD.	MFG.	FED'R'L.	TABLE NO.*
	A	A	A	K42	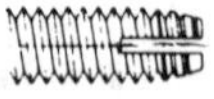	G	G	CS ALT. #2	K47
	AB	AB	AB	K44, K45, K46	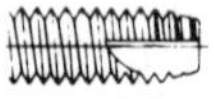	T	23	CG	K47
	B	B	B	K44, K45, K46		BF	BF	BF	16
	BP	BP	BP	K44, K45, K46		F	F	CF	K47
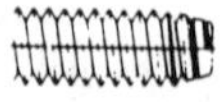	C	C	C	K43	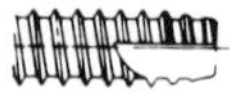	BT	25	BG	K48
	D	I	CS ALT. #1	K47	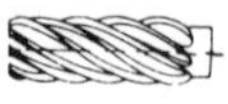	U	U	U	K49

* Table number refers to table in which recommended hole sizes for a particular thread type is given.

Table K34 . **USA Standard Threads and Points for Thread Forming Self-Tapping Screws** (USAS B18.6.4-1966)

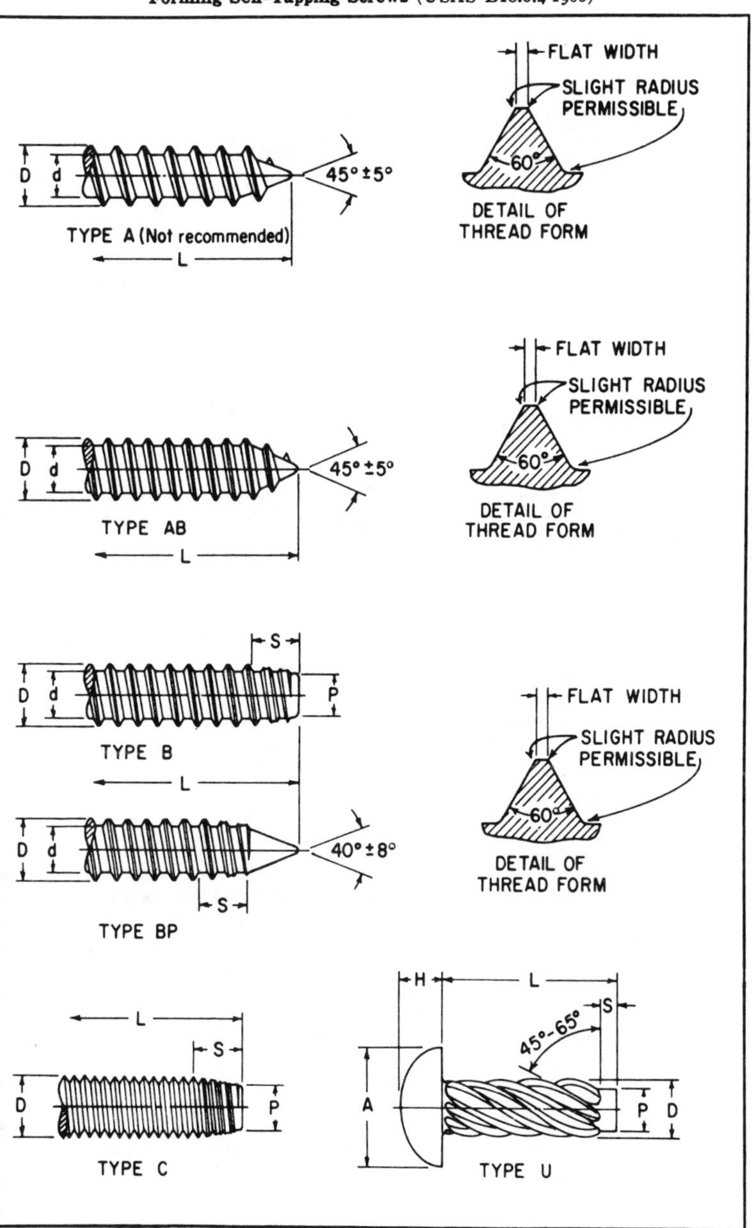

See Tables K37, K38 and K39 for thread data.

Table K35. **USA Standard Threads and Points for Thread Cutting Self-Tapping Screws** (USAS B18.6.4-1966)

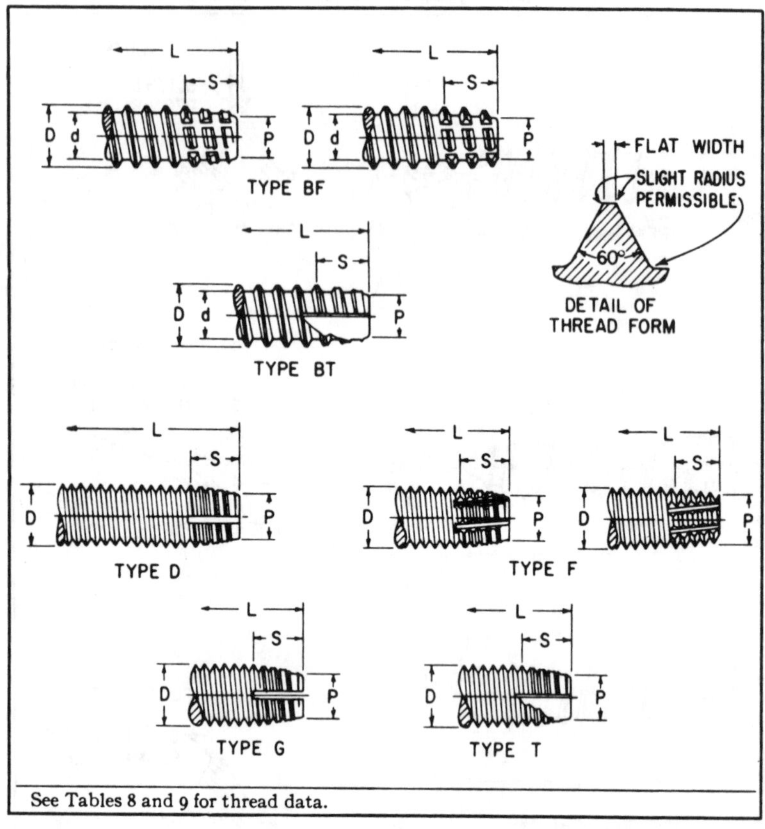

See Tables 8 and 9 for thread data.

Cross Recesses. — Type I cross recess has a large center opening, tapered wings, and blunt bottom, with all edges relieved or rounded. Type IA cross recess has a large center opening, wide straight wings, and blunt bottom, with all edges relieved or rounded. Type II consists of two intersecting slots with parallel sides converging to a slightly truncated apex at the bottom of the recess.

Table K36. **USA Standard Cross Recesses for Self-Tapping Screws** (USAS B18.6.4-1966)

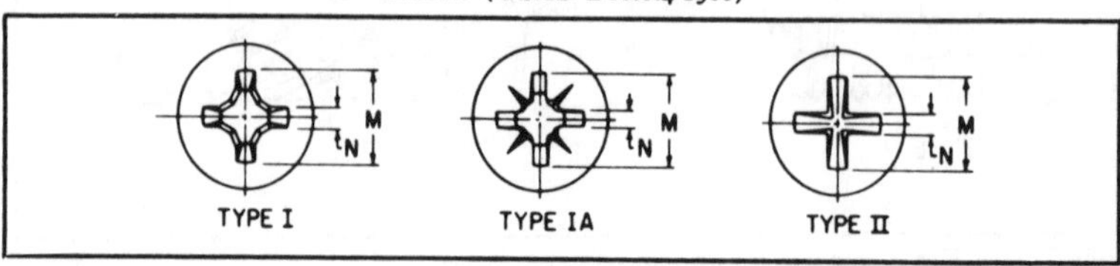

Table K37. USA Standard Thread and Point Dimensions for Types AB, A and U Thread Forming Tapping Screws (USAS B18.6.4-1966)

Type AB (Formerly BA)

Nominal Size or Basic Screw Diameter		Threads per inch	D Major Diameter Max.	D Major Diameter Min.	d Minor Diameter Max.	d Minor Diameter Min.	L Minimum Practical Screw Lengths 90° Heads	L Minimum Practical Screw Lengths Csk. Heads
0	0.0600	48	0.060	0.057	0.036	0.033	3/32	7/64
1	0.0730	42	0.075	0.072	0.049	0.046	1/8	9/64
2	**0.0860**	**32**	**0.088**	**0.084**	**0.064**	**0.060**	9/64	11/64
3	0.0990	28	0.101	0.097	0.075	0.071	11/64	3/16
4	**0.1120**	**24**	**0.114**	**0.110**	**0.086**	**0.082**	3/16	7/32
5	0.1250	20	0.130	0.126	0.094	0.090	3/16	1/4
6	**0.1380**	**20**	**0.139**	**0.135**	**0.104**	**0.099**	7/32	17/64
7	0.1510	19	0.154	0.149	0.115	0.109	17/64	5/16
8	**0.1640**	**18**	**0.166**	**0.161**	**0.122**	**0.116**	9/32	21/64
10	**0.1900**	**16**	**0.189**	**0.183**	**0.141**	**0.135**	21/64	3/8
12	0.2160	14	0.215	0.209	0.164	0.157	3/8	13/32
1/4	**0.2500**	**14**	**0.246**	**0.240**	**0.192**	**0.185**	13/32	15/32
5/16	0.3125	12	0.315	0.308	0.244	0.236	17/32	19/32

Type A

Nominal Size or Basic Screw Diameter		Threads per inch	D Major Diameter Max.	D Major Diameter Min.	d Minor Diameter Max.	d Minor Diameter Min.	L† These Lengths or Shorter — Use Type AB 90° Heads	L† These Lengths or Shorter — Use Type AB Csk. Heads
0	0.0600	40	0.060	0.057	0.042	0.039	1/8	3/16
1	0.0730	32	0.075	0.072	0.051	0.048	1/8	3/16
2	0.0860	32	0.088	0.084	0.061	0.056	5/32	3/16
3	0.0990	28	0.101	0.097	0.076	0.071	3/16	7/32
4	0.1120	24	0.114	0.110	0.083	0.078	3/16	1/4
5	0.1250	20	0.130	0.126	0.095	0.090	3/16	1/4
6	0.1380	18	0.141	0.136	0.102	0.096	1/4	5/16
7	0.1510	16	0.158	0.152	0.114	0.108	5/16	3/8
8	0.1640	15	0.168	0.162	0.123	0.116	3/8	7/16
10	0.1900	12	0.194	0.188	0.133	0.126	3/8	1/2
12	0.2160	11	0.221	0.215	0.162	0.155	7/16	9/16
14	0.2420	10	0.254	0.248	0.185	0.178	1/2	5/8
16	0.2680	10	0.280	0.274	0.197	0.189	9/16	3/4
18	0.2940	9	0.306	0.300	0.217	0.209	5/8	13/16
20	0.3200	9	0.333	0.327	0.234	0.226	11/16	13/16
24	0.3720	9	0.390	0.383	0.291	0.282	3/4	1

Type U Metallic Drive Screws

Nom. Size	No. of Starts	Out. Dia. Max.	Out. Dia. Min.	Pilot Dia. Max.	Pilot Dia. Min.	Nom. Size	No. of Starts	Out. Dia. Max.	Out. Dia. Min.	Pilot Dia. Max.	Pilot Dia. Min.
00	6	0.060	0:057	0.049	0.046	8	8	0.167	0.162	0.136	0.132
0	6	0.075	0.072	0.063	0.060	10	8	0.182	0.177	0.150	0.146
2	8	0.100	0.097	0.083	0.080	12	8	0.212	0.206	0.177	0.173
4	7	0.116	0.112	0.096	0.092	14	9	0.242	0.236	0.202	0.198
6	7	0.140	0.136	0.116	0.112	5/16	11	0.315	0.309	0.272	0.267
7	8	0.154	0.150	0.126	0.122	3/8	12	0.378	0.371	0.334	0.329

All dimensions are given in inches.
Sizes shown in bold face type are preferred. Type A screws no longer recommended.
† For screws of these nominal lengths and shorter use Type AB screws.
For Types A and AB screws no extrusion of excess metal beyond apex of the point resulting from thread rolling is permissible; a slight rounding or truncation of the point is desirable.
The width of flat at crest of thread shall not exceed 0.004 inch for sizes up to and including No. 8, and 0.006 inch for larger sizes of Types A and AB screws.

Table K38. **USA Standard Thread and Point Dimensions for Types B and BP Thread Forming Tapping Screws** (USAS B18.6.4-1966)

Nominal Size[1] or Basic Screw Diameter	Thds. per inch	D		d		P[2]		S[3,4]				L					
								Point Taper Length				Type B				Type BP	
		Major Diameter		Minor Diameter		Point Diameter		For Short Screws		For Long Screws		Determinant Length for Point Taper[3]		Minimum Practical Screw Lengths		Determinant and Minimum Practical Lengths[4]	
		Max.	Min.	Max.	Min.	Max.	Min.	Max.	Min.	Max.	Min.	90° Heads	Csk. Heads	90° Heads	Csk. Heads	90° Heads	Csk. Head
0 0.0600	48	0.060	0.057	0.036	0.033	0.031	0.027	0.042	0.031	0.052	0.042	5/64	1/8	5/64	3/32	5/32	13/64
1 0.0730	42	0.075	0.072	0.049	0.046	0.044	0.040	0.048	0.036	0.060	0.048	5/64	5/32	5/64	1/8	11/64	1/4
2 0.0860	32	0.088	0.084	0.064	0.060	0.058	0.054	0.062	0.047	0.078	0.062	7/64	3/16	7/64	5/32	13/64	9/32
3 0.0990	28	0.101	0.097	0.075	0.071	0.068	0.063	0.071	0.054	0.089	0.071	9/64	7/32	9/64	3/16	1/4	21/64
4 0.1120	24	0.114	0.110	0.086	0.082	0.079	0.074	0.083	0.063	0.104	0.083	3/16	1/4	9/64	3/16	5/16	3/8
5 0.1250	20	0.130	0.126	0.094	0.090	0.087	0.082	0.100	0.075	0.125	0.100	3/16	1/4	5/32	3/16	21/64	25/64
6 0.1380	20	0.139	0.135	0.104	0.099	0.095	0.089	0.100	0.075	0.125	0.100	1/4	5/16	11/64	1/4	25/64	15/32
7 0.1510	19	0.154	0.149	0.115	0.109	0.105	0.099	0.105	0.079	0.132	0.105	5/16	3/8	3/16	1/4	15/32	17/32
8 0.1640	18	0.166	0.161	0.122	0.116	0.112	0.106	0.111	0.083	0.139	0.111	5/16	7/16	3/16	1/4	31/64	39/64
10 0.1900	16	0.189	0.183	0.141	0.135	0.130	0.123	0.125	0.094	0.156	0.125	3/8	1/2	15/64	5/16	9/16	11/16
12 0.2160	14	0.215	0.209	0.164	0.157	0.152	0.145	0.143	0.107	0.179	0.143	7/16	9/16	9/32	3/8	21/32	25/32
1/4 0.2500	14	0.246	0.240	0.192	0.185	0.179	0.171	0.143	0.107	0.179	0.143	1/2	5/8	9/32	3/8	3/4	7/8
5/16 0.3125	12	0.315	0.308	0.244	0.236	0.230	0.222	0.167	0.125	0.208	0.167	1/2	5/8	5/16	7/16	53/64	61/64
3/8 0.3750	12	0.380	0.371	0.309	0.299	0.293	0.285	0.167	0.125	0.208	0.167	1/2	5/8	5/16	7/16	29/32	1 1/32
7/16 0.4375	10	0.440	0.431	0.359	0.349	0.343	0.335	0.200	0.150	0.250	0.200	5/8	3/4	15/32	5/8	1 7/64	1 15/64
1/2 0.5000	10	0.504	0.495	0.423	0.413	0.407	0.399	0.200	0.150	0.250	0.200	5/8	3/4	15/32	5/8	1 3/16	1 5/16

All dimensions are given in inches. See Table K34 for thread diagrams.
Tapered threads shall have unfinished crests.
The width of flat at crest of thread shall not exceed 0.004 inch for sizes up to and including No. 8, and 0.006 inch for larger sizes.

[1] Where specifying nominal size in decimals, zeros in the fourth decimal place shall be omitted.

[2] The tabulated values apply to screw blanks before roll threading.

[3] Type B screws of these nominal lengths and shorter shall have point taper length specified above for short screws. Longer lengths shall have point taper length specified for long screws.

[4] Type BP screws of these nominal lengths shall have point taper length specified above for short screws. Longer lengths shall have point taper length specified for long screws.

Table K39. USA Standard Thread and Point Dimensions for Type C Thread Forming Tapping Screws (USAS B18.6.4-1966)

Nominal Size[1] or Basic Screw Diameter		Threads per inch	D		P[2]		S[3]				L			
			Major Diameter		Point Diameter		Point Taper Length				Determinant Length for Point Taper[3]		Minimum Practical Screw Lengths	
							For Short Screws		For Long Screws					
			Max.	Min.	Max.	Min.	Max.	Min.	Max.	Min.	90° Heads	Csk. Heads	90° Heads	Csk. Heads
2	0.0860	56	0.0860	0.0820	0.067	0.061	0.062	0.045	0.080	0.062	9/64	3/16	1/8	5/32
2	0.0860	64	0.0860	0.0822	0.070	0.064	0.055	0.039	0.070	0.055	1/8	11/64	1/8	5/32
3	0.0990	48	0.0990	0.0946	0.077	0.070	0.073	0.052	0.094	0.073	11/64	7/32	1/8	3/16
3	0.0990	56	0.0990	0.0950	0.080	0.074	0.062	0.045	0.080	0.062	9/64	3/16	1/8	3/16
4	0.1120	40	0.1120	0.1072	0.086	0.077	0.088	0.062	0.112	0.088	13/64	1/4	1/8	3/16
4	0.1120	48	0.1120	0.1076	0.090	0.083	0.073	0.052	0.094	0.073	11/64	7/32	1/8	3/16
5	0.1250	40	0.1250	0.1202	0.099	0.090	0.088	0.062	0.112	0.088	13/64	9/32	1/8	3/16
5	0.1250	44	0.1250	0.1204	0.101	0.093	0.080	0.057	0.102	0.080	3/16	1/4	1/8	3/16
6	0.1380	32	0.1380	0.1326	0.106	0.095	0.109	0.078	0.141	0.109	1/4	5/16	3/16	1/4
6	0.1380	40	0.1380	0.1332	0.112	0.103	0.088	0.062	0.112	0.088	13/64	17/64	3/16	1/4
8	0.1640	32	0.1640	0.1586	0.132	0.121	0.109	0.078	0.141	0.109	1/4	21/64	3/16	1/4
8	0.1640	36	0.1640	0.1590	0.135	0.125	0.097	0.069	0.125	0.097	15/64	19/64	3/16	1/4
10	0.1900	24	0.1900	0.1834	0.147	0.133	0.146	0.104	0.188	0.146	11/32	27/64	15/64	5/16
10	0.1900	32	0.1900	0.1846	0.158	0.147	0.109	0.078	0.141	0.109	1/4	11/32	15/64	5/16
12	0.2160	24	0.2160	0.2094	0.173	0.159	0.146	0.104	0.188	0.146	11/32	7/16	17/64	3/8
12	0.2160	28	0.2160	0.2098	0.179	0.167	0.125	0.089	0.161	0.125	19/64	25/64	17/64	3/8
1/4	0.2500	20	0.2500	0.2428	0.198	0.181	0.175	0.125	0.225	0.175	13/32	33/64	17/64	3/8
1/4	0.2500	28	0.2500	0.2438	0.213	0.201	0.125	0.089	0.161	0.125	19/64	13/32	17/64	3/8
5/16	0.3125	18	0.3125	0.3043	0.255	0.236	0.194	0.139	0.250	0.194	29/64	19/32	9/32	7/16
5/16	0.3125	24	0.3125	0.3059	0.269	0.255	0.146	0.104	0.188	0.146	11/32	15/32	9/32	7/16
3/8	0.3750	16	0.3750	0.3660	0.310	0.289	0.219	0.156	0.281	0.219	1/2	47/64	9/32	7/16
3/8	0.3750	24	0.3750	0.3684	0.332	0.318	0.146	0.104	0.188	0.146	11/32	1/2	9/32	7/16

All dimensions in inches. See Table K34 for thread diagrams.
Tapered threads shall have unfinished crests.

[1] Where specifying nominal size in decimals, zeros in the fourth decimal place shall be omitted.

[2] The tabulated values apply to screw blanks before roll threading.

[3] Screws of these nominal lengths and shorter shall have point taper length specified above for short screws. Longer lengths shall have point taper length specified for long screws. Lengths for 90 deg heads equal 8 times the pitch of the thread, rounded upward to nearest 1/64 inch and for countersunk heads equal 8 times the pitch of the thread plus the maximum head side height (undercut head style), rounded upward to the nearest 1/64 inch.

Table K40. **USA Standard Thread and Point Dimensions for Types BF and BT* Thread Cutting Tapping Screws** (USAS B18.6.4-1966)

Nominal Size[1] or Basic Screw Diameter		Threads per inch	D Major Diameter		d Minor Diameter		P[2] Point Diameter		S[3] Point Taper Length For Short Screws		S[3] Point Taper Length For Long Screws		L Determinant Length for Point Taper[3]		L Minimum Practical Screw Lengths	
			Max.	Min.	Max.	Min.	Max.	Min.	Max.	Min.	Max.	Min.	90° Heads	Csk. Heads	90° Heads	Csk. Heads
0	0.0600	48	0.060	0.057	0.036	0.033	0.031	0.027	0.042	0.031	0.052	0.042	5/64	1/8	5/64	3/32
1	0.0730	42	0.075	0.072	0.049	0.046	0.044	0.040	0.048	0.036	0.060	0.048	5/64	5/32	5/64	1/8
2	0.0860	32	0.088	0.084	0.064	0.060	0.058	0.054	0.062	0.047	0.078	0.062	7/64	3/16	7/64	5/32
3	0.0990	28	0.101	0.097	0.075	0.071	0.068	0.063	0.071	0.054	0.089	0.071	9/64	7/32	9/64	3/16
4	0.1120	24	0.114	0.110	0.086	0.082	0.079	0.074	0.083	0.063	0.104	0.083	3/16	1/4	9/64	3/16
5	0.1250	20	0.130	0.126	0.094	0.090	0.087	0.082	0.100	0.075	0.125	0.100	3/16	1/4	5/32	3/16
6	0.1380	20	0.139	0.135	0.104	0.099	0.095	0.089	0.100	0.075	0.125	0.100	1/4	5/16	11/64	1/4
7	0.1510	19	0.154	0.149	0.115	0.109	0.105	0.099	0.105	0.079	0.132	0.105	5/16	3/8	3/16	1/4
8	0.1640	18	0.166	0.161	0.122	0.116	0.112	0.106	0.111	0.083	0.139	0.111	5/16	7/16	3/16	1/4
10	0.1900	16	0.189	0.183	0.141	0.135	0.130	0.123	0.125	0.094	0.156	0.125	3/8	1/2	15/64	5/16
12	0.2160	14	0.215	0.209	0.164	0.157	0.152	0.145	0.143	0.107	0.179	0.143	7/16	9/16	9/32	3/8
1/4	0.2500	14	0.246	0.240	0.192	0.185	0.179	0.171	0.143	0.107	0.179	0.143	1/2	5/8	9/32	3/8
5/16	0.3125	12	0.315	0.308	0.244	0.236	0.230	0.222	0.167	0.125	0.208	0.167	1/2	5/8	5/16	7/16
3/8	0.3750	12	0.380	0.371	0.309	0.299	0.293	0.285	0.167	0.125	0.208	0.167	1/2	5/8	5/16	7/16
7/16	0.4375	10	0.440	0.431	0.359	0.349	0.343	0.335	0.200	0.150	0.250	0.200	5/8	3/4	15/32	5/8
1/2	0.5000	10	0.504	0.495	0.423	0.413	0.407	0.399	0.200	0.150	0.250	0.200	5/8	3/4	15/32	5/8

All dimensions are given in inches. See Table K35 for thread diagrams.

* Otherwise designated "Type 25".

Points of screws shall be tapered and fluted or slotted. Details of taper and flute design shall be optional with manufacturer provided the screws meet the performance requirements. Flutes or slots shall extend through first full form thread except for Type BF screws on which the length of flutes may be one pitch (thread) short of first full form thread.

The width of flat at crest of thread shall not exceed 0.004 inch for sizes up to and including No. 8, and 0.006 inch for larger sizes.

[1] Where specifying nominal size in decimals, zeros in the fourth decimal place shall be omitted.

[2] The tabulated values apply to screw blanks before roll threading.

[3] Screws of these nominal lengths and shorter shall have point taper length specified above for short screws. Longer lengths shall have point taper length specified for long screws.

Table K41. **USA Standard Thread and Point Dimensions for Types D*, F, G, and T† Thread Cutting Tapping Screws** (USAS B18.6.4-1966)

Nominal Size[1] or Basic Screw Diameter		Threads per inch	D Major Diameter		P[2] Point Diameter		S[3] Point Taper Length For Short Screws		S[3] Point Taper Length For Long Screws		L Determinant Length for Point Taper[3]		L Minimum Practical Screw Lengths	
			Max.	Min.	Max.	Min.	Max.	Min.	Max.	Min.	90° Heads	Csk. Heads	90° Heads	Csk. Heads
2	0.0860	56	0.0860	0.0820	0.067	0.061	0.062	0.045	0.080	0.062	9/64	3/16	1/8	5/32
2	0.0860	64	0.0860	0.0822	0.070	0.064	0.055	0.039	0.070	0.055	1/8	11/64	1/8	5/32
3	0.0990	48	0.0990	0.0946	0.077	0.070	0.073	0.052	0.094	0.073	11/64	7/32	1/8	3/16
3	0.0990	56	0.0990	0.0950	0.080	0.074	0.062	0.045	0.080	0.062	9/64	3/16	1/8	3/16
4	0.1120	40	0.1120	0.1072	0.086	0.077	0.088	0.062	0.112	0.088	13/64	1/4	1/8	3/16
4	0.1120	48	0.1120	0.1076	0.090	0.083	0.073	0.052	0.094	0.073	11/64	7/32	1/8	3/16
5	0.1250	40	0.1250	0.1202	0.099	0.090	0.088	0.062	0.112	0.088	13/64	9/32	1/8	3/16
5	0.1250	44	0.1250	0.1204	0.101	0.093	0.080	0.057	0.102	0.080	3/16	1/4	1/8	3/16
6	0.1380	32	0.1380	0.1326	0.106	0.095	0.109	0.078	0.141	0.109	1/4	5/16	3/16	1/4
6	0.1380	40	0.1380	0.1332	0.112	0.103	0.088	0.062	0.112	0.088	13/64	17/64	3/16	1/4
8	0.1640	32	0.1640	0.1586	0.132	0.121	0.109	0.078	0.141	0.109	1/4	21/64	3/16	1/4
8	0.1640	36	0.1640	0.1590	0.135	0.125	0.097	0.069	0.125	0.097	15/64	19/64	3/16	1/4
10	0.1900	24	0.1900	0.1834	0.147	0.133	0.146	0.104	0.188	0.146	11/32	27/64	15/64	5/16
10	0.1900	32	0.1900	0.1846	0.158	0.147	0.109	0.078	0.141	0.109	1/4	11/32	15/64	5/16
12	0.2160	24	0.2160	0.2094	0.173	0.159	0.146	0.104	0.188	0.146	11/32	7/16	17/64	3/8
12	0.2160	28	0.2160	0.2098	0.179	0.167	0.125	0.089	0.161	0.125	19/64	25/64	17/64	3/8
1/4	0.2500	20	0.2500	0.2428	0.198	0.181	0.175	0.125	0.225	0.175	13/32	33/64	17/64	3/8
1/4	0.2500	28	0.2500	0.2438	0.213	0.201	0.125	0.089	0.161	0.125	19/64	13/32	17/64	3/8
5/16	0.3125	18	0.3125	0.3043	0.255	0.236	0.194	0.139	0.250	0.194	29/64	19/32	9/32	7/16
5/16	0.3125	24	0.3125	0.3059	0.269	0.255	0.146	0.104	0.188	0.146	11/32	15/32	9/32	7/16
3/8	0.3750	16	0.3750	0.3660	0.310	0.289	0.219	0.156	0.281	0.219	1/2	47/64	9/32	7/16
3/8	0.3750	24	0.3750	0.3684	0.332	0.318	0.146	0.104	0.188	0.146	11/32	1/2	9/32	7/16

All dimensions are given in inches. See Table K35 for thread diagrams.

Points of screws shall be tapered and fluted or slotted. Details of taper and flute design shall be optional with manufacturer provided screws meet the performance requirements. Flutes or slots shall extend through first full form thread except for Type F screws on which the length of flutes may be one pitch (thread) short of first full form thread.

[1] Where specifying nominal size in decimals, zeros in the fourth decimal place shall be omitted.

[2] The tabulated values apply to screw blanks before roll threading.

[3] Screws of these nominal lengths and shorter shall have point taper length specified above for short screws. Longer lengths shall have point taper length specified for long screws. Lengths for 90 deg heads equal 8 times the pitch of the thread, rounded upward to nearest 1/64 inch and for countersunk heads equal 8 times the pitch of the thread plus the maximum head side height (undercut head style), rounded upward to the nearest 1/64 inch.

* Otherwise designated "Type 1". † Otherwise designated "Type 23".

Table K42. **Approximate Hole Sizes for Type A Steel Thread Forming Screws***

In Steel, Stainless Steel, Monel Metal, Brass, and Aluminum Sheet Metal				
Screw Size	Metal Thickness	Hole Required: Pierced or Extruded	Hole Required: Drilled or Clean Punched	Drill Size
4	.015	....	.086	44
	.018	....	.086	44
	.024	.098	.094	42
	.030	.098	.094	42
	.036	.098	.098	40
6	.015	....	.104	37
	.018	....	.104	37
	.024	.111	.104	37
	.030	.111	.104	37
	.036	.111	.106	36
7	.015	....	.116	32
	.018	....	.116	32
	.024	.120	.116	32
	.030	.120	.116	32
	.036	.120	.116	32
	.048	.120	.120	31
8	.018	....	.125	1/8
8	.024	.136	.125	1/8
	.030	.136	.125	1/8
	.036	.136	.125	1/8
	.048	.136	.128	30
10	.018	....	.136	29
	.024	.157	.136	29
	.030	.157	.136	29
	.036	.157	.136	29
	.048	.157	.149	25
12	.024	....	.161	20
	.030	.185	.161	20
	.036	.185	.161	20
	.048	.185	.161	20
14	.024	....	.185	13
	.030	.209	.189	12
	.036	.209	.191	11
	.048	.209	.196	9

In Plywood (Resin Impregnated)					
Screw Size	Hole Required	Drill Size	Min. Mat'l Thickness	Penetration in Blind Holes: Min.	Penetration in Blind Holes: Max.
4	.098	40	3/16	1/4	3/4
6	.110	35	3/16	1/4	3/4
7	.128	30	1/4	5/16	3/4
8	.140	28	1/4	5/16	3/4
10	.169	18	5/16	3/8	1
12	.189	12	5/16	3/8	1
14	.228	1	7/16	1/2	1

In Asbestos Compositions					
Screw Size	Hole Required	Drill Size	Min. Mat'l Thickness	Penetration in Blind Holes: Min.	Penetration in Blind Holes: Max.
4	.093	42	3/16	1/4	3/4
6	.106	36	3/16	1/4	3/4
7	.125	1/8	1/4	5/16	3/4
8	.136	29	1/4	5/16	3/4
10	.161	20	5/16	3/8	1
12	.185	13	5/16	3/8	1
14	.213	3	7/16	1/2	1

See footnote at bottom of Table K43. Type A is not recommended, use Type AB.

Table K43. **Approximate Hole Sizes for Type C Steel Thread Forming Screws***

In Sheet Steel											
Screw Size	Metal Thickness	Hole Required	Drill Size	Screw Size	Metal Thickness	Hole Required	Drill Size	Screw Size	Metal Thickness	Hole Required	Drill Size
4–40	.037	.093	42	10–24	.037	.154	23	1/4–20	.037	.221	2
	.048	.093	42		.048	.161	20		.048	.221	2
	.062	.096	41		.062	.166	19		.062	.228	1
	.075	.0995	39		.075	.1695	18		.075	.234	A
	.105	.101	38		.105	.173	17		.105	.234	A
	.134	.101	38		.134	.177	16		.134	.236	6mm
6–32	.037	.113	33	10–32	.037	.1695	18	1/4–28	.037	.224	5.7mm
	.048	.116	32		.048	.1695	18		.048	.228	1
	.062	.116	32		.062	.1695	18		.062	.232	5.9mm
	.075	.122	3.1mm		.075	.173	17		.075	.234	A
	.105	.125	1/8		.105	.177	16		.105	.238	B
	.134	.125	1/8		.134	.177	16		.134	.238	B
8–32	.037	.136	29	12–24	.037	.189	12	5/16–18	.037	.290	L
	.048	.144	27		.048	.1935	10		.048	.290	L
	.062	.144	27		.062	.1935	10		.062	.290	L
	.075	.147	26		.075	.199	8		.075	.295	M
	.105	.1495	25		.105	.199	8		.105	.295	M
	.134	.1495	25		.134	.199	8		.134	.295	M

All dimensions in inches except drill sizes. *** Since conditions differ widely, it may be necessary to vary the hole size to suit a particular application.**

Table K44. **Approximate Pierced or Extruded Hole Sizes for Types AB, B and BP Steel Thread Forming Screws***

Screw Size	Metal Thickness	Pierced or Extruded Hole Required	Screw Size	Metal Thickness	Pierced or Extruded Hole Required	Screw Size	Metal Thickness	Pierced or Extruded Hole Required
In Steel, Stainless Steel, Monel Metal, and Brass Sheet Metal								
4	.015	.086	7	.024	.120	10	.030	.157
	.018	.086		.030	.120		.036	.157
	.024	.098		.036	.120		.048	.157
	.030	.098		.048	.120	12	.024	.185
	.036	.098	8	.018	.136		.030	.185
6	.015	.111		.024	.136		.036	.185
	.018	.111		.030	.136		.048	.185
	.024	.111		.036	.136	1/4	.030	.209
	.030	.111		.048	.136		.036	.209
	.036	.111	10	.018	.157		.048	.209
7	.018	.120		.024	.157	..	....	....
In Aluminum Alloy Sheet Metal								
4	.024	.086	6	.048	.111	8	.036	.136
	.030	.086	7	.024	.120		.048	.136
	.036	.086		.030	.120	10	.024	.157
	.048	.086		.036	.120		.030	.157
6	.024	.111		.048	.120		.036	.157
	.030	.111	8	.024	.136		.048	.157
	.036	.111		.030	.136	..	....	....

All dimensions are given in inches except whole number screw sizes.
See footnotes at bottom of page.

Table K45. **Approximate Drilled Hole Sizes for Types AB, B and BP Steel Thread Forming Screws***

Screw Size	Hole Required	Drill Size	Min. Mat'l Thickness	Penetration in Blind Holes Min.	Penetration in Blind Holes Max.	Screw Size	Hole Required	Drill Size	Min. Mat'l Thickness	Penetration in Blind Holes Min.	Penetration in Blind Holes Max.
In Plywood (Resin Impregnated)						In Asbestos Compositions					
2	.073	49	1/8	3/16	1/2	2	.076	48	1/8	3/16	1/2
4	.099	39	3/16	1/4	5/8	4	.101	38	3/16	1/4	5/8
6	.125	1/8	3/16	1/4	5/8	6	.120	31	3/16	1/4	5/8
7	.136	29	3/16	1/4	3/4	7	.136	29	1/4	5/16	3/4
8	.144	27	3/16	1/4	3/4	8	.147	26	5/16	3/8	3/4
10	.173	17	1/4	5/16	1	10	.166	19	5/16	3/8	1
12	.193	10	5/16	3/8	1	12	.196	9	5/16	3/8	1
1/4	.228	I	5/16	3/8	1	1/4	.228	I	7/16	1/2	1
In Aluminum, Magnesium, Zinc, Brass, and Bronze Castings						In Phenol Formaldehyde Plastics					
2	.078	47	...	1/8	...	2	.078	47	...	...	...
4	.104	37	...	3/16	...	4	.099	39	...	...	...
6	.128	30	...	1/4	...	6	.128	30	...	...	...
7	.144	27	...	1/4	...	7	.136	29	...	...	...
8	.152	24	...	1/4	...	8	.149	25	...	...	...
10	.177	16	...	1/4	...	10	.177	16	...	...	...
12	.199	8	...	9/32	...	12	.199	8	...	...	...
1/4	.234	15/64	...	5/16	...	1/4	.234	15/64	...	...	...
In Cellulose Acetate and Nitrate, and Acrylic and Styrene Resins											
2	.078	47	...	3/16	...	8	.144	27	...	5/16	...
4	.093	42	...	1/4	...	10	.169	18	...	5/16	...
6	.120	31	...	1/4	...	12	.191	11	...	3/8	...
7	.128	30	...	1/4	...	1/4	.221	2	...	3/8	...

All dimensions are given in inches except whole number screw and drill sizes.
* Since conditions differ widely, it may be necessary to vary the hole size to suit a particular application. Data below double line in Table K45 apply to Types B and BP only.

Table K46. **Approximate Drilled or Clean-Punched Hole Sizes for Types AB, B and BP Steel Thread Forming Screws***

Screw Size	Metal Thickness	Hole Required	Drill Size
In Steel, Stainless Steel, Monel Metal, and Brass Sheet Metal			
2	.015	.063	52
	.018	.063	52
	.024	.067	51
	.030	.070	50
	.036	.073	49
	.048	.073	49
	.060	.076	48
4	.015	.086	44
	.018	.086	44
	.024	.089	43
	.030	.093	42
	.036	.093	42
	.048	.096	41
	.060	.099	39
	.075	.101	38
6	.015	.104	37
	.018	.104	37
	.024	.106	36
	.030	.106	36
	.036	.110	35
	.048	.111	34
	.060	.116	32
	.075	.120	31
	.105	.128	30
In Aluminum Alloy Sheet Metal			
2	.024	.063	52
	.030	.063	52
	.036	.063	52
	.048	.067	51
	.060	.070	50
4	.030	.086	44
	.036	.086	44
	.048	.086	44
	.060	.089	43
	.075	.089	43
	.105	.093	42
6	.030	.104	37
	.036	.104	37
	.048	.104	37
	.060	.106	36
	.075	.110	35
	.105	.111	34
	.128 to .250	.120	31
7	.030	.113	33
	.036	.113	33
	.048	.116	32

Screw Size	Metal Thickness	Hole Required	Drill Size
In Steel, Stainless Steel, Monel Metal, and Brass Sheet Metal			
7	.018	.116	32
	.024	.116	32
	.030	.116	32
	.036	.116	32
	.048	.120	31
	.060	.128	30
	.075	.136	29
	.105	.140	28
8	.024	.125	1/8
	.030	.125	1/8
	.036	.125	1/8
	.048	.128	30
	.060	.136	29
	.075	.140	28
	.105	.149	25
	.125	.149	25
	.135	.152	24
10	.024	.144	27
	.030	.144	27
	.036	.147	26
	.048	.152	24
	.060	.152	24
	.075	.157	22
	.105	.161	20
In Aluminum Alloy Sheet Metal			
7	.060	.120	31
	.075	.128	30
	.105	.136	29
	.128 to .250	.136	29
8	.030	.116	32
	.036	.120	31
	.048	.128	30
	.060	.136	29
	.075	.140	28
	.105	.147	26
	.125	.147	26
	.135	.149	25
	.162 to .375	.152	24
10	.036	.144	27
	.048	.144	27
	.060	.144	27
	.075	.147	26
	.105	.147	26
	.125	.154	23
	.135	.154	23

Screw Size	Metal Thickness	Hole Required	Drill Size
In Steel, Stainless Steel, Monel Metal, and Brass Sheet Metal			
10	.125	.169	18
	.135	.169	18
	.164	.173	17
12	.024	.166	19
	.030	.166	19
	.036	.166	19
	.048	.169	18
	.060	.177	16
	.075	.182	14
	.105	.185	13
	.125	.196	9
	.135	.196	9
	.164	.201	7
1/4	.030	.194	10
	.036	.194	10
	.048	.194	10
	.060	.199	8
	.075	.204	6
	.105	.209	4
	.125	.228	1
	.135	.228	1
	.164	.234	15/64
	.187	.234	15/64
	.194	.234	15/64
In Aluminum Alloy Sheet Metal			
10	.164	.159	21
	.200 to .375	.166	19
12	.048	.161	20
	.060	.166	19
	.075	.173	17
	.105	.180	15
	.125	.182	14
	.135	.182	14
	.164	.189	12
	.200 to .375	.196	9
1/4	.060	.199	8
	.075	.201	7
	.105	.204	6
	.125	.209	4
	.135	.209	4
	.164	.213	3
	.187	.213	3
	.194	.221	2
	.200 to .375	.228	1

All dimensions are given in inches except whole number screw and drill sizes.

* Since conditions differ widely, it may be necessary to vary the hole size to suit a particular application. Hole sizes for metal thicknesses above .075 inch are for Types B and BP only.

Table K47. **Approximate Hole Sizes for Types D, F, G, and T Steel Thread Cutting Screws***

Screw Size	Stock Thickness										
	.050	.060	.083	.109	.125	.140	3/16	1/4	5/16	3/8	1/2
	Hole Sizes† in Steel										
2-56	.0730	.0730	.0730	.0730	.0760	.0760	.....	.....	.....	.....	.....
3-48	.0810	.0810	.0820	.0860	.0860	.0860	.0890	.....	.....	.....	.....
4-40	.0890	.0890	.0935	.0960	.0980	.0980	.1015	.....	.....	.....	.....
5-40	.1060	.1060	.1060	.1065	.1094	.1100	.1160	.1160	.....	.....	.....
6-32	.1100	.1130	.1160	.1160	.1160	.1200	.1250	.1250	.....	.....	.....
8-32	.1360	.1405	.1405	.1440	.1440	.1470	.1495	.1495	.1495	.....	.....
10-24	.1520	.1540	.1610	.1610	.1660	.1695	.1730	.1730	.1730	.1730	.....
10-32	.1590	.1660	.1660	.1695	.1695	.1695	.1770	.1770	.1770	.1770	.....
12-24	.....	.1800	.1820	.1875	.1910	.1910	.1990	.1990	.1990	.1990	.1990
1/4-20	.....	.....	.2130	.2188	.2210	.2210	.2280	.2280	.2280	.2280	.2280
1/4-28	.....	.....	.2210	.2280	.2280	.2340	.2344	.2344	.2344	.2344	.2344
5/16-18	.....	.....	.....	.2770	.2770	.2813	.2900	.2900	.2900	.2900	.2900
5/16-24	.....	.....	.....	.2900	.2900	.2900	.2950	.2950	.2950	.2950	.2950
3/8-16	.....	.....	.....	.....	.3390	.3390	.3480	.3580	.3580	.3580	.3580
3/8-24	.....	.....	.....	.....	.3480	.3480	.3580	.3580	.3580	.3580	.3580
	Hole Sizes† in Aluminum										
2-56	.0700	.0730	.0730	.0730	.0730	.0730	.....	.....	.....	.....	.....
3-48	.0781	.0810	.0820	.0820	.0820	.0860	.0860	.....	.....	.....	.....
4-40	.0890	.0890	.0890	.0935	.0935	.0938	.0980	.....	.....	.....	.....
5-40	.1015	.1015	.1040	.1040	.1065	.1065	.1100	.1130	.....	.....	.....
6-32	.1094	.1094	.1110	.1130	.1160	.1160	.1200	.1250	.....	.....	.....
8-32	.1360	.1360	.1360	.1405	.1405	.1440	.1470	.1495	.1495	.....	.....
10-24	.1495	.1520	.1540	.1570	.1590	.1610	.1660	.1719	.1730	.1730	.....
10-32	.1610	.1610	.1610	.1660	.1660	.1660	.1719	.1770	.1770	.1770	.....
12-24	.....	.1770	.1800	.1820	.1850	.1875	.1910	.1990	.1990	.1990	.1990
1/4-20	.....	.....	.2055	.2090	.2130	.2130	.2210	.2280	.2280	.2280	.2280
1/4-28	.....	.....	.2188	.2210	.2210	.2210	.2280	.2344	.2344	.2344	.2344
5/16-18	.....	.....	.....	.2660	.2720	.2720	.2810	.2900	.2900	.2900	.2900
5/16-24	.....	.....	.....	.2810	.2812	.2812	.2900	.2950	.2950	.2950	.2950
3/8-16	.....	.....	.....	.....	.3281	.3320	.3390	.3480	.3480	.3480	.3480
3/8-24	.....	.....	.....	.....	.3438	.3438	.3480	.3580	.3580	.3580	.3580
	Hole Sizes† in Cast Iron										
2-56	.0760	.0760	.0760	.0781	.0781	.0781	.....	.....	.....	.....	.....
3-48	.0890	.0890	.0890	.0890	.0890	.0935	.0935	.....	.....	.....	.....
4-40	.0995	.0995	.1015	.1015	.1015	.1015	.1040	.....	.....	.....	.....
5-40	.1110	.1110	.1130	.1130	.1160	.1160	.1160	.1160	.....	.....	.....
6-32	.1200	.1200	.1250	.1250	.1250	.1250	.1285	.1285	.....	.....	.....
8-32	.1470	.1495	.1495	.1495	.1495	.1495	.1540	.1540	.1540	.....	.....
10-24	.1695	.1695	.1719	.1730	.1730	.1730	.1770	.1770	.1770	.1770	.....
10-32	.1730	.1730	.1770	.1770	.1770	.1770	.1800	.1800	.1800	.1800	.....
12-24	.....	.1960	.1990	.1990	.1990	.1990	.2031	.2040	.2040	.2040	.2040
1/4-20	.....	.....	.2280	.2280	.2280	.2280	.2344	.2344	.2344	.2344	.2344
1/4-28	.....	.....	.2340	.2344	.2344	.2344	.2380	.2380	.2380	.2380	.2380
5/16-18	.....	.....	.....	.2900	.2900	.2900	.2950	.2950	.2950	.2950	.2950
5/16-24	.....	.....	.....	.2950	.2950	.2950	3020	.3020	.3020	.3020	.3020
3/8-16	.....	.....	.....	.....	.3480	.3480	.3480	.3480	.3480	.3480	.3480
3/8-24	.....	.....	.....	.....	.3580	.3580	.3580	.3580	.3580	.3580	.3580

All dimensions are given in inches except the whole number screw sizes.

* Since conditions differ widely, it may be necessary to vary the hole size to suit a particular application.

† Hole sizes listed are standard drill sizes.

Table K47 (*Continued*). **Approximate Hole Sizes for Types D, F, G, and T Steel Thread Cutting Screws***

Screw Size	Stock Thickness										
	.050	.060	.083	.109	.125	.140	3/16	1/4	5/16	3/8	1/2
	Hole Sizes† in Zinc and Aluminum Die Castings										
2-56	.0730	.0730	.0760	.0760	.0760	.0760	.....	.....	.....	.....	.....
3-48	.0820	.0820	.0820	.0860	.0890	.0890	.0890	.....	.....	.....	.....
4-40	.0960	.0960	.0960	.0960	.0995	.0995	.0995	.....	.....	.....	.....
5-40	.1060	.1060	.1060	.1100	.1100	.1100	.1110	.1130	.....	.....	.....
6-32	.1160	.1200	.1200	.1200	.1200	.1200	.1200	.1200	.....	.....	.....
8-32	.1440	.1440	.1440	.1440	.1470	.1470	.1470	.1495	.1495	.....	.....
10-24	.1610	.1660	.1660	.1660	.1660	.1660	.1695	.1695	.1719	.1719	.....
10-32	.1695	.1695	.1719	.1719	.1719	.1719	.1719	.1730	.1730	.1770	.....
12-24	.....	.1800	.1910	.1910	.1910	.1935	.1935	.1960	.1960	.1990	.1990
1/4-20	.....	.....	.2188	.2188	.2210	.2210	.2210	.2280	.2280	.2280	.2280
1/4-28	.....	.....	.2280	.2280	.2280	.2280	.2280	.2340	.2340	.2344	.2344
5/16-18	.....	.....	.....	.2770	.2810	.2810	.2812	.2812	.2900	.2900	.2900
5/16-24	.....	.....	.....	.2900	.2900	.2900	.2900	.2900	.2950	.2950	.2950
3/8-16	.....	.....	.....	.....	.3390	.3390	.3390	.3438	.3438	.3480	.3480
3/8-24	.....	.....	.....	.....	.3480	.3480	.3480	.3580	.3580	.3580	.3580

In Phenol Formaldehyde Plastics					In Acrylic and Other Resins				
Screw Size	Hole Required	Drill Size	Depth of Penetration		Screw Size	Hole Required	Drill Size	Depth of Penetration	
			Min.	Max.				Min.	Max.
2-56	.0781	5/64	7/32	3/8	2-56	.076	48	7/32	3/8
3-48	.089	43	7/32	3/8	3-48	.086	44	7/32	3/8
4-40	.098	40	1/4	5/16	4-40	.093	42	1/4	5/16
5-40	.113	33	1/4	7/16	5-40	.110	35	1/4	7/16
6-32	.116	32	1/4	5/16	6-32	.116	32	1/4	5/16
8-32	.144	27	5/16	1/2	8-32	.144	27	5/16	1/2
10-24	.161	20	3/8	1/2	10-24	.161	20	3/8	1/2
10-32	.166	19	3/8	1/2	10-32	.166	19	3/8	1/2
1/4-20	.228	1	3/8	5/8	1/4-20	.228	1	3/8	1

For footnotes see bottom of table on previous page.

Table K48. **Approximate Hole Sizes for Types BF and BT Steel Thread Cutting Screws***

Stock Thickness	Screw Size										
	2-32	3-28	4-24	5-20	6-20	8-18	10-16	12-14	1/4-14	5/16-12	3/8-12
	Hole Sizes† in Zinc and Aluminum Die Castings										
.060	.0730	.0860	.....	.....	.....	.....	.....	.....	.....	.....	.....
.083	.0730	.0860	.....	.....	.....	.....	.....	.....	.....	.....	.....
.109	.0760	.0860	.0980	.1110	.....	.....	.....	.....	.....	.....	.....
.125	.0760	.0860	.0995	.1110	.1200	.1490	.1660	.1910	.2210	.2810	.344
.140	.0760	.0890	.0995	.1130	.1200	.1490	.1660	.1910	.2210	.2810	.344
3/16	.....	.0890	.0995	.1130	.1200	.1490	.1660	.1910	.2210	.2810	.344
1/4	.....	.....	.1015	.1160	.1250	.1520	.1695	.1960	.2280	.2810	.344
5/16	.....	.....	.....	.....	.1250	.1520	.1719	.1960	.2280	.2900	.348
3/8	.....	.....	.....	.....	.....	.....	.1719	.1960	.2280	.2900	.348

In Phenol Formaldehyde Plastics					In Acrylic and Other Resins				
Screw Size	Hole Required	Drill Size	Depth of Penetration		Screw Size	Hole Required	Drill Size	Depth of Penetration	
			Min.	Max.				Min.	Max.
2-32	.0781	5/64	3/32	1/4	2-32	.076	48	3/32	1/4
3-28	.089	43	1/8	5/16	3-28	.089	43	1/8	5/16
4-24	.104	37	1/8	5/16	4-24	.0995	39	1/8	5/16
5-20	.116	32	3/16	3/8	5-20	.113	33	3/16	3/8
6-20	.125	1/8	3/16	3/8	6-20	.120	31	3/16	3/8
8-18	.147	26	1/4	1/2	8-18	.144	27	1/4	1/2
10-16	.1695	18	5/16	5/8	10-16	.166	19	5/16	5/8
12-14	.1935	10	3/8	5/8	12-14	.189	12	3/8	5/8
1/4-14	.228	1	3/8	3/4	1/4-14	.221	2	3/8	3/4

For footnotes see bottom of table on previous page.

Table K49. Approximate Hole Sizes for Type U Hardened Steel Metallic Drive Screws

In Ferrous and Non-Ferrous Castings, Sheet Metals, Plastics, Plywood (Resin-Impregnated) and Fiber					
Screw Size	Hole Size	Drill Size	Screw Size	Hole Size	Drill Size
00	.052	55	8	.144	27
0	.067	51	10	.161	20
2	.086	44	12	.191	11
4	.104	37	14	.221	2
6	.120	31	5/16	.295	M
7	.136	29	3/8	.358	T

All dimensions are given in inches except whole number screw and drill sizes and letter drill sizes.

Self-tapping Thread Inserts. — Self-tapping screw thread inserts are essentially hard bushings with internal and external threads. The internal threads conform to Unified and American standard classes 2B and 3B, depending on the type of insert used. The external thread has cutting edges on the end that provide the self-tapping feature. These inserts may be used in magnesium, aluminum, cast iron, zinc, plastics, and other materials. Self-tapping inserts are made of case-hardened carbon steel, stainless steel, and brass, the brass type being designed specifically for installation in wood.

Screw Thread Inserts. — Screw thread inserts are helically formed coils of diamond-shaped stainless steel or phosphor bronze wire that screw into a threaded hole to form a mating internal thread for a screw or stud. These inserts provide a convenient means of repairing stripped-out threads and are also used to provide stronger threads in soft materials such as aluminum, zinc die castings, wood, magnesium, etc., than can be obtained by direct tapping of the base metal involved.

According to the Heli-Coil Corp., conventional design practice in specifying boss diameters or edge distances can usually be applied since the major diameter of a

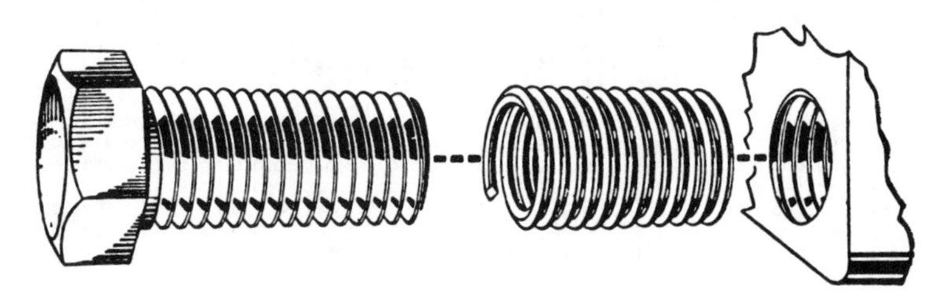

hole tapped to receive a thread insert is not much larger than the major diameter of thread the insert provides. For example, the major diameter of the tapped hole to receive a 1/4-28 cap screw is 0.2904 inch while a conventional tapped hole for a 1/4-28 cap screw has a major diameter of 0.250 inch.

Screw thread inserts are available in thread sizes from 4-40 to 1 1/2-6 inch National and Unified Coarse Thread Series and in 6-40 to 1 1/2-12 sizes in the fine-thread series. When used in conjunction with appropriate taps and gages, screw thread inserts will meet requirements of 2, 2B, 3 and 3B thread classes.

Table K50A. American Standard Type A Plain Washers — Preferred Sizes (ASA B27.2-1965)**

Nominal Washer Size***		Series	Inside Diameter			Outside Diameter			Thickness		
			Basic	Tolerance		Basic	Tolerance		Basic	Max.	Min.
				Plus	Minus		Plus	Minus			
—	—		0.078	0.000	0.005	0.188	0.000	0.005	0.020	0.025	0.016
—	—		0.094	0.000	0.005	0.250	0.000	0.005	0.020	0.025	0.016
—	—		0.125	0.008	0.005	0.312	0.008	0.005	0.032	0.040	0.025
No. 6	0.138		0.156	0.008	0.005	0.375	0.015	0.005	0.049	0.065	0.036
No. 8	0.164		0.188	0.008	0.005	0.438	0.015	0.005	0.049	0.065	0.036
No. 10	0.190		0.219	0.008	0.005	0.500	0.015	0.005	0.049	0.065	0.036
3/16	0.188		0.250	0.015	0.005	0.562	0.015	0.005	0.049	0.065	0.036
No. 12	0.216		0.250	0.015	0.005	0.562	0.015	0.005	0.065	0.080	0.051
1/4	0.250	N	0.281	0.015	0.005	0.625	0.015	0.005	0.065	0.080	0.051
1/4	0.250	W	0.312	0.015	0.005	0.734*	0.015	0.007	0.065	0.080	0.051
5/16	0.312	N	0.344	0.015	0.005	0.688	0.015	0.007	0.065	0.080	0.051
5/16	0.312	W	0.375	0.015	0.005	0.875	0.030	0.007	0.083	0.104	0.064
3/8	0.375	N	0.406	0.015	0.005	0.812	0.015	0.007	0.065	0.080	0.051
3/8	0.375	W	0.438	0.015	0.005	1.000	0.030	0.007	0.083	0.104	0.064
7/16	0.438	N	0.469	0.015	0.005	0.922	0.015	0.007	0.065	0.080	0.051
7/16	0.438	W	0.500	0.015	0.005	1.250	0.030	0.007	0.083	0.104	0.064
1/2	0.500	N	0.531	0.015	0.005	1.062	0.030	0.007	0.095	0.121	0.074
1/2	0.500	W	0.562	0.015	0.005	1.375	0.030	0.007	0.109	0.132	0.086
9/16	0.562	N	0.594	0.015	0.005	1.156*	0.030	0.007	0.095	0.121	0.074
9/16	0.562	W	0.625	0.015	0.005	1.469*	0.030	0.007	0.109	0.132	0.086
5/8	0.625	N	0.656	0.030	0.007	1.312	0.030	0.007	0.095	0.121	0.074
5/8	0.625	W	0.688	0.030	0.007	1.750	0.030	0.007	0.134	0.160	0.108
3/4	0.750	N	0.812	0.030	0.007	1.469	0.030	0.007	0.134	0.160	0.108
3/4	0.750	W	0.812	0.030	0.007	2.000	0.030	0.007	0.148	0.177	0.122
7/8	0.875	N	0.938	0.030	0.007	1.750	0.030	0.007	0.134	0.160	0.108
7/8	0.875	W	0.938	0.030	0.007	2.250	0.030	0.007	0.165	0.192	0.136
1	1.000	N	1.062	0.030	0.007	2.000	0.030	0.007	0.134	0.160	0.108
1	1.000	W	1.062	0.030	0.007	2.500	0.030	0.007	0.165	0.192	0.136
1 1/8	1.125	N	1.250	0.030	0.007	2.250	0.030	0.007	0.134	0.160	0.108
1 1/8	1.125	W	1.250	0.030	0.007	2.750	0.030	0.007	0.165	0.192	0.136
1 1/4	1.250	N	1.375	0.030	0.007	2.500	0.030	0.007	0.165	0.192	0.136
1 1/4	1.250	W	1.375	0.030	0.007	3.000	0.030	0.007	0.165	0.192	0.136
1 3/8	1.375	N	1.500	0.030	0.007	2.750	0.030	0.007	0.165	0.192	0.136
1 3/8	1.375	W	1.500	0.045	0.010	3.250	0.045	0.010	0.180	0.213	0.153
1 1/2	1.500	N	1.625	0.030	0.007	3.000	0.030	0.007	0.165	0.192	0.136
1 1/2	1.500	W	1.625	0.045	0.010	3.500	0.045	0.010	0.180	0.213	0.153
1 5/8	1.625		1.750	0.045	0.010	3.750	0.045	0.010	0.180	0.213	0.153
1 3/4	1.750		1.875	0.045	0.010	4.000	0.045	0.010	0.180	0.213	0.153
1 7/8	1.875		2.000	0.045	0.010	4.250	0.045	0.010	0.180	0.213	0.153
2	2.000		2.125	0.045	0.010	4.500	0.045	0.010	0.180	0.213	0.153
2 1/4	2.250		2.375	0.045	0.010	4.750	0.045	0.010	0.220	0.248	0.193
2 1/2	2.500		2.625	0.045	0.010	5.000	0.045	0.010	0.238	0.280	0.210
2 3/4	2.750		2.875	0.065	0.010	5.250	0.065	0.010	0.259	0.310	0.228
3	3.000		3.125	0.065	0.010	5.500	0.065	0.010	0.284	0.327	0.249

All dimensions are in inches.

* The 0.734-inch, 1.156-inch, and 1.469-inch outside diameters avoid washers which could be used in coin operated devices.

** Preferred sizes are for the most part from series previously designated "Standard Plate" and "SAE." Where common sizes existed in the two series, the SAE size is designated "N" (narrow) and the Standard Plate "W" (wide). These sizes as well as all other sizes of Type A Plain Washers are to be ordered by ID, OD, and thickness dimensions.

*** Nominal washer sizes are intended for use with comparable nominal screw or bolt sizes.

Additional selected sizes of Type A Plain Washers are shown in Table K50B.

Table K50B. **American Standard Type A Plain Washers — Additional Selected Sizes** (ASA B27.2-1965)

Inside Diameter			Outside Diameter			Thickness		
Basic	Tolerance		Basic	Tolerance		Basic	Max.	Min.
	Plus	Minus		Plus	Minus			
0.094	0.000	0.005	0.219	0.000	0.005	0.020	0.025	0.016
0.125	0.000	0.005	0.250	0.000	0.005	0.022	0.028	0.017
0.156	0.008	0.005	0.312	0.008	0.005	0.035	0.048	0.027
0.172	0.008	0.005	0.406	0.015	0.005	0.049	0.065	0.036
0.188	0.008	0.005	0.375	0.015	0.005	0.049	0.065	0.036
0.203	0.008	0.005	0.469	0.015	0.005	0.049	0.065	0.036
0.219	0.008	0.005	0.438	0.015	0.005	0.049	0.065	0.036
0.234	0.008	0.005	0.531	0.015	0.005	0.049	0.065	0.036
0.250	0.015	0.005	0.500	0.015	0.005	0.049	0.065	0.036
0.266	0.015	0.005	0.625	0.015	0.005	0.049	0.065	0.036
0.312	0.015	0.005	0.875	0.015	0.007	0.065	0.080	0.051
0.375	0.015	0.005	0.734*	0.015	0.007	0.065	0.080	0.051
0.375	0.015	0.005	1.125	0.015	0.007	0.065	0.080	0.051
0.438	0.015	0.005	0.875	0.030	0.007	0.083	0.104	0.064
0.438	0.015	0.005	1.375	0.030	0.007	0.083	0.104	0.064
0.500	0.015	0.005	1.125	0.030	0.007	0.083	0.104	0.064
0.500	0.015	0.005	1.625	0.030	0.007	0.083	0.104	0.064
0.562	0.015	0.005	1.250	0.030	0.007	0.109	0.132	0.086
0.562	0.015	0.005	1.875	0.030	0.007	0.109	0.132	0.086
0.625	0.015	0.005	1.375	0.030	0.007	0.109	0.132	0.086
0.625	0.015	0.005	2.125	0.030	0.007	0.134	0.160	0.108
0.688	0.030	0.007	1.469*	0.030	0.007	0.134	0.160	0.108
0.688	0.030	0.007	2.375	0.030	0.007	0.165	0.192	0.136
0.812	0.030	0.007	1.750	0.030	0.007	0.148	0.177	0.122
0.812	0.030	0.007	2.875	0.030	0.007	0.165	0.192	0.136
0.938	0.030	0.007	2.000	0.030	0.007	0.165	0.192	0.136
0.938	0.030	0.007	3.375	0.045	0.010	0.180	0.213	0.153
1.062	0.030	0.007	2.250	0.030	0.007	0.165	0.192	0.136
1.062	0.045	0.010	3.875	0.045	0.010	0.238	0.280	0.210
1.250	0.030	0.007	2.500	0.030	0.007	0.165	0.192	0.136
1.375	0.030	0.007	2.750	0.030	0.007	0.165	0.192	0.136
1.500	0.045	0.010	3.000	0.045	0.010	0.180	0.213	0.153
1.625	0.045	0.010	3.250	0.045	0.010	0.180	0.213	0.153
1.688	0.045	0.010	3.500	0.045	0.010	0.180	0.213	0.153
1.812	0.045	0.010	3.750	0.045	0.010	0.180	0.213	0.153
1.938	0.045	0.010	4.000	0.045	0.010	0.180	0.213	0.153
2.062	0.045	0.010	4.250	0.045	0.010	0.180	0.213	0.153

All dimensions are in inches.
* The 0.734-inch and 1.469-inch outside diameters avoid washers which could be used in coin operated devices.
The above sizes are to be ordered by ID, OD, and thickness dimensions.
Preferred Sizes of Type A Plain Washers are shown in Table K50A.

American Standard Plain Washers. — The Type A plain washers were originally developed in a light, medium, heavy and extra heavy series. These series have been discontinued and the washers are now designated by their nominal dimensions.

The Type B plain washers are available in a narrow, regular and wide series with proportions designed to distribute the load over larger areas of lower strength materials.

Plain washers are made of ferrous or non-ferrous metal, plastic or other material as specified. The tolerances indicated in the tables are intended for metal washers only.

Table K51. **American Standard Type B Plain Washers** (ASA B27.2-1965)

Nominal Washer Size**		Series†	Inside Diameter			Outside Diameter			Thickness		
			Basic	Tolerance		Basic	Tolerance		Basic	Max.	Min.
				Plus	Minus		Plus	Minus			
No. 0	0.060	N	0.068	0.000	0.005	0.125	0.000	0.005	0.025	0.028	0.022
		R	0.068	0.000	0.005	0.188	0.000	0.005	0.025	0.028	0.022
		W	0.068	0.000	0.005	0.250	0.000	0.005	0.025	0.028	0.022
No. 1	0.073	N	0.084	0.000	0.005	0.156	0.000	0.005	0.025	0.028	0.022
		R	0.084	0.000	0.005	0.219	0.000	0.005	0.025	0.028	0.022
		W	0.084	0.000	0.005	0.281	0.000	0.005	0.032	0.036	0.028
No. 2	0.086	N	0.094	0.000	0.005	0.188	0.000	0.005	0.025	0.028	0.022
		R	0.094	0.000	0.005	0.250	0.000	0.005	0.032	0.036	0.028
		W	0.094	0.000	0.005	0.344	0.000	0.005	0.032	0.036	0.028
No. 3	0.099	N	0.109	0.000	0.005	0.219	0.000	0.005	0.025	0.028	0.022
		R	0.109	0.000	0.005	0.312	0.000	0.005	0.032	0.036	0.028
		W	0.109	0.008	0.005	0.406	0.008	0.005	0.040	0.045	0.036
No. 4	0.112	N	0.125	0.000	0.005	0.250	0.000	0.005	0.032	0.036	0.028
		R	0.125	0.008	0.005	0.375	0.008	0.005	0.040	0.045	0.036
		W	0.125	0.008	0.005	0.438	0.008	0.005	0.040	0.045	0.036
No. 5	0.125	N	0.141	0.000	0.005	0.281	0.000	0.005	0.032	0.036	0.028
		R	0.141	0.008	0.005	0.406	0.008	0.005	0.040	0.045	0.036
		W	0.141	0.008	0.005	0.500	0.008	0.005	0.040	0.045	0.036
No. 6	0.138	N	0.156	0.000	0.005	0.312	0.000	0.005	0.032	0.036	0.028
		R	0.156	0.008	0.005	0.438	0.008	0.005	0.040	0.045	0.036
		W	0.156	0.008	0.005	0.562	0.008	0.005	0.040	0.045	0.036
No. 8	0.164	N	0.188	0.008	0.005	0.375	0.008	0.005	0.040	0.045	0.036
		R	0.188	0.008	0.005	0.500	0.008	0.005	0.040	0.045	0.036
		W	0.188	0.008	0.005	0.625	0.015	0.005	0.063	0.071	0.056
No. 10	0.190	N	0.203	0.008	0.005	0.406	0.008	0.005	0.040	0.045	0.036
		R	0.203	0.008	0.005	0.562	0.008	0.005	0.040	0.045	0.036
		W	0.203	0.008	0.005	0.734*	0.015	0.007	0.063	0.071	0.056
No. 12	0.216	N	0.234	0.008	0.005	0.438	0.008	0.005	0.040	0.045	0.036
		R	0.234	0.008	0.005	0.625	0.015	0.005	0.063	0.071	0.056
		W	0.234	0.008	0.005	0.875	0.015	0.007	0.063	0.071	0.056
¼	0.250	N	0.281	0.015	0.005	0.500	0.015	0.005	0.063	0.071	0.056
		R	0.281	0.015	0.005	0.734*	0.015	0.007	0.063	0.071	0.056
		W	0.281	0.015	0.005	1.000	0.015	0.007	0.063	0.071	0.056
5/16	0.312	N	0.344	0.015	0.005	0.625	0.015	0.005	0.063	0.071	0.056
		R	0.344	0.015	0.005	0.875	0.015	0.007	0.063	0.071	0.056
		W	0.344	0.015	0.005	1.125	0.015	0.007	0.063	0.071	0.056
3/8	0.375	N	0.406	0.015	0.005	0.734*	0.015	0.007	0.063	0.071	0.056
		R	0.406	0.015	0.005	1.000	0.015	0.007	0.063	0.071	0.056
		W	0.406	0.015	0.005	1.250	0.030	0.007	0.100	0.112	0.090
7/16	0.438	N	0.469	0.015	0.005	0.875	0.015	0.007	0.063	0.071	0.056
		R	0.469	0.015	0.005	1.125	0.015	0.007	0.063	0.071	0.056
		W	0.469	0.015	0.005	1.469*	0.030	0.007	0.100	0.112	0.090

All dimensions are in inches.

* The 0.734-inch and 1.469-inch outside diameters avoid washers which could be used in coin operated devices.

** Nominal washer sizes are intended for use with comparable nominal screw or bolt sizes.

† N indicates Narrow; R, Regular, and W, Wide Series.

Inside and outside diameters shall be concentric within at least the inside diameter tolerance.

Washers shall be flat within 0.005-inch for basic outside diameters up to and including 0.875-inch, and within 0.010 inch for larger outside diameters.

Table K51 (*Concluded*). **American Standard Type B Plain Washers** (ASA B27.2-1965)

Nominal Washer Size**		Series†	Inside Diameter			Outside Diameter			Thickness		
			Basic	Tolerance		Basic	Tolerance		Basic	Max.	Min.
				Plus	Minus		Plus	Minus			
½	0.500	N	0.531	0.015	0.005	1.000	0.015	0.007	0.063	0.071	0.056
		R	0.531	0.015	0.005	1.250	0.030	0.007	0.100	0.112	0.090
		W	0.531	0.015	0.005	1.750	0.030	0.007	0.100	0.112	0.090
9/16	0.562	N	0.594	0.015	0.005	1.125	0.015	0.007	0.063	0.071	0.056
		R	0.594	0.015	0.005	1.469*	0.030	0.007	0.100	0.112	0.090
		W	0.594	0.015	0.005	2.000	0.030	0.007	0.100	0.112	0.090
5/8	0.625	N	0.656	0.030	0.007	1.250	0.030	0.007	0.100	0.112	0.090
		R	0.656	0.030	0.007	1.750	0.030	0.007	0.100	0.112	0.090
		W	0.656	0.030	0.007	2.250	0.030	0.007	0.160	0.174	0.146
¾	0.750	N	0.812	0.030	0.007	1.375	0.030	0.007	0.100	0.112	0.090
		R	0.812	0.030	0.007	2.000	0.030	0.007	0.100	0.112	0.090
		W	0.812	0.030	0.007	2.500	0.030	0.007	0.160	0.174	0.146
7/8	0.875	N	0.938	0.030	0.007	1.469*	0.030	0.007	0.100	0.112	0.090
		R	0.938	0.030	0.007	2.250	0.030	0.007	0.160	0.174	0.146
		W	0.938	0.030	0.007	2.750	0.030	0.007	0.160	0.174	0.146
1	1.000	N	1.062	0.030	0.007	1.750	0.030	0.007	0.100	0.112	0.090
		R	1.062	0.030	0.007	2.500	0.030	0.007	0.160	0.174	0.146
		W	1.062	0.030	0.007	3.000	0.030	0.007	0.160	0.174	0.146
1⅛	1.125	N	1.188	0.030	0.007	2.000	0.030	0.007	0.100	0.112	0.090
		R	1.188	0.030	0.007	2.750	0.030	0.007	0.160	0.174	0.146
		W	1.188	0.030	0.007	3.250	0.030	0.007	0.160	0.174	0.146
1¼	1.250	N	1.312	0.030	0.007	2.250	0.030	0.007	0.160	0.174	0.146
		R	1.312	0.030	0.007	3.000	0.030	0.007	0.160	0.174	0.146
		W	1.312	0.045	0.010	3.500	0.045	0.010	0.250	0.266	0.234
1⅜	1.375	N	1.438	0.030	0.007	2.500	0.030	0.007	0.160	0.174	0.146
		R	1.438	0.030	0.007	3.250	0.030	0.007	0.160	0.174	0.146
		W	1.438	0.045	0.010	3.750	0.045	0.010	0.250	0.266	0.234
1½	1.500	N	1.562	0.030	0.007	2.750	0.030	0.007	0.160	0.174	0.146
		R	1.562	0.045	0.010	3.500	0.045	0.010	0.250	0.266	0.234
		W	1.562	0.045	0.010	4.000	0.045	0.010	0.250	0.266	0.234
1⅝	1.625	N	1.750	0.030	0.007	3.000	0.030	0.007	0.160	0.174	0.146
		R	1.750	0.045	0.010	3.750	0.045	0.010	0.250	0.266	0.234
		W	1.750	0.045	0.010	4.250	0.045	0.010	0.250	0.266	0.234
1¾	1.750	N	1.875	0.030	0.007	3.250	0.030	0.007	0.160	0.174	0.146
		R	1.875	0.045	0.010	4.000	0.045	0.010	0.250	0.266	0.234
		W	1.875	0.045	0.010	4.500	0.045	0.010	0.250	0.266	0.234
1⅞	1.875	N	2.000	0.045	0.010	3.500	0.045	0.010	0.250	0.266	0.234
		R	2.000	0.045	0.010	4.250	0.045	0.010	0.250	0.266	0.234
		W	2.000	0.045	0.010	4.750	0.045	0.010	0.250	0.266	0.234
2	2.000	N	2.125	0.045	0.010	3.750	0.045	0.010	0.250	0.266	0.234
		R	2.125	0.045	0.010	4.500	0.045	0.010	0.250	0.266	0.234
		W	2.125	0.045	0.010	5.000	0.045	0.010	0.250	0.266	0.234

All dimensions are in inches.

* The 1.469-inch outside diameter avoids washers which could be used in coin operated devices.

** Nominal washer sizes are intended for use with comparable nominal screw or bolt sizes.

† N indicates Narrow; R, Regular; and W, Wide Series.

Inside and outside diameters shall be concentric within at least the inside diameter tolerance.

Washers shall be flat within 0.005-inch for basic outside diameters up through 0.875-inch and within 0.010-inch for larger outside diameters.

For 2¼-, 2½-, 2¾-, and 3-inch sizes see ASA B27.2-1965.

Table K52. American Standard Helical Spring Lock Washers (ASA B27.1-1965)

Nominal Size	Outside Diameter, Max.*				Inside Diameter		Light		Regular		Heavy		Extra Duty	
	Light	Regular	Heavy	Extra Duty	Min.	Max.	Width	Thickness	Width	Thickness	Width	Thickness	Width	Thickness
0.086 (No. 2)	0.162	0.172	0.182	0.208	0.088	0.094	0.030	0.015	0.035	0.020	0.040	0.025	0.053	0.027
0.099 (No. 3)	0.185	0.195	0.209	0.239	0.101	0.107	0.035	0.020	0.040	0.025	0.047	0.031	0.062	0.034
0.112 (No. 4)	0.199	0.209	0.223	0.253	0.115	0.121	0.035	0.020	0.040	0.025	0.047	0.031	0.062	0.034
0.125 (No. 5)	0.222	0.236	0.252	0.300	0.128	0.134	0.040	0.025	0.047	0.031	0.055	0.040	0.079	0.045
0.138 (No. 6)	0.236	0.250	0.266	0.314	0.141	0.148	0.040	0.025	0.047	0.031	0.055	0.040	0.079	0.045
0.164 (No. 8)	0.277	0.293	0.307	0.375	0.168	0.175	0.047	0.031	0.055	0.040	0.062	0.047	0.096	0.057
0.190 (No. 10)	0.320	0.334	0.350	0.434	0.194	0.202	0.055	0.040	0.062	0.047	0.070	0.056	0.112	0.068
0.216 (No. 12)	0.361	0.377	0.391	0.497	0.221	0.229	0.062	0.047	0.070	0.056	0.077	0.063	0.130	0.080
1/4	0.485	0.489	0.491	0.535	0.255	0.263	0.107	0.047	0.109	0.062	0.110	0.077	0.132	0.084
5/16	0.570	0.586	0.596	0.622	0.318	0.328	0.117	0.056	0.125	0.078	0.130	0.097	0.143	0.108
3/8	0.673	0.683	0.691	0.741	0.382	0.393	0.136	0.070	0.141	0.094	0.145	0.115	0.170	0.123
7/16	0.775	0.779	0.787	0.839	0.446	0.459	0.154	0.085	0.156	0.109	0.160	0.133	0.186	0.143
1/2	0.871	0.873	0.883	0.939	0.509	0.523	0.170	0.099	0.171	0.125	0.176	0.151	0.204	0.162
9/16	0.967	0.971	0.981	1.041	0.572	0.587	0.186	0.113	0.188	0.141	0.193	0.170	0.223	0.182
5/8	1.075	1.079	1.093	1.157	0.636	0.653	0.201	0.126	0.203	0.156	0.210	0.189	0.242	0.202
11/16	1.170	1.176	1.192	1.258	0.700	0.718	0.216	0.138	0.219	0.172	0.227	0.207	0.260	0.221
3/4	1.269	1.271	1.291	1.361	0.763	0.783	0.233	0.153	0.234	0.188	0.244	0.226	0.279	0.241
13/16	1.365	1.367	1.391	1.463	0.826	0.847	0.249	0.168	0.250	0.203	0.262	0.246	0.298	0.261
7/8	1.460	1.464	1.494	1.576	0.890	0.912	0.264	0.179	0.266	0.219	0.281	0.266	0.322	0.285
15/16	1.552	1.560	1.594	1.688	0.954	0.978	0.277	0.191	0.281	0.234	0.298	0.284	0.345	0.308
1	1.645	1.661	1.705	1.799	1.017	1.042	0.289	0.202	0.297	0.250	0.319	0.306	0.366	0.330
1 1/16	1.734	1.756	1.808	1.910	1.080	1.107	0.301	0.213	0.312	0.266	0.338	0.326	0.389	0.352
1 1/8	1.825	1.853	1.909	2.019	1.144	1.172	0.314	0.224	0.328	0.281	0.356	0.345	0.411	0.375
1 3/16	1.910	1.950	2.008	2.124	1.208	1.237	0.324	0.234	0.344	0.297	0.373	0.364	0.431	0.396
1 1/4	1.999	2.045	2.113	2.231	1.271	1.302	0.336	0.244	0.359	0.312	0.393	0.384	0.452	0.417
1 5/16	2.083	2.141	2.211	2.335	1.334	1.366	0.346	0.254	0.375	0.328	0.410	0.403	0.472	0.438
1 3/8	2.169	2.239	2.311	2.439	1.398	1.432	0.356	0.264	0.391	0.344	0.427	0.422	0.491	0.458
1 7/16	2.254	2.334	2.406	2.540	1.462	1.497	0.366	0.273	0.406	0.359	0.442	0.440	0.509	0.478
1 1/2	2.336	2.430	2.502	2.638	1.525	1.561	0.375	0.282	0.422	0.375	0.458	0.458	0.526	0.496
1 5/8	2.493	.....	2.693	.....	1.650	1.686	0.391	0.344	.....	.....	0.491	0.458	.....	.....
1 3/4	2.618	.....	2.818	.....	1.775	1.811	0.391	0.344	.....	.....	0.491	0.458	.....	.....
1 7/8	2.747	.....	2.943	.....	1.900	1.936	0.393	0.384	.....	.....	0.491	0.458	.....	.....
2	2.872	.....	3.068	.....	2.025	2.061	0.393	0.384	.....	.....	0.491	0.458	.....	.....
2 1/4	3.190	.....	3.388	.....	2.275	2.311	0.427	0.422	.....	.....	0.526	0.496	.....	.....
2 1/2	3.440	.....	3.638	.....	2.525	2.561	0.427	0.422	.....	.....	0.526	0.496	.....	.....
2 3/4	3.752	.....	3.888	.....	2.775	2.811	0.458	0.458	.....	.....	0.526	0.496	.....	.....
3	4.002	.....	4.138	.....	3.025	3.061	0.458	0.458	.....	.....	0.526	0.496	.....	.....

* The maximum outside diameters specified allow for commercial tolerances on cold-drawn wire.

Table K53. **American Standard Hi-Collar* Helical Spring Lock Washers** (ASA B27.1-1965)

Nominal Washer Size		Inside Diameter		Outside Diameter	Washer Section	
					Width	Thickness
		Min.	Max.	Max.**	Min.	Min.
No. 4	0.112	0.115	0.121	0.173	0.022	0.022
No. 5	0.125	0.128	0.134	0.202	0.030	0.030
No. 6	0.138	0.141	0.148	0.216	0.030	0.030
No. 8	0.164	0.168	0.175	0.267	0.042	0.047
No. 10	0.190	0.194	0.202	0.294	0.042	0.047
1/4	0.250	0.255	0.263	0.365	0.047	0.078
5/16	0.312	0.318	0.328	0.460	0.062	0.093
3/8	0.375	0.382	0.393	0.553	0.076	0.125
7/16	0.438	0.446	0.459	0.647	0.090	0.140
1/2	0.500	0.509	0.523	0.737	0.103	0.172
5/8	0.625	0.636	0.653	0.923	0.125	0.203
3/4	0.750	0.763	0.783	1.111	0.154	0.218
7/8	0.875	0.890	0.912	1.296	0.182	0.234
1	1.000	1.017	1.042	1.483	0.208	0.250
1 1/8	1.125	1.144	1.172	1.669	0.236	0.313
1 1/4	1.250	1.271	1.302	1.799	0.236	0.313
1 3/8	1.375	1.398	1.432	2.041	0.292	0.375
1 1/2	1.500	1.525	1.561	2.170	0.292	0.375
1 3/4	1.750	1.775	1.811	2.602	0.383	0.469
2	2.000	2.025	2.061	2.852	0.383	0.469
2 1/4	2.250	2.275	2.311	3.352	0.508	0.508
2 1/2	2.500	2.525	2.561	3.602	0.508	0.508
2 3/4	2.750	2.775	2.811	4.102	0.633	0.633
3	3.000	3.025	3.061	4.352	0.633	0.633

* For use with 1960 Series Socket Head Cap Screws.
** The maximum outside diameters specified allow for the commercial tolerances on cold-drawn wire.

American Standard Helical Spring Lock and Tooth Lock Washers (ASA B27.1-1965). — This standard covers helical spring lock washers of carbon steel; corrosion resistant steel, Types 302, 305 and 420; aluminum-zinc alloy; phosphor-bronze; silicon-bronze; and K-Monel; in various series. It also covers tooth lock washers of carbon steel having internal teeth, external teeth, and both internal and external teeth, of two constructions, designated as Type A and Type B.

Helical spring lock washers: These washers are intended for general industrial application. They have the dual function of: (1) spring takeup devices to compensate for developed looseness and the loss of tension between component parts of an assembly; and (2) hardened thrust bearings to facilitate assembly and disassembly of bolted fastenings by decreasing the frictional resistance between the bolted surface and the bearing face of the bolt head or nut.

They are specified or designated by nominal size and series as, for example, 1/4 inch Regular, 1/4 inch Extra Duty, or 1/4 inch Hi-Collar. Nominal washer sizes are intended for use with comparable nominal screw or bolt sizes.

Carbon steel helical spring lock washers are available in five series as given in Tables 1 and 2. Preference should be given to the Regular, Extra Duty, and Hi-Collar series.

Aluminum-zinc alloy helical spring lock washers are available in the Regular series.

Phosphor-bronze, silicon-bronze, and K-Monel helical spring lock washers are available in the Regular and Light series. Preference should be given to the Regular series.

When carbon steel helical spring lock washers are to be hot-dipped galvanized for use with hot-dipped galvanized bolts or screws, they are to be coiled to limits 0.020 inch in excess of those specified in Tables 1 and 2 for minimum inside diameter and maximum outside diameter. Galvanizing on washers under ¼ inch nominal size is not recommended.

Tooth lock washers: These washers are intended for general industrial application and serve to lock fasteners, such as bolts and nuts, to the component parts of an assembly, or increase the friction between the fasteners and the assembly.

Table K54. **American Standard Internal-External Tooth Lock Washers** (ASA B27.1-1965)

All dimensions are given in inches except whole numbers under "Size"

TYPE A

TYPE B

Size	A Inside Diameter Max.	A Inside Diameter Min.	B Outside Diameter Max.	B Outside Diameter Min.	C Thickness Max.	C Thickness Min.
No. 4	.123	.115	.475	.460	.021	.016
	.123	.115	.510 .610	.495 .580	.021	.017
No. 6	.150	.141	.510 .610 .690	.495 .580 .670	.028	.023
No. 8	.176	.168	.610 .690 .760	.580 .670 .740	.034	.028
No. 10	.204	.195	.610	.580	.034	.028
	.204	.195	.690 .760 .900	.670 .740 .880	.040	.032
No. 12	.231	.221	.690 .760 .900	.670 .725 .880	.040	.032
	.231	.221	.985	.965	.045	.037
¼	.267	.256	.760 .900	.725 .880	.040	.032
	.267	.256	.985 1.070	.965 1.045	.045	.037

Size	A Inside Diameter Max.	A Inside Diameter Min.	B Outside Diameter Max.	B Outside Diameter Min.	C Thickness Max.	C Thickness Min.
5/16	.332	.320	.900	.865	.040	.032
	.332	.320	.985	.965	.045	.037
	.332	.320	1.070 1.155	1.045 1.130	.050	.042
⅜	.398	.384	.985	.965	.045	.037
	.398	.384	1.070 1.155 1.260	1.045 1.130 1.220	.050	.042
7/16	.464	.448	1.070 1.155	1.045 1.130	.050	.042
	.464	.448	1.260 1.315	1.220 1.290	.055	.047
½	.530	.512	1.260 1.315	1.220 1.290	.055	.047
	.530	.512	1.410	1.380	.060	.052
	.530	.512	1.620	1.590	.067	.059
9/16	.596	.576	1.315	1.290	.055	.047
	.596	.576	1.430	1.380	.060	.052
	.596	.576	1.620 1.830	1.590 1.797	.067	.059
⅝	.663	.640	1.410	1.380	.060	.052
	.663	.640	1.620 1.830 1.975	1.590 1.797 1.935	.067	.059

Table K55. American Standard Internal and External Tooth Lock Washers (ASA B27.1-1965)

Internal Tooth Lock Washers

	Size	#2	#3	#4	#5	#6	#8	#10	#12	¼	5⁄16	⅜	7⁄16	½	9⁄16	⅝	11⁄16	¾	13⁄16	⅞	1	1⅛	1¼
A	Max	0.095	0.109	0.123	0.136	0.150	0.176	0.204	0.231	0.267	0.332	0.398	0.464	0.530	0.596	0.663	0.728	0.795	0.861	0.927	1.060	1.192	1.325
	Min	0.089	0.102	0.115	0.129	0.141	0.168	0.195	0.221	0.256	0.320	0.384	0.448	0.512	0.576	0.640	0.704	0.769	0.832	0.894	1.019	1.144	1.275
B	Max	0.200	0.232	0.270	0.280	0.295	0.340	0.381	0.410	0.478	0.610	0.692	0.789	0.900	0.985	1.071	1.166	1.245	1.315	1.410	1.637	1.830	1.975
	Min	0.175	0.215	0.255	0.245	0.275	0.325	0.365	0.394	0.460	0.594	0.670	0.740	0.867	0.957	1.045	1.130	1.220	1.290	1.364	1.590	1.799	1.921
C	Max	0.015	0.019	0.019	0.021	0.021	0.023	0.025	0.025	0.028	0.034	0.040	0.040	0.045	0.045	0.050	0.050	0.055	0.055	0.060	0.067	0.067	0.067
	Min	0.010	0.012	0.015	0.017	0.017	0.018	0.020	0.020	0.023	0.028	0.032	0.032	0.037	0.037	0.042	0.042	0.047	0.047	0.052	0.059	0.059	0.059

External Tooth Lock Washers

	Size	#2	#3	#4	#5	#6	#8	#10	#12	¼	5⁄16	⅜	7⁄16	½	9⁄16	⅝	11⁄16	¾	13⁄16	⅞	1	1⅛	1¼
A	Max	....	....	0.123	....	0.150	0.176	0.204	0.231	0.267	0.332	0.398	0.464	0.530	0.596	0.663	0.728	0.795	0.861	0.927	1.060	....	....
	Min	....	....	0.115	....	0.141	0.168	0.195	0.221	0.256	0.320	0.384	0.448	0.513	0.576	0.641	0.704	0.768	0.833	0.897	1.025	....	....
B	Max	....	....	0.260	....	0.320	0.381	0.410	0.475	0.510	0.610	0.694	0.760	0.900	0.985	1.070	1.155	1.260	1.315	1.410	1.620	....	....
	Min	....	....	0.245	....	0.305	0.365	0.395	0.460	0.494	0.588	0.670	0.740	0.880	0.960	1.045	1.130	1.220	1.290	1.380	1.590	....	....
C	Max	....	....	0.019	....	0.022	0.023	0.025	0.028	0.028	0.034	0.040	0.040	0.045	0.045	0.050	0.050	0.055	0.055	0.060	0.067	....	....
	Min	....	....	0.015	....	0.016	0.018	0.020	0.023	0.023	0.028	0.032	0.032	0.037	0.037	0.042	0.042	0.047	0.047	0.052	0.059	....	....

Heavy Internal Tooth Lock Washers

	Size	¼	5⁄16	⅜	7⁄16	½	9⁄16	⅝	¾	⅞
A	Max	0.267	0.332	0.398	0.464	0.530	0.596	0.663	0.795	0.927
	Min	0.256	0.320	0.384	0.448	0.512	0.576	0.640	0.768	0.894
B	Max	0.536	0.607	0.748	0.858	0.924	1.034	1.135	1.265	1.447
	Min	0.500	0.590	0.700	0.800	0.880	0.990	1.100	1.240	1.400
C	Max	0.045	0.050	0.050	0.067	0.067	0.067	0.067	0.084	0.084
	Min	0.035	0.040	0.042	0.050	0.055	0.055	0.059	0.070	0.075

Countersunk External Tooth Lock Washers*

	Size	#4	#6	#8	#10	#12	¼	#16	5⁄16	⅜	7⁄16	½
A	Max	0.123	0.150	0.177	0.205	0.231	0.267	0.287	0.333	0.398	0.463	0.529
	Min	0.113	0.140	0.167	0.195	0.220	0.255	0.273	0.318	0.383	0.448	0.512
C	Max	0.019	0.021	0.021	0.025	0.025	0.025	0.028	0.028	0.034	0.045	0.045
	Min	0.015	0.017	0.017	0.020	0.020	0.020	0.023	0.023	0.028	0.037	0.037
D	Max	0.065	0.092	0.105	0.099	0.128	0.128	0.147	0.192	0.255	0.270	0.304
	Min	0.050	0.082	0.088	0.083	0.118	0.113	0.137	0.165	0.242	0.260	0.294

All dimensions are given in inches. * Starting with #4, approx. O.D.'s. are: 0.213, 0.289, 0.322, 0.354, 0.421, 0.454, 0.505, 0.599, 0.765, 0.867 and 0.976.

Table K56. **British Standard Single Coil Rectangular Section Spring Washers; Metric Series — Types B and BP** (BS 4464: 1969)

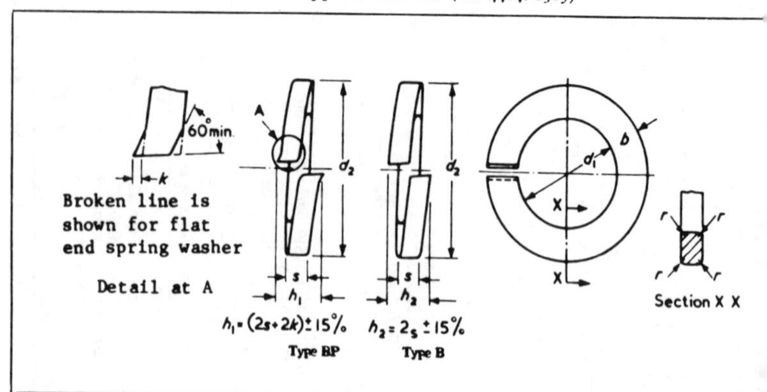

Nom. Size & Thread Diam., d	Inside Diam., d_1		Width, b	Thick-ness, s	Outside Diam., d_2 Max	Radius, r Max	k (Type BP Only)
	Max	Min					
M1.6	1.9	1.7	0.7 ± 0.1	0.4 ± 0.1	3.5	0.15	...
M2	2.3	2.1	0.9 ± 0.1	0.5 ± 0.1	4.3	0.15	...
(M2.2)	2.5	2.3	1.0 ± 0.1	0.6 ± 0.1	4.7	0.2	...
M2.5	2.8	2.6	1.0 ± 0.1	0.6 ± 0.1	5.0	0.2	...
M3	3.3	3.1	1.3 ± 0.1	0.8 ± 0.1	6.1	0.25	...
(M3.5)	3.8	3.6	1.3 ± 0.1	0.8 ± 0.1	6.6	0.25	0.15
M4	4.35	4.1	1.5 ± 0.1	0.9 ± 0.1	7.55	0.3	0.15
M5	5.35	5.1	1.8 ± 0.1	1.2 ± 0.1	9.15	0.4	0.15
M6	6.4	6.1	2.5 ± 0.15	1.6 ± 0.1	11.7	0.5	0.2
M8	8.55	8.2	3 ± 0.15	2 ± 0.1	14.85	0.65	0.3
M10	10.6	10.2	3.5 ± 0.2	2.2 ± 0.15	18.0	0.7	0.3
M12	12.6	12.2	4 ± 0.2	2.5 ± 0.15	21.0	0.8	0.4
(M14)	14.7	14.2	4.5 ± 0.2	3 ± 0.15	24.1	1.0	0.4
M16	16.9	16.3	5 ± 0.2	3.5 ± 0.2	27.3	1.15	0.4
(M18)	19.0	18.3	5 ± 0.2	3.5 ± 0.2	29.4	1.15	0.4
M20	21.1	20.3	6 ± 0.2	4 ± 0.2	33.5	1.3	0.4
(M22)	23.3	22.4	6 ± 0.2	4 ± 0.2	35.7	1.3	0.4
M24	25.3	24.4	7 ± 0.25	5 ± 0.2	39.8	1.65	0.5
(M27)	28.5	27.5	7 ± 0.25	5 ± 0.2	43.0	1.65	0.5
M30	31.5	30.5	8 ± 0.25	6 ± 0.25	48.0	2.0	0.8
(M33)	34.6	33.5	10 ± 0.25	6 ± 0.25	55.1	2.0	0.8
M36	37.6	36.5	10 ± 0.25	6 ± 0.25	58.1	2.0	0.8
(M39)	40.8	39.6	10 ± 0.25	6 ± 0.25	61.3	2.0	0.8
M42	43.8	42.6	12 ± 0.25	7 ± 0.25	68.3	2.3	0.8
(M45)	46.8	45.6	12 ± 0.25	7 ± 0.25	71.3	2.3	0.8
M48	50.0	48.8	12 ± 0.25	7 ± 0.25	74.5	2.3	0.8
(M52)	54.1	52.8	14 ± 0.25	8 ± 0.25	82.6	2.65	1.0
M56	58.1	56.8	14 ± 0.25	8 ± 0.25	86.6	2.65	1.0
(M60)	62.3	60.9	14 ± 0.25	8 ± 0.25	90.8	2.65	1.0
M64	66.3	64.9	14 ± 0.25	8 ± 0.25	93.8	2.65	1.0
(M68)	70.5	69.0	14 ± 0.25	8 ± 0.25	99.0	2.65	1.0

All dimensions are given in millimeters. Sizes shown in parentheses are non-preferred, and are not usually stock sizes.

Table K57. British Standard Double Coil Rectangular Section Spring Washers; Metric Series — Type D (BS 4464: 1969)

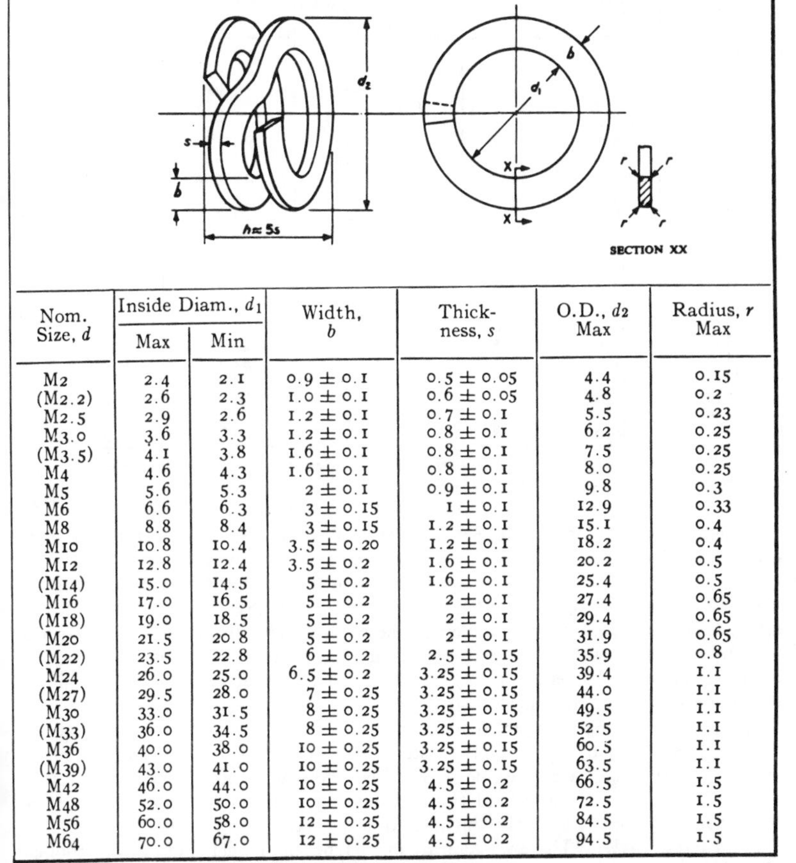

Nom. Size, d	Inside Diam., d_1		Width, b	Thick-ness, s	O.D., d_2 Max	Radius, r Max
	Max	Min				
M2	2.4	2.1	0.9 ± 0.1	0.5 ± 0.05	4.4	0.15
(M2.2)	2.6	2.3	1.0 ± 0.1	0.6 ± 0.05	4.8	0.2
M2.5	2.9	2.6	1.2 ± 0.1	0.7 ± 0.1	5.5	0.23
M3.0	3.6	3.3	1.2 ± 0.1	0.8 ± 0.1	6.2	0.25
(M3.5)	4.1	3.8	1.6 ± 0.1	0.8 ± 0.1	7.5	0.25
M4	4.6	4.3	1.6 ± 0.1	0.8 ± 0.1	8.0	0.25
M5	5.6	5.3	2 ± 0.1	0.9 ± 0.1	9.8	0.3
M6	6.6	6.3	3 ± 0.15	1 ± 0.1	12.9	0.33
M8	8.8	8.4	3 ± 0.15	1.2 ± 0.1	15.1	0.4
M10	10.8	10.4	3.5 ± 0.20	1.2 ± 0.1	18.2	0.4
M12	12.8	12.4	3.5 ± 0.2	1.6 ± 0.1	20.2	0.5
(M14)	15.0	14.5	5 ± 0.2	1.6 ± 0.1	25.4	0.5
M16	17.0	16.5	5 ± 0.2	2 ± 0.1	27.4	0.65
(M18)	19.0	18.5	5 ± 0.2	2 ± 0.1	29.4	0.65
M20	21.5	20.8	5 ± 0.2	2 ± 0.1	31.9	0.65
(M22)	23.5	22.8	6 ± 0.2	2.5 ± 0.15	35.9	0.8
M24	26.0	25.0	6.5 ± 0.2	3.25 ± 0.15	39.4	1.1
(M27)	29.5	28.0	7 ± 0.25	3.25 ± 0.15	44.0	1.1
M30	33.0	31.5	8 ± 0.25	3.25 ± 0.15	49.5	1.1
(M33)	36.0	34.5	8 ± 0.25	3.25 ± 0.15	52.5	1.1
M36	40.0	38.0	10 ± 0.25	3.25 ± 0.15	60.5	1.1
(M39)	43.0	41.0	10 ± 0.25	3.25 ± 0.15	63.5	1.1
M42	46.0	44.0	10 ± 0.25	4.5 ± 0.2	66.5	1.5
M48	52.0	50.0	10 ± 0.25	4.5 ± 0.2	72.5	1.5
M56	60.0	58.0	12 ± 0.25	4.5 ± 0.2	84.5	1.5
M64	70.0	67.0	12 ± 0.25	4.5 ± 0.2	94.5	1.5

All dimensions are given in millimeters. Sizes shown in parentheses are non-preferred, and are not usually stock sizes. The free height of double coil washers before compression is normally approximately five times the thickness but, if required, washers with other free heights may be obtained by arrangement with manufacturer.

British Standard Single Coil Square Section Spring Washers; Metric Series — Type A — 1 (BS 4464:1969)

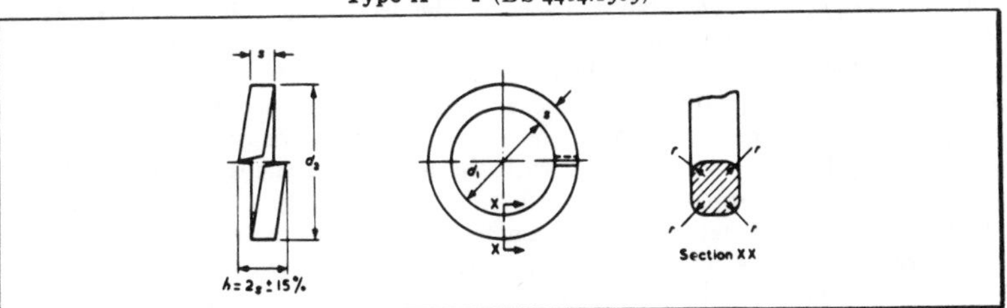

Table K58. **British Standard Single Coil Square Section Spring Washers; Metric Series — Type A — 2** (BS 4464: 1969)

Nom. Size, d	Inside Diam., d_1		Thickness & Width, s	O.D., d_2 Max	Radius, r Max
	Max	Min			
M3	3.3	3.1	1 ± 0.1	5.5	0.3
(M3.5)	3.8	3.6	1 ± 0.1	6.0	0.3
M4	4.35	4.1	1.2 ± 0.1	6.95	0.4
M5	5.35	5.1	1.5 ± 0.1	8.55	0.5
M6	6.4	6.1	1.5 ± 0.1	9.6	0.5
M8	8.55	8.2	2 ± 0.1	12.75	0.65
M10	10.6	10.2	2.5 ± 0.15	15.9	0.8
M12	12.6	12.2	2.5 ± 0.15	17.9	0.8
(M14)	14.7	14.2	3 ± 0.2	21.1	1.0
M16	16.9	16.3	3.5 ± 0.2	24.3	1.15
(M18)	19.0	18.3	3.5 ± 0.2	26.4	1.15
M20	21.1	20.3	4.5 ± 0.2	30.5	1.5
(M22)	23.3	22.4	4.5 ± 0.2	32.7	1.5
M24	25.3	24.4	5 ± 0.2	35.7	1.65
(M27)	28.5	27.5	5 ± 0.2	38.9	1.65
M30	31.5	30.5	6 ± 0.2	43.9	2.0
(M33)	34.6	33.5	6 ± 0.2	47.0	2.0
M36	37.6	36.5	7 ± 0.25	52.1	2.3
(M39)	40.8	39.6	7 ± 0.25	55.3	2.3
M42	43.8	42.6	8 ± 0.25	60.3	2.65
(M45)	46.8	45.6	8 ± 0.25	63.3	2.65
M48	50.0	48.8	8 ± 0.25	66.5	2.65

All dimensions are given in millimeters. Sizes shown in parentheses are non-preferred and are not usually stock sizes.

British Standard for Metric Series Metal Washers. — BS 4320:1968 specifies bright and black metal washers for general engineering purposes.

Bright metal washers: These washers are made from either CS4 cold rolled strip steel (BS 1449:Part 3B) or from CZ 108 brass strip (BS 265), both in the hard condition. However, by mutual agreement between purchaser and supplier, washers may be made available with the material in any other condition, or they may be made from another material, or may be coated with a protective or decorative finish to some appropriate British Standard. Washers are reasonably flat and free from burrs and are normally supplied unchamfered. They may, however, have a 30-degree chamfer on one edge of the external diameter. These washers are made available in two size categories, normal and large diameter, and in two thicknesses, normal (Form A or C), and light (Form B or D). The thickness of a light range washer ranges from ½ to ⅔ the thickness of a normal range washer.

Black metal washers: These washers are made from mild steel, and can be supplied in three size categories designated normal, large, and extra large diameters. The normal diameter series are intended for bolts ranging from M5 to M68 (Form E washers); the large diameter series for bolts ranging from M8 to M39 (Form F washers) and the extra large series for bolts from M5 to M39 (Form G washers). A protective finish can be specified by the purchaser in accordance with any appropriate British Standard.

Washer designations: The Standard specifies the details that should be given when ordering or placing an inquiry for washers. They are the general description, namely, bright or black washers; the nominal size of the bolt or screw involved, for example, M5; the designated form, for example, Form A or Form E; the dimensions of any chamfer required on bright washers; the number of the Standard (BS 4320), and coating information if required, with the number of the appropriate British Standard and the coating thickness needed. As an example, in the use of this information, the designation for a chamfered, normal diameter series washer of normal range thickness to suit a 12-mm diameter bolt would be: Bright washers M12 (Form A) chamfered to BS 4320.

Table K59. **British Standard Bright Metal Washers — Metric Series** (BS 4320:1968)

NORMAL DIAMETER SIZES

Nominal Size of Bolt or Screw	Inside Diameter			Outside Diameter			Thickness Form A (Normal Range)			Thickness Form B (Light Range)		
	Nom	Max	Min	Nom	Max	Min	Nom	Max	Min	Nom	Max	Min
M 1.0	1.1	1.25	1.1	2.5	2.5	2.3	0.3	0.4	0.2	....	....	....
M 1.2	1.3	1.45	1.3	3.0	3.0	2.8	0.3	0.4	0.2	....	....	....
(M 1.4)	1.5	1.65	1.5	3.0	3.0	2.8	0.3	0.4	0.2	....	....	....
M 1.6	1.7	1.85	1.7	4.0	4.0	3.7	0.3	0.4	0.2	....	....	....
M 2.0	2.2	2.35	2.2	5.0	5.0	4.7	0.3	0.4	0.2	....	....	....
(M 2.2)	2.4	2.55	2.4	5.0	5.0	4.7	0.5	0.6	0.4	....	....	....
M 2.5	2.7	2.85	2.7	6.5	6.5	6.2	0.5	0.6	0.4	....	....	....
M 3	3.2	3.4	3.2	7	7	6.7	0.5	0.6	0.4	....	....	....
(M 3.5)	3.7	3.9	3.7	7	7	6.7	0.5	0.6	0.4	....	....	....
M 4	4.3	4.5	4.3	9	9	8.7	0.8	0.9	0.7	....	....	....
(M 4.5)	4.8	5.0	4.8	9	9	8.7	0.8	0.9	0.7	....	....	....
M 5	5.3	5.5	5.3	10	10	9.7	1.0	1.1	0.9	....	....	....
M 6	6.4	6.7	6.4	12.5	12.5	12.1	1.6	1.8	1.4	0.8	0.9	0.7
(M 7)	7.4	7.7	7.4	14	14	13.6	1.6	1.8	1.4	0.8	0.9	0.7
M 8	8.4	8.7	8.4	17	17	16.6	1.6	1.8	1.4	1.0	1.1	0.9
M 10	10.5	10.9	10.5	21	21	20.5	2.0	2.2	1.8	1.25	1.45	1.05
M 12	13.0	13.4	13.0	24	24	23.5	2.5	2.7	2.3	1.6	1.80	1.40
(M 14)	15.0	15.4	15.0	28	28	27.5	2.5	2.7	2.3	1.6	1.8	1.4
M 16	17.0	17.4	17.0	30	30	29.5	3.0	3.3	2.7	2.0	2.2	1.8
(M 18)	19.0	19.5	19.0	34	34	33.2	3.0	3.3	2.7	2.0	2.2	1.8
M 20	21	21.5	21	37	37	36.2	3.0	3.3	2.7	2.0	2.2	1.8
(M 22)	23	23.5	23	39	39	38.2	3.0	3.3	2.7	2.0	2.2	1.8
M 24	25	25.5	25	44	44	43.2	4.0	4.3	3.7	2.5	2.7	2.3
(M 27)	28	28.5	28	50	50	49.2	4.0	4.3	3.7	2.5	2.7	2.3
M 30	31	31.6	31	56	56	55.0	4.0	4.3	3.7	2.5	2.7	2.3
(M 33)	34	34.6	34	60	60	59.0	5.0	5.6	4.4	3.0	3.3	2.7
M 36	37	37.6	37	66	66	65.0	5.0	5.6	4.4	3.0	3.3	2.7
(M 39)	40	40.6	40	72	72	71.0	6.0	6.6	5.4	3.0	3.3	2.7

LARGE DIAMETER SIZES

Nominal Size of Bolt or Screw	Inside Diameter			Outside Diameter			Thickness Form C (Normal Range)			Thickness Form D (Light Range)		
	Nom	Max	Min	Nom	Max	Min	Nom	Max	Min	Nom	Max	Min
M 4	4.3	4.5	4.3	10.0	10.0	9.7	0.8	0.9	0.7	....	....	....
M 5	5.3	5.5	5.3	12.5	12.5	12.1	1.0	1.1	0.9	....	....	....
M 6	6.4	6.7	6.4	14	14	13.6	1.6	1.8	1.4	0.8	0.9	0.7
M 8	8.4	8.7	8.4	21	21	20.5	1.6	1.8	1.4	1.0	1.1	0.9
M 10	10.5	10.9	10.5	24	24	23.5	2.0	2.2	1.8	1.25	1.45	1.05
M 12	13.0	13.4	13.0	28	28	27.5	2.5	2.7	2.3	1.6	1.8	1.4
(M 14)	15.0	15.4	15	30	30	29.5	2.5	2.7	2.3	1.6	1.8	1.4
M 16	17.0	17.4	17	34	34	33.2	3.0	3.3	2.7	2.0	2.2	1.8
(M 18)	19.0	19.5	19	37	37	36.2	3.0	3.3	2.7	2.0	2.2	1.8
M 20	21	21.5	21	39	39	38.2	3.0	3.3	2.7	2.0	2.2	1.8
(M 22)	23	23.5	23	44	44	43.2	3.0	3.3	2.7	2.0	2.2	1.8
M 24	25	25.5	25	50	50	49.2	4.0	4.3	3.7	2.5	2.7	2.3
(M 27)	28	28.5	28	56	56	55	4.0	4.3	3.7	2.5	2.7	2.3
M 30	31	31.6	31	60	60	59	4.0	4.3	3.7	2.5	2.7	2.3
(M 33)	34	34.6	34	66	66	65	5.0	5.6	4.4	3.0	3.3	2.7
M 36	37	37.6	37	72	72	71	5.0	5.6	4.4	3.0	3.3	2.7
(M 39)	40	40.6	40	77	77	76	6.0	6.6	5.4	3.0	3.3	2.7

All dimensions are given in millimeters.
Nominal bolt or screw sizes shown in parentheses are non-preferred.

Table K60. British Standard Black Metal Washers — Metric Series (BS 4320:1968)

Nom Bolt or Screw Size	Inside Diameter			Outside Diameter			Thickness		
	Nom	Max	Min	Nom	Max	Min	Nom	Max	Min
NORMAL DIAMETER SIZES (Form E)									
M 5	5.5	5.8	5.5	10.0	10.0	9.2	1.0	1.2	0.8
M 6	6.6	7.0	6.6	12.5	12.5	11.7	1.6	1.9	1.3
(M 7)	7.6	8.0	7.6	14.0	14.0	13.2	1.6	1.9	1.3
M 8	9.0	9.4	9.0	17	17	16.2	1.6	1.9	1.3
M 10	11.0	11.5	11.0	21	21	20.2	2.0	2.3	1.7
M 12	14	14.5	14	24	24	23.2	2.5	2.8	2.2
(M 14)	16	16.5	16	28	28	27.2	2.5	2.8	2.2
M 16	18	18.5	18	30	30	29.2	3.0	3.6	2.4
(M 18)	20	20.6	20	34	34	32.8	3.0	3.6	2.4
M 20	22	22.6	22	37	37	35.8	3.0	3.6	2.4
(M 22)	24	24.6	24	39	39	37.8	3.0	3.6	2.4
M 24	26	26.6	26	44	44	42.8	4	4.6	3.4
(M 27)	30	30.6	30	50	50	48.8	4	4.6	3.4
M 30	33	33.8	33	56	56	54.5	4	4.6	3.4
(M 33)	36	36.8	36	60	60	58.5	5	6.0	4.0
M 36	39	39.8	39	66	66	64.5	5	6.0	4.0
(M 39)	42	42.8	42	72	72	70.5	6	7.0	5.0
M 42	45	45.8	45	78	78	76.5	7	8.2	5.8
(M 45)	48	48.8	48	85	85	83	7	8.2	5.8
M 48	52	53	52	92	92	90	8	9.2	6.8
(M 52)	56	57	56	98	98	96	8	9.2	6.8
M 56	62	63	62	105	105	103	9	10.2	7.8
(M 60)	66	67	66	110	110	108	9	10.2	7.8
M 64	70	71	70	115	115	113	9	10.2	7.8
(M 68)	74	75	74	120	120	118	10	11.2	8.8
LARGE DIAMETER SIZES (Form F)									
M 8	9	9.4	9.0	21	21	20.2	1.6	1.9	1.3
M 10	11	11.5	11	24	24	23.2	2	2.3	1.7
M 12	14	14.5	14	28	28	27.2	2.5	2.8	2.2
(M 14)	16	16.5	16	30	30	29.2	2.5	2.8	2.2
M 16	18	18.5	18	34	34	32.8	3	3.6	2.4
(M 18)	20	20.6	20	37	37	35.8	3	3.6	2.4
M 20	22	22.6	22	39	39	37.8	3	3.6	2.4
(M 22)	24	24.6	24	44	44	42.8	3	3.6	2.4
M 24	26	26.6	26	50	50	48.8	4	4.6	3.4
(M 27)	30	30.6	30	56	56	54.5	4	4.6	3.4
M 30	33	33.8	33	60	60	58.5	4	4.6	3.4
(M 33)	36	36.8	36	66	66	64.5	5	6.0	4
M 36	39	39.8	39	72	72	70.5	5	6.0	4
(M 39)	42	42.8	42	77	77	75.5	6	7	5
EXTRA LARGE DIAMETER SIZES (Form G)									
M 5	5.5	5.8	5.5	15	15	14.2	1.6	1.9	1.3
M 6	6.6	7.0	6.6	18	18	17.2	2	2.3	1.7
(M 7)	7.6	8.0	7.6	21	21	20.2	2	2.3	1.7
M 8	9	9.4	9.0	24	24	23.2	2	2.3	1.7
M 10	11	11.5	11.0	30	30	29.2	2.5	2.8	2.2
M 12	14	14.5	14.0	36	36	34.8	3	3.6	2.4
(M 14)	16	16.5	16.0	42	42	40.8	3	3.6	2.4
M 16	18	18.5	18	48	48	46.8	4	4.6	3.4
(M 18)	20	20.6	20	54	54	52.5	4	4.6	3.4
M 20	22	22.6	22	60	60	58.5	5	6.0	4
(M 22)	24	24.6	24	66	66	64.5	5	6.0	4
M 24	26	26.6	26	72	72	70.5	6	7	5
(M 27)	30	30.6	30	81	81	79	6	7	5
M 30	33	33.8	33	90	90	88	8	9.2	6.8
(M 33)	36	36.8	36	99	99	97	8	9.2	6.8
M 36	39	39.8	39	108	108	106	10	11.2	8.8
(M 39)	42	42.8	42	117	117	115	10	11.2	8.8

All dimensions are given in millimeters.
Nominal bolt or screw sizes shown in parentheses are non-preferred.

Section L

Tables— *Page*

Table L1. BRITISH STANDARD WHITWORTH (B.S.W.)
Tapping Drill Sizes and Percentage Depth of Thread

Nominal Size	
1/8	3/32 in. (97); 2·4 mm. (95); 2·5 mm. (83); 2·55 mm. (77); 2·6 mm. (71)
3/16	3·4 mm. (100); 3·5 mm. (93); 3·6 mm. (86); 3·7 mm. (78); 3·8 mm. (71)
1/4	3/16 in. (98); 4·9 mm. (89): 5·0 mm. (83); 5·1 mm. (77); 13/64 in. (73)
5/16	6·2 mm. (96); 1/4 in. (88); 6·4 mm. (85); 6·5 mm. (79)
3/8	7·5 mm. (100); 7·7 mm. (90); 7·9 mm. (80); 5/16 in. (78)
7/16	8·8 mm. (100); 9·0 mm. (91); 9·2 mm. (82); 9·3 mm. (78)
1/2	10·0 mm. (100); 10·3 mm. (89); 13/32 in. (88); 10·4 mm. (85); 10·5 mm. (81)
9/16	11·6 mm. (99); 11·8 mm. (92); 15/32 in. (88); 12·0 mm. (84); 12·1 mm. (81)
5/8	13 mm. (97); 33/64 in. (94); 17/32 in. (80); 13·5 mm. (80)
11/16	37/64 in. (94); 15 mm. (83); 19/32 in. (80)
3/4	5/8 in. (98); 16 mm. (94); 41/64 in. (86); 16·25 mm. (86)
7/8	47/64 in. (99); 19 mm. (89); 3/4 in. (88); 19·25 mm. (82)
1	27/32 in. (98); 21·5 mm. (96); 55/64 in. (88); 22 mm. (84)
1 1/8	24 mm. (98); 61/64 in. (94); 24·5 mm. (88); 31/32 in. (85); 24·75 mm. (82)
1 1/4	1 5/64 in. (94); 27·5 mm. (91); 1 3/32 in. (85); 28 mm. (81)
1 1/2	1 19/64 in. (95); 33 mm. (94); 1 5/16 in. (88); 33·5 mm. (85)
1 3/4	38 mm. (99); 1 1/2 in. (97); 1 33/64 in. (91); 1 17/32 in. (85); 39·0 mm. (84)
2	1 23/32 in. (99); 44 mm. (94); 1 3/4 in. (88); 44·5 mm. (87)

Table L2. BRITISH STANDARD FINE (B.S.F.)

Tapping Drill Sizes and Percentage Depth of Thread

Nominal Size	
$\frac{3}{16}$	3·8 mm. (95); 3·9 mm. (85); $\frac{5}{32}$ in. (78); 4 mm. (75)
$\frac{7}{32}$	4·4 mm. (100); 4·5 mm. (91); 4·6 mm. (82); 4·7 mm. (74)
$\frac{1}{4}$	$\frac{13}{64}$ in. (95); 5·2 mm. (92); 5·3 mm. (84); 5·4 mm. (76)
$\frac{9}{32}$	$\frac{15}{64}$ in. (95); 6 mm. (91); 6·1 mm. (84); 6·2 mm. (76)
$\frac{5}{16}$	6·5 mm. (97); 6·6 mm. (91); $\frac{17}{64}$ in. (81); 6·8 mm. (77)
$\frac{3}{8}$	$\frac{5}{16}$ in. (97); 8·0 mm. (94); 8·1 mm. (88); 8·2 mm. (82); 8·3 mm. (75)
$\frac{7}{16}$	9·4 mm. (95); 9·5 mm. (89); $\frac{3}{8}$ in. (88); 9·6 mm. (83); 9·7 mm. (78)
$\frac{1}{2}$	$\frac{27}{64}$ in. (98); 10·9 mm. (89); 11·0 mm. (84); $\frac{7}{16}$ in. (78)
$\frac{9}{16}$	$\frac{31}{64}$ in. (98); 12·5 mm. (88); 12·6 mm. (83); $\frac{1}{2}$ in. (78)
$\frac{5}{8}$	$\frac{35}{64}$ in. (85); 14·0 mm. (81)
$\frac{11}{16}$	$\frac{39}{64}$ in. (85); 15·5 mm. (85)
$\frac{3}{4}$	16·5 mm. (94); $\frac{21}{32}$ in. (88); 16·75 mm. (85)
$\frac{13}{16}$	18 mm. (97); $\frac{23}{32}$ in. (88); 18·5 mm. (79)
$\frac{7}{8}$	$\frac{49}{64}$ in. (94); 19·5 mm. (92); 19·75 mm. (84); $\frac{25}{32}$ in. (80)
1	$\frac{7}{8}$ in. (98); 22·5 mm. (89); $\frac{57}{64}$ in. (86); 22·75 mm. (82)
$1\frac{1}{8}$	25·0 mm. (99); $\frac{63}{64}$ in. (99); 1 in. (88); 25·5 mm. (85)
$1\frac{1}{4}$	$1\frac{7}{64}$ in. (99); 28·5 mm. (90); $1\frac{1}{8}$ in. (88)
$1\frac{3}{8}$	$1\frac{7}{32}$ in. (98); 31·0 mm. (97); $1\frac{15}{64}$ in. (88); 31·5 mm. (84)
$1\frac{1}{2}$	$1\frac{11}{32}$ in. (98); 34·5 mm. (89); $1\frac{23}{64}$ in. (88)

Table L3. BRITISH STANDARD ELECTRICAL CONDUIT THREAD

Tapping Drill Sizes and Percentage Depth of Thread

Nominal Size (ins.)	
$\frac{1}{2}$	10·9 mm. (99); 11·0 mm. (94); $\frac{7}{16}$ in. (88); 11·3 mm. (78); 11·4 mm. (72); $\frac{29}{64}$ in. (66); 11·6 mm. (61)
$\frac{5}{8}$	$\frac{9}{16}$ in. (88); 14·5 mm. (76); $\frac{37}{64}$ in. (66)
$\frac{3}{4}$	$\frac{43}{64}$ in. (98); $\frac{11}{16}$ in. (78); 17·5 mm. (76); $\frac{45}{64}$ in. (59); 18·0 mm. (52)
1	$\frac{59}{64}$ in. (98); 23·5 mm. (93); $\frac{15}{16}$ in. (78); 24·0 mm. (69); $\frac{61}{64}$ in. (59)
$1\frac{1}{4}$	$1\frac{11}{64}$ in. (98); 30·0 mm. (86); $1\frac{3}{16}$ in. (78); 30·5 mm. (61); $1\frac{13}{64}$ in. (59)
$1\frac{1}{2}$	36 mm. (90); $1\frac{27}{64}$ in. (85); 36·5 mm. (69); $1\frac{7}{16}$ in. (68); $1\frac{29}{64}$ in. (51)
2	48·5 mm. (99); $1\frac{59}{64}$ in. (84); 49 mm. (77); $1\frac{15}{16}$ in. (68); 49·5 mm. (56); $1\frac{61}{64}$ in. (51)
$2\frac{1}{2}$	$2\frac{27}{64}$ in. (84); $2\frac{7}{16}$ in. (68); 62·0 mm. (65); $2\frac{29}{64}$ in. (51)

Table L4. BRITISH STANDARD PARALLEL PIPE (B.S.P.)
Tapping Drill Sizes and Percentage Depth of Thread

Nominal Size	
$\frac{1}{8}$	8·6 mm. (97); $\frac{11}{32}$ in. (86); 8·8 mm. (80)
$\frac{1}{4}$	11·5 mm. (97); $\frac{29}{64}$ in. (96); 11·7 mm. (85); 11·8 mm. (80)
$\frac{3}{8}$	15·0 mm. (97); $\frac{19}{32}$ in. (92); 15·25 mm. (83)
$\frac{1}{2}$	$\frac{47}{64}$ in. (99); 19 mm. (84); $\frac{3}{4}$ in. (82)
$\frac{5}{8}$	$\frac{13}{16}$ in. (98); 21·0 mm. (82); $\frac{53}{64}$ in. (81)
$\frac{3}{4}$	$\frac{61}{64}$ in. (96); 24·25 mm. (94); 24·5 mm. (84); $\frac{31}{32}$ in. (79)
$\frac{7}{8}$	28·0 mm. (95); 1$\frac{7}{64}$ in. (87); 28·25 mm. (84)
1	30·5 mm. (92); 1$\frac{13}{64}$ in. (91); 30·75 mm. (85)
1$\frac{1}{4}$	39·0 mm. (99); 1$\frac{35}{64}$ in. (88); 39·5 mm. (82)
1$\frac{1}{2}$	45·0 mm. (95); 1$\frac{25}{32}$ in. (87)

Table L5. BRITISH ASSOCIATION (B.A.)
Tapping Drill Sizes and Percentage Depth of Thread

B.A. No.	
0	4·8 mm. (100); 4·9 mm. (92); 5·0 mm. (84); 5·1 mm. (75)
1	4·2 mm. (100); 4·3 mm. (92); 4·4 mm. (83); 4·5 mm. (74)
2	3·8 mm. (93); 3·9 mm. (83); $\frac{5}{32}$ in. (75); 4·0 mm. (72)
3	3·2 mm. (100); 3·3 mm. (91); 3·4 mm. (80)
4	2·85 mm. (95); 2·9 mm. (89); 2·95 mm. (82); 3·00 mm. (76)
5	2·50 mm. (99); 2·55 mm. (92); 2·60 mm. (84); 2·65 mm. (78)
6	2·15 mm. (100); 2·20 mm. (94); 2·25 mm. (86); 2·30 mm. (78)
7	1·95 mm. (95); 2·00 mm. (86); 2·05 mm. (78)
8	1·70 mm. (96); 1·75 mm. (87); 1·80 mm. (77)
9	1·45 mm. (96); 1·5 mm. (85); 1·55 mm. (74)
10	1·30 mm. (95); 1·35 mm. (83); 1·40 mm. (71)
11	1·15 mm. (95); $\frac{3}{64}$ in. (84); 1·20 mm. (82)
12	0·98 mm. (94); 1·00 mm. (88); 1·05 mm. (74)
13	0·90 mm. (100); 0·92 mm. (93); 0·95 mm. (83); 0·98 mm. (73)
14	0·72 mm. (100); 0·75 mm. (89); 0·78 mm. (79); $\frac{1}{32}$ in. (75)
15	0·65 mm. (100); 0·68 mm. (88); 0·70 mm. (81)

Table L6. **BRITISH STANDARD CYCLE (B.S.C.)**

Tapping Drill Sizes and Percentage Depth of Thread

Nominal Size	
(ins.)	
0·0825	1·60 mm. (100); 1·65 mm. (92); 1·70 mm. (82); 1·75 mm. (72); 1·8 mm. (61)
0·0905	1·80 mm. (100); 1·85 mm. (93); 1·90 mm. (83); 1·95 mm. (72); $\frac{5}{64}$ in. (65)
0·1025	2·10 mm. (100); 2·15 mm. (94); 2·20 mm. (84); 2·25 mm. (73); 2·30 mm. (63)
0·1145	2·45 mm. (95); 2·50 mm. (85); 2·55 mm. (74); 2·60 mm. (64)
$\frac{1}{8}$	2·50 mm. (100); 2·55 mm. (93); 2·60 mm. (85); 2·65 mm. (78); 2·70 mm. (70); 2·75 mm. (63)
0·1291	2·70 mm. (94); 2·75 mm. (86); $\frac{7}{64}$ in. (81); 2·80 mm. (78); 2·85 mm. (70); 2·90 mm. (62)
0·1423	2·95 mm. (99); 3·00 mm. (91); 3·10 mm. (76); $\frac{1}{8}$ in. (65)
$\frac{5}{32}$	$\frac{1}{8}$ in. (94); 3·20 mm. (91); 3·30 mm. (80); 3·40 mm. (68)
0·1583	3·40 mm. (92); 3·50 mm. (78); $\frac{9}{64}$ in. (67); 3·60 mm. (62)
0·1776	3·70 mm. (96); 3·80 mm. (84); 3·90 mm. (73); $\frac{5}{32}$ in. (64); 4·00 mm. (61)
$\frac{3}{16}$	$\frac{5}{32}$ in. (94); 4·00 mm. (90); 4·10 mm. (79); 4·2 mm. (67)
$\frac{7}{32}$	4·60 mm. (92); 4·70 mm. (82); $\frac{3}{16}$ in. (76); 4·80 mm. (73); 4·90 mm. (63)
$\frac{1}{4}$	5·30 mm. (100); 5·40 mm. (91); 5·50 mm. (82); $\frac{7}{32}$ in. (76); 5·6 mm. (72); 5·7 mm. (62)
$\frac{17}{64}$	5·70 mm. (100); 5·80 mm. (91); 5·90 mm. (81); $\frac{15}{64}$ in. (76); 6·00 mm. (72); 6·10 mm. (62)
$\frac{9}{32}$	6·10 mm. (100); 6·20 mm. (91); 6·30 mm. (81); $\frac{1}{4}$ in. (76); 6·40 mm. (72); 6·50 mm. (62)
$\frac{5}{16}$	6·90 mm. (99); 7·00 mm. (90); 7·10 mm. (81); $\frac{9}{32}$ in. (76); 7·20 mm. (71); 7·30 mm. (61)
$\frac{3}{8}$	8·50 mm. (99); 8·60 mm. (89); 8·70 mm. (79); $\frac{11}{32}$ in. (76); 8·80 mm. (70); 8·90 mm. (60)
$\frac{7}{16}$	10·10 mm. (97); 10·20 mm. (88); 10·30 mm. (78); $\frac{13}{32}$ in. (76); 10·40 mm. (69); 10·50 mm. (59)
$\frac{1}{2}$	11·7 mm. (96); 11·8 mm. (86); 11·9 mm. (77); $\frac{15}{32}$ in. (76); 12·0 mm. (67); 12·1 mm. (58)
$\frac{9}{16}$	$\frac{17}{32}$ in. (76); 13·5 mm. (75)
$\frac{5}{8}$	15·0 mm. (84); $\frac{19}{32}$ in. (76)
$\frac{11}{16}$	16·5 mm. (92); $\frac{21}{32}$ in. (76)
$\frac{3}{4}$	$\frac{23}{32}$ in. (76); 18·5 mm. (53)
$\frac{7}{8}$	$\frac{27}{32}$ in. (70); 21·5 mm. (65)
$\frac{31}{32}$	$\frac{15}{16}$ in. (88); 24·0 mm. (67)
1	24·5 mm. (80); $\frac{31}{32}$ in. (70)
$1\frac{1}{8}$	$1\frac{3}{32}$ in. (76); 28·0 mm. (55)
1·290	$1\frac{1}{4}$ in. (90); 32·0 mm. (68); $1\frac{17}{64}$ in. (55)
1·370	$1\frac{21}{64}$ in. (94); 34·0 mm. (71); $1\frac{11}{32}$ in. (57)
1·450	36·0 mm. (80); $1\frac{27}{64}$ in. (68)
$1\frac{9}{16}$	$1\frac{17}{32}$ in. (71); 39·0 mm. (60)
$1\frac{5}{8}$	$1\frac{19}{32}$ in. (71); 40·5 mm. (69); $1\frac{39}{64}$ in. (57)

Table L7. BRITISH STANDARD METRIC SERIES

Tapping Drill Sizes and Percentage Depth of Thread

Nominal Size (mm.)	
6	4·9 mm. (97); 5·0 mm. (88); 5·1 mm. (79); $\frac{13}{64}$ in. (74); 5·2 mm. (70); 5·3 mm. (62)
7	5·9 mm. (97); $\frac{15}{64}$ in. (92); 6·0 mm. (88); 6·1 mm. (79); 6·2 mm. (70); 6·3 mm. (62)
8	6·6 mm. (99); 6·7 mm. (92); $\frac{17}{64}$ in. (88); 6·8 mm. (84); 6·9 mm. (77); 7·0 mm. (70); 7·1 mm. (63)
9	7·6 mm. (99); 7·7 mm. (92); 7·8 mm. (84); 7·9 mm. (77); $\frac{5}{16}$ in. (75); 8·0 mm. (70); 8·1 mm. (63)
10	8·3 mm. (100); $\frac{21}{64}$ in. (98); 8·4 mm. (94); 8·5 mm. (88); 8·6 mm. (82); 8·7 mm. (76); $\frac{11}{32}$ in. (74); 8·8 mm. (70); 8·9 mm. (65)
11	9·3 mm. (100); 9·4 mm. (94); 9·5 mm. (88); $\frac{3}{8}$ in. (87); 9·6 mm. (82); 9·7 mm. (76); 9·8 mm. (70); 9·9 mm. (65); $\frac{25}{64}$ in. (63)
12	10·0 mm. (100); 10·1 mm. (96); 10·2 mm. (91); 10·3 mm. or $\frac{13}{32}$ in. (85); 10·4 mm. (80); 10·5 mm. (76); 10·6 mm. (70); 10·7 mm. or $\frac{27}{64}$ in. (65)
14	11·8 mm. (97); $\frac{15}{32}$ in. (92); 12·0 mm. (88); 12·2 mm. (79); $\frac{31}{64}$ in. (75); $\frac{1}{2}$ in. (57)
16	$\frac{35}{64}$ in. (93); 14·0 mm. (88); $\frac{9}{16}$ in. (75); 14·5 mm. (66); $\frac{37}{64}$ in. (58)
18	$\frac{39}{64}$ in. (89); 15·5 mm. (88); $\frac{5}{8}$ in. (75); 16·0 mm. (70); $\frac{41}{64}$ in. (61); 16·5 mm. (53)
20	$\frac{11}{16}$ in. (89); 17·5 mm. (88); $\frac{45}{64}$ in. (75); 18·0 mm. (70); $\frac{23}{32}$ in. (61); 18·5 mm. (53)
22	$\frac{49}{64}$ in. (90); 19·5 mm. (88); $\frac{25}{32}$ in. (76); 20·0 mm. (70); $\frac{51}{64}$ in. (62); 20·5 mm. (53)
24	$\frac{13}{16}$ in. (99); 21·0 mm. (88); $\frac{53}{64}$ in. (87); $\frac{27}{32}$ in. (75); 21·5 mm. (73); $\frac{55}{64}$ in. (65)
27	$\frac{15}{16}$ in. (93); 24·0 mm. (88); $\frac{61}{64}$ in. (82); 24·5 mm. (73); $\frac{31}{32}$ in. (70); 25·0 mm. (59)
30	$1\frac{1}{32}$ in. (96); 26·5 mm. (88); $1\frac{1}{16}$ in. (76); $1\frac{5}{64}$ in. (66); $1\frac{3}{32}$ in. (56); 28·0 mm. (50)
33	$1\frac{5}{32}$ in. (92); 29·5 mm. (88); 30 mm. (75); $1\frac{3}{16}$ in. (65); 30·5 mm. (63); $1\frac{7}{32}$ in. (51)
36	31·5 mm. (99); $1\frac{1}{4}$ in. (93); 32 mm. (88); 32·5 mm. (83); $1\frac{9}{32}$ in. (76); 33 mm. (66)
39	34·5 mm. (99); $1\frac{3}{8}$ in. (90); 35·0 mm. (88); $1\frac{13}{32}$ in. (72); 36·5 mm. (55)
42	$1\frac{29}{64}$ in. (99); $1\frac{15}{32}$ in. (92); $1\frac{31}{64}$ in. (84); $1\frac{1}{2}$ in. (76); $1\frac{33}{64}$ in. (68)
45	$1\frac{37}{64}$ in. (96); $1\frac{19}{32}$ in. (88); $1\frac{5}{8}$ in. (73); $1\frac{21}{32}$ in. (57)
48	$1\frac{43}{64}$ in. (97); $1\frac{11}{16}$ in. (90); $1\frac{23}{32}$ in. (76); $1\frac{3}{4}$ in. (62); $1\frac{49}{64}$ in. (56)
52	$1\frac{53}{64}$ in. (98); $1\frac{27}{32}$ in. (91); $1\frac{55}{64}$ in. (84); $1\frac{7}{8}$ in. (77); $1\frac{29}{32}$ in. (63)
56	$1\frac{31}{32}$ in. (96); 2 in. (83): $2\frac{1}{32}$ in. (71); $2\frac{1}{16}$ in. (58); $2\frac{5}{64}$ in. (51)
60	$2\frac{1}{8}$ in. (96); $2\frac{5}{32}$ in. (84); $2\frac{3}{16}$ in. (71); $2\frac{7}{32}$ in. (58)

able L8. FRENCH, SWISS AND ORIGINAL S.I. METRIC SERIES
Tapping Drill Sizes and Percentage Depth of Thread

Nominal Size	
(mm.)	
6	4·6 mm. (99); 4·7 mm. (92); $\frac{3}{16}$ in. (88); 4·8 mm. (85); 4·9 mm. (78); 5·0 mm. (71); 5·1 mm. (64); $\frac{13}{64}$ in. (60); 5·2 mm. (57)
7	5·6 mm. (99); 5·7 mm. (92); 5·8 mm. (85); 5·9 mm. (78); $\frac{15}{64}$ in. (74); 6·0 mm. (71); 6·1 mm. (64); 6·2 mm. (57)
8	$\frac{1}{4}$ in. (94); 6·4 mm. (91); 6·5 mm. (85); 6·6 mm. (80); 6·7 mm. (74); $\frac{17}{64}$ in. (71); 6·8 mm. (68); 6·9 mm. (63); 7·0 mm. (57)
9	7·4 mm. (91); 7·5 mm. (85); $\frac{19}{64}$ in. (83); 7·6 mm. (80); 7·7 mm. (74); 7·8 mm. (68); 7·9 mm. (63); $\frac{5}{16}$ in. (60); 8·0 mm. (57)
10	$\frac{5}{16}$ in. (98); 8·0 mm. (95); 8·1 mm. (90); 8·2 mm. (85); 8·3 mm. (80); $\frac{21}{64}$ in. (79); 8·4 mm. (76); 8·5 mm. (71); 8·6 mm. (66); 8·7 mm. (62); $\frac{11}{32}$ in. (60)
11	9·0 mm. (95); 9·1 mm. (90); $\frac{23}{64}$ in. (89); 9·2 mm. (85); 9·3 mm (80); 9·4 mm. (76); 9·5 mm. (71); $\frac{3}{8}$ in. (70); 9·6 mm. (66); 9·7 mm. (62); 9·8 mm. (57)
12	$\frac{3}{8}$ in. (99); 9·6 mm. (98); 9·7 mm. (94); 9·8 mm. (89); 9·9 mm. (85); $\frac{25}{64}$ in. (84); 10·0 mm. (81); 10·1 mm. (77); 10·2 mm. (73); 10·3 mm. (69); $\frac{13}{32}$ in. (68); 10·4 mm. (65); 10·5 mm. (61)
14	11·6 mm. (85); 11·8 mm. (78); $\frac{15}{32}$ in. (74); 12·0 mm. (71); 12·2 mm. (64); $\frac{31}{64}$ in (60)
16	13·5 mm. (89); $\frac{35}{64}$ in. (75); 14·0 mm. (71); $\frac{9}{16}$ in. (60); 14·5 mm. (53)
18	15·0 mm. (85); $\frac{39}{64}$ in. (72); 15·5 mm. (71); $\frac{5}{8}$ in. (60); 16·0 mm. (56)
20	17·0 mm. (85); $\frac{11}{16}$ in. (72); 17·5 mm. (71); $\frac{45}{64}$ in. (60); 18·0 mm. (56)
22	19·0 mm. (85); $\frac{49}{64}$ in. (73); 19·5 mm. (71); $\frac{25}{32}$ in. (61); 20·0 mm. (56); $\frac{51}{64}$ in. (50)
24	20·5 mm. (83); $\frac{13}{16}$ in. (80); 21·0 mm. (71); $\frac{53}{64}$ in. (70); $\frac{27}{32}$ in. (60); 21·5 mm. (59); $\frac{55}{64}$ in. (52)
27	$\frac{59}{64}$ in. (85); $\frac{15}{16}$ in. (75); 24·0 mm. (71); $\frac{61}{64}$ in. (66); 24·5 mm. (59); $\frac{31}{32}$ in. (56)
30	25 mm. (91); $1\frac{1}{32}$ in. (77); 26·5 mm. (71); $1\frac{1}{16}$ in. (61); $1\frac{5}{64}$ in. (53)
33	$1\frac{1}{8}$ in. (90); $1\frac{5}{32}$ in. (74); 29·5 mm. (71); 30 mm. (60); $1\frac{3}{16}$ in. (52); 30·5 mm. (51)
36	31 mm. (89); 31·5 mm. (80); $1\frac{1}{4}$ in. (75); 32 mm. (71); 32·5 mm. (67); $1\frac{9}{32}$ in. (61); 33 mm. (53)
39	$1\frac{11}{32}$ in. (87); 34·5 mm. (80); $1\frac{3}{8}$ in. (73); 35 mm. (71); $1\frac{13}{32}$ in. (58)
42	36·5 mm. (87); $1\frac{29}{64}$ in. (80); $1\frac{15}{32}$ in. (74); $1\frac{31}{64}$ in. (68); $1\frac{1}{2}$ in. (61); $1\frac{33}{64}$ in. (53)
45	39·5 mm. (87); $1\frac{37}{64}$ in. (77); $1\frac{19}{32}$ in. (71); $1\frac{5}{8}$ in. (59)
48	42 mm. (85); $1\frac{43}{64}$ in. (78); $1\frac{11}{16}$ in. (73); $1\frac{23}{32}$ in. (61); $1\frac{3}{4}$ in. (50)
52	46 mm. (85); $1\frac{53}{64}$ in. (79); $1\frac{27}{32}$ in. (74); $1\frac{55}{64}$ in. (68); $1\frac{7}{8}$ in. (62); $1\frac{29}{32}$ in. (51)
56	49·5 mm. (84); $1\frac{31}{32}$ in. (77); 2 in. (67); $2\frac{1}{32}$ in. (57)
60	$2\frac{7}{64}$ in. (83); $2\frac{1}{8}$ in. (77); $2\frac{5}{32}$ in. (68); $2\frac{3}{16}$ in. (57)

Table L9. GERMAN METRIC SERIES

Tapping Drill Sizes and Percentage Depth of Thread

Nominal Size	
(mm.)	
6	4·7 mm. (100); $\frac{3}{16}$ in. (95); 4·8 mm. (92); 4·9 mm. (85); 5·0 mm. (7 5·1 mm. (69); $\frac{13}{64}$ in. (65); 5·2 mm. (62)
7	5·7 mm. (100); 5·8 mm. (92); 5·9 mm. (85); $\frac{15}{64}$ in. (81); 6·0 mm. (7 6·1 mm. (69); 6·2 mm. (62)
8	6·4 mm. (99); 6·5 mm. (92); 6·6 mm. (86); 6·7 mm. (80); $\frac{17}{64}$ in. (7 6·8 mm. (74); 6·9 mm. (68); 7·0 mm. (62)
9	7·4 mm. (99); 7·5 mm. (92); $\frac{19}{64}$ in. (90); 7·6 mm. (86); 7·7 mm. (8 7·9 mm. (68); $\frac{5}{16}$ in. (66); 8·0 mm. (62)
10	8·1 mm. (98); 8·2 mm. (93); 8·3 mm. (87); $\frac{21}{64}$ in. (86); 8·4 mm. (8 8·5 mm. (77); 8·6 mm. (72); 8·7 mm. (67); $\frac{11}{32}$ in. (65); 8·8 mm. (6
11	9·1 mm. (89); $\frac{23}{64}$ in. (96); 9·2 mm. (93); 9·3 mm. (87); 9·4 mm. (8 9·5 mm. (77); $\frac{3}{8}$ in. (76); 9·6 mm. (72); 9·7 mm. (67); 9·8 mm. (
12	9·8 mm. (97); 9·9 mm. (92); $\frac{25}{64}$ in. (91); 10·0 mm. (88); 10·1 mm. (8 10·2 mm. (79); 10·3 mm. (75); $\frac{13}{32}$ in. (74); 10·4 mm. (70); 10·5 m (66); 10·6 mm. (62)
14	11·8 mm. (85); $\frac{15}{32}$ in. (81); 12·0 mm. (77); 12·2 mm. (69); $\frac{31}{64}$ in. (6 $\frac{1}{2}$ in. (50)
16	$\frac{35}{64}$ in. (82); 14·0 mm. (77); $\frac{9}{16}$ in. (66); 14·5 mm. (58); $\frac{37}{64}$ in. (51)
18	$\frac{39}{64}$ in. (78); 15·5 mm. (77); $\frac{5}{8}$ in. (66); 16·0 mm. (62); $\frac{41}{64}$ in. (54)
20	$\frac{11}{16}$ in. (78); 17·5 mm. (77); $\frac{45}{64}$ in. (66); 18·0 mm. (62); $\frac{23}{32}$ in. (54)
22	$\frac{49}{64}$ in. (79); 19·5 mm. (77); $\frac{25}{32}$ in. (67); 20·0 mm. (62); $\frac{51}{64}$ in. (55)
24	$\frac{13}{16}$ in. (87); 21·0 mm. (77); $\frac{53}{64}$ in. (76); $\frac{27}{32}$ in. (66); 21·5 mm. (64 $\frac{55}{64}$ in. (57)
27	$\frac{15}{16}$ in. (82); 24·0 mm. (77); $\frac{61}{64}$ in. (72); 24·5 mm. (64); $\frac{31}{32}$ in. (62 25·0 mm. (52)
30	$1\frac{1}{32}$ in. (85); 26·5 mm. (77); $1\frac{1}{16}$ in. (67); $1\frac{5}{64}$ in. (58)
33	$1\frac{5}{32}$ in. (81); 29·5 mm. (77); 30 mm. (66); $1\frac{3}{16}$ in. (57); 30·5 mm. (5
36	31·5 mm. (87); $1\frac{1}{4}$ in. (82); 32 mm. (77); 32·5 mm. (73); $1\frac{9}{32}$ in. (67 33 mm. (58)
39	34·5 mm. (87); $1\frac{3}{8}$ in. (79); 35·0 mm. (77); $1\frac{13}{32}$ in. (63)
42	$1\frac{29}{64}$ in. (87); $1\frac{15}{32}$ in. (81); $1\frac{31}{64}$ in. (74); $1\frac{1}{2}$ in. (67); $1\frac{33}{64}$ in. (60)
45	$1\frac{37}{64}$ in. (85); $1\frac{19}{32}$ in. (77); $1\frac{5}{8}$ in. (64); $1\frac{21}{32}$ in. (50)
48	$1\frac{43}{64}$ in. (85); $1\frac{11}{16}$ in. (79); $1\frac{23}{32}$ in. (67); $1\frac{3}{4}$ in. (55)
52	$1\frac{53}{64}$ in. (86); $1\frac{27}{32}$ in. (80); $1\frac{55}{64}$ in. (74); $1\frac{7}{8}$ in. (68); $1\frac{29}{32}$ in. (55)
56	$1\frac{31}{64}$ in. (85); 2 in. (73); $2\frac{1}{32}$ in. (62); $2\frac{1}{16}$ in. (51)
60	$2\frac{1}{8}$ in. (85); $2\frac{5}{32}$ in. (74); $2\frac{3}{16}$ in. (62); $2\frac{7}{32}$ in. (51)

Table L10.

UNIFIED AND AMERICAN NATIONAL COARSE

Tapping Drill Sizes and Percentage Basic Depth of Thread

Nominal Size	
No.	
1	1·4 mm. (93); 1·45 mm. (83); 1·5 mm. (73); 1·55 mm. (63)
2	$\frac{1}{16}$ in. or 1·6 mm. (100); 1·65 mm. (95); 1·7 mm. (86); 1·75 mm. (77); 1·8 mm. (68); 1·85 mm. (59)
3	1·85 mm. (100); 1·9 mm. (94); 1·95 mm. (86); $\frac{5}{64}$ in. (82); 2 mm. (78); 2·05 mm. (70); 2·1 mm. (63)
4	2·1 mm. (95); 2·15 mm. (88); 2·2 mm. (82); 2·25 mm. (75); 2·3 mm. (69); 2·35 mm. (62); $\frac{3}{32}$ in. (59)
5	$\frac{3}{32}$ in. (100); 2·4 mm. (98); 2·45 mm. (92); 2·5 mm. (88); 2·55 mm. (82); 2·6 mm. (75); 2·65 mm. (69); 2·7 mm. (62)
6	2·55 mm. (98); 2·6 mm. (93); 2·65 mm. (88); 2·7 mm. (83); 2·75 mm. (77); $\frac{7}{64}$ in. (75); 2·8 mm. (72); 2·85 mm. (67); 2·9 mm. (62)
8	3·2 mm. (99); 3·3 mm. (89); 3·4 mm. (78); 3·5 mm. (68); $\frac{9}{64}$ in. (61); 3·6 mm. (58)
10	3·6 mm. (95); 3·7 mm. (87); 3·8 mm. (79); 3·9 mm. (71); $\frac{5}{32}$ in. (66); 4 mm. (64)
12	4·2 mm. (99); 4.3 mm. (91); $\frac{11}{64}$ in. (86); 4·4 mm. (84); 4·5 mm. (76); 4·6 mm. (68); 4·7 mm. (61)
In.	
$\frac{1}{4}$	$\frac{3}{16}$ in. (100); 4·8 mm. (99); 4·9 mm. (93); 5·0 mm. (87); 5·1 mm. (80); $\frac{13}{64}$ in. (76); 5·2 mm. (74); 5·3 mm. (67); 5·4 mm. (61)
$\frac{5}{16}$	6·2 mm. (100); $\frac{1}{4}$ in. (92); 6·3 mm. (94); 6·4 mm. (89); 6·5 mm. (82); 6·6 mm. (77); 6·7 mm. (71); $\frac{17}{64}$ in. (69); 6·8 mm. (66); 6·9 mm. (60)
$\frac{3}{8}$	7·6 mm. (99); 7·7 mm. (94); 7·8 mm. (89); 7·9 mm. (84); $\frac{5}{16}$ in. (82); 8 mm. (78); 8·1 mm. (73); 8·2 mm. (68); 8·3 mm. (63); $\frac{21}{64}$ in. (61)
$\frac{7}{16}$	8·9 mm. (100); 9 mm. (95); 9·1 mm. (90); $\frac{23}{64}$ in. (89); 9·2 mm. (86); 9·3 mm. (81); 9·4 mm. (77); 9·5 mm. (72); $\frac{3}{8}$ in. (71); 9·6 mm. (68); 9·7 mm. (64)
$\frac{1}{2}$	$\frac{13}{32}$ in. (99); 10·4 mm. (96); 10·5 mm. (92); 10·6 mm. (88); 10·7 mm. or $\frac{27}{64}$ in. (83); 10·8 mm. (79); 10·9 mm. (75); 11 mm. (71); 11·1 mm. (67); 11·2 mm. (63)
$\frac{9}{16}$	11·7 mm. (100); $\frac{15}{32}$ in. or 11·9 mm. (92); 12·1 mm. (84); 12·3 mm. (77); 12·5 mm. (69); $\frac{1}{2}$ in. or 12·7 mm. (61)
$\frac{5}{8}$	$\frac{33}{64}$ in. or 13·1 mm. (98); 13·3 mm. (91); $\frac{17}{32}$ in. or 13·5 mm. (84); 13·7 mm. (77); $\frac{35}{64}$ in. or 13·9 mm. (70)
$\frac{3}{4}$	16 mm. (98); $\frac{41}{64}$ in. (90); 16·5 mm. (82); $\frac{21}{32}$ in. (77); 17 mm. (66); $\frac{43}{64}$ in. (64)
$\frac{7}{8}$	19 mm. (93); $\frac{3}{4}$ in. (92); $\frac{49}{64}$ in. (80); 19·5 mm. (79); $\frac{25}{32}$ in. (69); 20 mm. (64)

[*Continued on page* 298

Table L10
continued.

UNIFIED AND AMERICAN NATIONAL COARSE
Tapping Drill Sizes and Percentage Basic Depth of Thread

Nominal Size	
No.	
1	21·5 mm. (100); $\frac{55}{64}$ in. (92); 22 mm. (87); $\frac{7}{8}$ in. (82); 22·5 mm. (75 $\frac{57}{64}$ in. (71); 23 mm. (62)
$1\frac{1}{8}$	$\frac{61}{64}$ in. (97); 24·5 mm. (94); $\frac{31}{32}$ in. (89); $\frac{63}{64}$ in. or 25 mm. (80); 1 in. (71 25·5 mm. (69); $1\frac{1}{64}$ in. (61)
$1\frac{1}{4}$	$1\frac{5}{64}$ in. (98); 27·5 mm. (96); $1\frac{3}{32}$ in. (89); $1\frac{7}{64}$ in. (80); 28·5 mm. (73 $1\frac{1}{8}$ in. (71); 29 mm. (62)
$1\frac{3}{8}$	$1\frac{11}{64}$ in. (99); 30 mm. (95); $1\frac{3}{16}$ in. (92); 30·5 mm. (85); $1\frac{13}{64}$ in. (84 $1\frac{7}{32}$ in. (77); 31·0 mm. (76); $1\frac{15}{64}$ in. (69); 31·5 mm. (66)
$1\frac{1}{2}$	$1\frac{19}{64}$ in. (99); 33 mm. (98); $1\frac{5}{16}$ in. (92); 33·5 mm. (89); $1\frac{21}{64}$ in. (84 34 mm. (79); $1\frac{11}{32}$ in. (77); 34·5 mm. (69); $1\frac{3}{8}$ in. (61)
$1\frac{3}{4}$	$1\frac{33}{64}$ in. or 38·5 mm. (96); $1\frac{17}{32}$ in. (90); 39 mm. (87); $1\frac{9}{16}$ in. (76); 40 mn (71); $1\frac{19}{32}$ in. or 40·5 mm. (64)
2	44 mm. (98); $1\frac{3}{4}$ in. (92); 45 mm. (84); $1\frac{25}{32}$ in. (80); 45·5 mm. (77 $1\frac{13}{16}$ in. or 46 mm. (69); 46·5 mm. (62)

Table L11.

UNIFIED AND AMERICAN NATIONAL FINE

Tapping Drill Sizes and Percentage Basic Depth of Thread

Nominal Size	
No.	
0	1·15 mm. (96); $\frac{3}{64}$ in. (85); 1·2 mm. (83); 1·25 mm. (70); 1·3 mm. (57)
1	1·45 mm. (94); 1.5 mm. (82); 1.55 mm. (71); $\frac{1}{16}$ in. (62); 1·6 mm. (59)
2	1·7 mm. (100); 1·75 mm. (89); 1·8 mm. (79); 1·85 mm. (69); 1·9 mm. (58)
3	1·95 mm. (100); $\frac{5}{64}$ in. (95); 2 mm. (92); 2·05 mm. (83); 2·1 mm. (74); 2·15 mm. (65)
4	2·2 mm. (99); 2·25 mm. (92); 2·3 mm. (84); 2·35 mm. (76); $\frac{3}{32}$ in. (71); 2·4 mm. (68); 2·45 mm. (60)
5	2·5 mm. (96); 2·55 mm. (89); 2·6 mm. (82); 2·65 mm. (75); 2·7 mm. (68); 2·75 mm. (61)
6	2·75 mm. (98); $\frac{7}{64}$ in. (93); 2·8 mm. (91); 2·85 mm. (84); 2·9 mm. (78); 2·95 mm. (71); 3 mm. (65)
8	3·3 mm. (100); 3·4 mm. (88); 3·5 mm. (77); $\frac{9}{64}$ in. (69); 3·6 mm. (65)
10	3·9 mm. (95); $\frac{5}{32}$ in. (88); 4·0 mm. (85); 4·1 mm. (75); 4·2 mm. (64)
12	$\frac{11}{64}$ in. (100); 4·4 mm. (98); 4·5 mm. (89); 4·6 mm. (80); 4·7 mm. (71); $\frac{3}{16}$ in. (65); 4·8 mm. (62)
In.	
$\frac{1}{4}$	5·3 mm. (94); 5·4 mm. (85); 5·5 mm. (76); $\frac{7}{32}$ in. (71); 5·6 mm. (67); 5·7 mm. (58)
$\frac{5}{16}$	6·7 mm. (95); $\frac{17}{64}$ in. (91); 6·8 mm. (87); 6·9 mm. (80); 7·0 mm. (72); 7·1 mm. (64); $\frac{9}{32}$ in. (61)
$\frac{3}{8}$	8·3 mm. (94); $\frac{21}{64}$ in. (92); 8·4 mm. (87); 8·5 mm. (79); 8·6 mm. (71); 8·7 mm. (63); $\frac{11}{32}$ in. (61)
$\frac{7}{16}$	9·6 mm. (97); 9·7 mm. (91); 9·8 mm. (84); 9·9 mm. (78); $\frac{25}{64}$ in. (74); 10 mm. (71); 10·1 mm. (65); 10·2 mm. (59)
$\frac{1}{2}$	11·2 mm. (96); 11·3 mm. (90); 11·4 mm. (83); 11·5 mm. (77); $\frac{29}{64}$ in. (76); 11·6 mm. (71); 11·7 mm. (64)
$\frac{9}{16}$	12·6 mm. (97); $\frac{1}{2}$ in. or 12·7 mm. (92); 12·8 mm. (86); 12·9 mm. (80); 13 mm. (74); $\frac{33}{64}$ in. or 13·1 mm. (69); 13·2 mm. (63)
$\frac{5}{8}$	14·25 mm. (94); $\frac{9}{16}$ in. (92); 14·5 mm. (79); $\frac{37}{64}$ in. (69); 14·75 mm. (65)
$\frac{3}{4}$	$\frac{43}{64}$ in. (100); 17·25 mm. (92); $\frac{11}{16}$ in. (82); 17·5 mm. (80); 17·75 mm. (67); $\frac{45}{64}$ in. (61)
$\frac{7}{8}$	20 mm. (100); $\frac{51}{64}$ in. (89); 20·5 mm. (78); $\frac{13}{16}$ in. (71); 21 mm. (55)
1 ANF	$\frac{59}{64}$ in. (89); 23·5 mm. (85); $\frac{15}{16}$ in. (71); 24 mm. (63)
1 UNF	$\frac{29}{32}$ in. or 23 mm. (92); 23·5 mm. (73); $\frac{15}{16}$ in. (61)
$1\frac{1}{8}$	$1\frac{1}{32}$ in. (92); 26·5 mm. (80); $1\frac{3}{64}$ in. (77); $1\frac{1}{16}$ in. or 27 mm. (61)
$1\frac{1}{4}$	$1\frac{5}{32}$ in. (92); 29·5 mm. (87); $1\frac{11}{64}$ in. (77); 30 mm. (67); $1\frac{3}{16}$ in. (61)
$1\frac{3}{8}$	32·5 mm. (94); $1\frac{9}{32}$ in. (92); $1\frac{19}{64}$ in. (77); 33 mm. (74); $1\frac{5}{16}$ in. (61)
$1\frac{1}{2}$	35·5 mm. (100); $1\frac{13}{32}$ in. (92); 36 mm. (81); $1\frac{27}{64}$ in. (77); $1\frac{7}{16}$ in. or 36·5 mm. (61)

Table L12. ISO METRIC COARSE PITCH SERIES
Tapping Drill Sizes and Percentage Basic Depth of Thread

Nominal Size	
M1	0·72 mm. (91); 0·75 mm. (82); 0·78 mm. (72); 0·8 mm. (65)
M1·1	0·82 mm. (91); 0·85 mm. (82); 0·88 mm. (72); 0·9 mm. (65)
M1·2	0·92 mm. (91); 0·95 mm. (82); 0·98 mm. (72); 1 mm. (65)
M1·4	1·05 mm. (95); 1·1 mm. (82); 1·15 mm. (68)
M1·6	1·2 mm. (93); 1·25 mm. (82); 1·3 mm. (70); 1·35 mm. (58)
M1·8	1·4 mm. (93); 1·45 mm. (82); 1·5 mm. (70); 1·55 mm. (58)
M2	1·55 mm. (92); 1·6 mm. (82); 1·65 mm. (71); 1·7 mm. (61)
M2·2	1·7 mm. (91); 1·75 mm. (82); 1·8 mm. (73); 1·85 mm. (64)
M2·5	2 mm. (91); 2·05 mm. (82); 2·1 mm. (73); 2·15 mm. (64)
M3	2·4 mm. (98); 2·45 mm. (90); 2·5 mm. (82); 2·55 mm. (73); 2·6 mm. (6
M3·5	2·8 mm. (95); 2·85 mm. (88); 2·9 mm. (82); 2·95 mm. (75); 3 mm. (6
M4	3·2 mm. (93); 3·3 mm. (82); 3·4 mm. (70); 3·5 mm. (58)
M4·5	3·6 mm. (98); 3·7 mm. (87); 3·8 mm. (76); 3·9 mm. (65)
M5	4·1 mm. (92); 4·2 mm. (82); 4·3 mm. (71); 4·4 mm. (61)
M6	4·8 mm. (98); 4·9 mm. (90); 5 mm. (82); 5·1 mm. (73); 5·2 mm. (65)
M7	5·8 mm. (98); 5·9 mm. (90); 6 mm. (82); 6·1 mm. (73); 6·2 mm. (65)
M8	6·5 mm. (98); 6·6 mm. (91); 6·7 mm. (85); 6·8 mm. (78); 6·9 mm. (72 7 mm. (65)
M9	7·5 mm. (98); 7·6 mm. (91); 7·7 mm. (85); 7·8 mm. (78); 7·9 mm. (72 8 mm. (65)
M10	8·2 mm. (98); 8·3 mm. (92); 8·4 mm. (87); 8·5 mm. (82); 8·6 mm. (76 8·7 mm. (71); 8·8 mm. (65); 8·9 mm. (60)
M11	9·2 mm. (98); 9·3 mm. (92); 9·4 mm. (87); 9·5 mm. (82); 9·6 mm. (76 9·7 mm. (71); 9·8 mm. (65); 9·9 mm. (60)
M12	9·9 mm. (98); 10 mm. (93); 10·1 mm. (89); 10·2 mm. (84); 10·3 mm. (79 10·4 mm. (75); 10·5 mm. (70); 10·6 mm. (65); 10·7 mm. (60)
M14	11·6 mm. (98); 11·8 mm. (90); 12 mm. (82); 12·2 mm. (74); 12·4 mm. (6
M16	13·6 mm. (98); 13·8 mm. (90); 14 mm. (82); 14·2 mm. (74); 14·4 mm. (6
M18	15 mm. (98); 15·25 mm. (90); 15·5 mm. (82); 15·75 mm. (73); 16 mm. (6
M20	17 mm. (98); 17·25 mm. (90); 17·5 mm. (82); 17·75 mm. (73); 18 mm. (6
M22	19 mm. (98); 19·25 mm. (90); 19·5 mm. (82); 19·75 mm. (73); 20 mm. (6
M24	20·5 mm. (95); 20·75 mm. (88); 21 mm. (82); 21·25 mm. (75); 21·5 mn (68); 21·75 mm. (61)
M27	23·5 mm. (95); 23·75 mm. (88); 24 mm. (82); 24·25 mm. (75); 24·5 mn (68); 24·75 mm. (61)
M30	26 mm. (93); 26·5 mm. (82); 27 mm. (70)
M33	29 mm. (93); 29·5 mm. (82); 30 mm. (70)
M36	31·5 mm. (92); 32 mm. (82); 32·5 mm. (71); 33 mm. (61)
M39	34·5 mm. (92); 35 mm. (82); 35·5 mm. (71); 36 mm. (61)

Table L13. ISO METRIC FINE PITCH SERIES (1·5 mm. pitch)

Tapping Drill Sizes and Percentage Depth of Thread

Nominal Size	
mm.	
12	10·2 mm. (98); 10·3 mm. (92); 10·4 mm. (87); 10·5 mm. (82); 10·6 mm. (76); 10·7 mm. (70); 10·8 mm. (65); 10·9 mm. (60)
14	12·2 mm. (98); 12·3 mm. (92); 12·4 mm. (87); 12·5 mm. (82); 12·6 mm. (76); 12·7 mm. (70); 12·8 mm. (65); 12·9 mm. (60)
15	13·2 mm. (98); 13·3 mm. (92); 13·4 mm. (87); 13·5 mm. (82); 13·6 mm. (76); 13·7 mm. (70); 13·8 mm. (65); 13·9 mm. (60)
16	14·25 mm. (95); 14·5 mm. (82); 14·75 mm. (68)
17	15·25 mm. (95); 15·5 mm. (82); 15·75 mm. (68)
18	16·25 mm. (95); 16·5 mm. (82); 16·75 mm. (68)
20	18·25 mm. (95); 18·5 mm. (82); 18·75 mm. (68)
22	20·25 mm. (95); 20·5 mm. (82); 20·75 mm. (68)
24	22·25 mm. (95); 22·5 mm. (82); 22·75 mm. (68)
25	23·25 mm. (95); 23·5 mm. (82); 23·75 mm. (68)
27	25·5 mm. (82)
28	26·5 mm. (82)
30	28·5 mm. (82)
32	30·5 mm. (82)
33	31·5 mm. (82)
35	33·5 mm. (82)
36	34·5 mm. (82)
39	37·5 mm. (82)
40	38·5 mm. (82)
42	40·5 mm. (82)
45	43·5 mm. (82)
48	46·5 mm. (82)
50	48·5 mm. (82)
52	50·5 mm. (82)

Table L14. ISO METRIC FINE PITCH SERIES
(2 mm. pitch)
Tapping Drill Sizes and Percentage Depth of Thread

Nominal Size	
mm.	
18	15·75 mm. (92); 16 mm. (82); 16·25 mm. (71); 16·5 mm. (61)
20	17·75 mm. (92); 18 mm. (82); 18·25 mm. (71); 18·5 mm. (61)
22	19·75 mm. (92); 20 mm. (82); 20·25 mm. (71); 20·5 mm. (61)
24	21·75 mm. (92); 22 mm. (82); 22·25 mm. (71); 22·5 mm. (61)
27	24·75 mm. (92); 25 mm. (82); 25·5 mm. (61)
30	28 mm. (82); 28·5 mm. (61)
33	31 mm. (82); 31·5 mm. (61)
36	34 mm. (82); 34·5 mm. (61)
39	37 mm. (82); 37·5 mm. (61)
42	40 mm. (82); 40·5 mm. (61)
45	43 mm. (82); 43·5 mm. (61)
48	46 mm. (82); 46·5 mm. (61)
52	50 mm. (82); 50·5 mm. (61)
56	54 mm. (82); 54·5 mm. (61)
60	58 mm. (82); 58·5 mm. (61)
64	62 mm. (82); 62·5 mm. (61)
68	66 mm. (82); 66·5 mm. (61)

Table L15. ISO METRIC FINE PITCH SERIES
(3 mm. pitch)
Tapping Drill Sizes and Percentage Depth of Thread

Nominal Size	
mm.	
30	26·5 mm. (95); 27 mm. (82); 27·5 mm. (68)
33	29·5 mm. (95); 30 mm. (82); 30·5 mm. (68)
36	32·5 mm. (95); 33 mm. (82); 33·5 mm. (68)
39	35·5 mm. (95); 36 mm. (82); 36·5 mm. (68)
40	36·5 mm. (95); 37 mm. (82); 37·5 mm. (68)
42	38·5 mm. (95); 39 mm. (82); 39·5 mm. (68)
45	41·5 mm. (95); 42 mm. (82); 42·5 mm. (68)
48	43·5 mm. (95); 45 mm. (82); 45·5 mm. (68)
52	48·5 mm. (95); 49 mm. (82); 49·5 mm. (68)
56	52·5 mm. (95); 53 mm. (82); 53·5 mm. (68)
60	56·5 mm. (95); 57 mm. (82); 57·5 mm. (68)
64	60·5 mm. (95); 61 mm. (82); 61·5 mm. (68)
68	64·5 mm. (95); 65 mm. (82); 65·5 mm. (68)

Table L16. ISO METRIC FINE PITCH SERIES (4 mm. pitch)

Tapping Drill Sizes and Percentage Depth of Thread

Nominal Size	
mm.	
42	37·5 mm. (92); 38 mm. (82); 38·5 mm. (71); 39 mm. (61)
45	40·5 mm. (92); 41 mm. (82); 41·5 mm. (71); 42 mm. (61)
48	43·5 mm. (92); 44 mm. (82); 44·5 mm. (71); 45 mm. (61)
52	47·5 mm. (92); 48 mm. (82); 48·5 mm. (71); 49 mm. (61)
56	51·5 mm. (92); 52 mm. (82); 52·5 mm. (71); 53 mm. (61)
60	55·5 mm. (92); 56 mm. (82); 56·5 mm. (71); 57 mm. (61)
64	59·5 mm. (92); 60 mm. (82); 60·5 mm. (71); 61 mm. (61)
68	63·5 mm. (92); 64 mm. (82); 64·5 mm. (71); 65 mm. (61)

Table L17. ISO METRIC FINE PITCH SERIES (0·35, 0·5 and 0·75 mm. pitch)

Tapping Drill Sizes and Percentage Depth of Thread

Nominal Size	Pitch	
mm.	mm.	
3	0·35	2·6 mm. (93); 2·65 mm. (82); 2·7 mm. (70); 2·75 mm. (58)
3·5	0·35	3·1 mm. (93); 3·2 mm. (70)
4	0·5	3·4 mm. (98); 3·5 mm. (82); 3·6 mm. (65)
4·5	0·5	3·9 mm. (98); 4 mm. (82); 4·1 mm. (65)
5	0·5	4·4 mm. (98); 4·5 mm. (82); 4·6 mm. (65)
5·5	0·5	4·9 mm. (98); 5 mm. (82); 5·1 mm. (65)
6	0·75	5·1 mm. (98); 5·2 mm. (87); 5·3 mm. (76); 5·4 mm. (65)
7	0·75	6·1 mm. (98); 6·2 mm. (87); 6·3 mm. (76); 6·4 mm. (65)
8	0·75	7·1 mm. (98); 7·2 mm. (87); 7·3 mm. (76); 7·4 mm. (65)
9	0·75	8·1 mm. (98); 8·2 mm. (87); 8·3 mm. (76); 8·4 mm. (65)
10	0·75	9·1 mm. (98); 9·2 mm. (87); 9·3 mm. (76); 9·4 mm. (65)
11	0·75	10·1 mm. (98); 10·2 mm. (87); 10·3 mm. (76); 10·4 mm. (65)

Table L18. ISO METRIC FINE PITCH SERIES (1·0 and 1·25 mm. pitch)

Tapping Drill Sizes and Percentage Depth of Thread

Nominal Size	Pitch	
mm.	mm.	
8	1	6·8 mm. (98); 6·9 mm. (90); 7 mm. (82); 7·1 mm. (73); 7·2 mm. (65)
9	1	7·8 mm. (98); 7·9 mm. (90); 8 mm. (82); 8·1 mm. (73); 8·2 mm. (65)
10	1	8·8 mm. (98); 8·9 mm. (90); 9 mm. (82); 9·1 mm. (73); 9·2 mm. (65)
10	1·25	8·5 mm. (98); 8·6 mm. (91); 8·7 mm. (85); 8·8 mm. (79); 8·9 mm. (71); 9 mm. (65)
11	1	9·8 mm. (98); 9·9 mm. (90); 10 mm. (82); 10·1 mm. (73); 10·2 mm. (65)
12	1	10·8 mm. (98); 10·9 mm. (90); 11 mm. (82); 11·1 mm. (73); 11·2 mm. (65)
12	1·25	10·5 mm. (98); 10·6 mm. (91); 10·7 mm. (85); 10·8 mm. (79); 10·9 mm. (71); 11 mm. (65)
14	1	12·8 mm. (98); 12·9 mm. (90); 13 mm. (82); 13·1 mm. (73); 13·2 mm. (65)
14	1·25	12·5 mm. (98); 12·6 mm. (91); 12·7 mm. (85); 12·8 mm. (79); 12·9 mm. (71); 13 mm. (65)
15	1	13·8 mm. (98); 13·9 mm. (90); 14 mm. (82); 14·25 mm. (61)
16	1	15 mm. (82); 15·25 mm. (61)
17	1	16 mm. (82); 16·25 mm. (61)
18	1	17 mm. (82); 17·25 mm. (61)
20	1	19 mm. (82); 19·25 mm. (61)
22	1	21 mm. (82); 21·25 mm. (61)
24	1	23 mm. (82); 23·25 mm. (61)
25	1	24 mm. (82); 24·25 mm. (61)
27	1	26 mm. (82); 26·25 mm. (61)
28	1	27 mm. (82); 27·25 mm. (61)
30	1	29 mm. (82); 29·25 mm. (61)

Table L19. AMERICAN PIPE THREAD

Tapping Drill Sizes

Size	Taper Thread		Parallel Thread
	With Reamer	Without Reamer	
$\frac{1}{16}$ in.	6·1 mm.	6·25 mm.	$\frac{1}{4}$ in.
$\frac{1}{8}$	$\frac{21}{64}$ in.	8·4 mm.	$\frac{11}{32}$
$\frac{1}{4}$	$\frac{27}{64}$	$\frac{7}{16}$ in.	$\frac{7}{16}$
$\frac{3}{8}$	$\frac{9}{16}$	$\frac{9}{16}$	$\frac{37}{64}$
$\frac{1}{2}$	$\frac{11}{16}$	$\frac{45}{64}$	$\frac{23}{32}$
$\frac{3}{4}$	$\frac{57}{64}$	$\frac{29}{32}$	$\frac{59}{64}$
1	$1\frac{1}{8}$	$1\frac{9}{64}$	$1\frac{5}{32}$
$1\frac{1}{4}$	$1\frac{15}{32}$	$1\frac{31}{64}$	$1\frac{1}{2}$
$1\frac{1}{2}$	$1\frac{23}{32}$	$1\frac{47}{64}$	$1\frac{3}{4}$
2	$2\frac{3}{16}$	$2\frac{13}{64}$	$2\frac{7}{32}$
$2\frac{1}{2}$	$2\frac{19}{32}$	$2\frac{5}{8}$	$2\frac{21}{32}$

Table L20. PROGRESSIVE DRILL SIZES

in.	mm.	Wire No. or Letter	in.	in.	mm.	Wire No. or Letter	in.
0·0126	0·32			0·0244	0·62		
0·0135		80		0·0250		72	
0·0138	0·35			0·0256	0·650		
0·0145		79		0·0260		71	
0·0150	0·38			0·0268	0·68		
0·0156			$\frac{1}{64}$	0·0276	0·700		
0·0157	0·400			0·0280		70	
0·0160		78		0·0283	0·72		
0·0165	0·42			0·0292		69	
0·0177	0·450			0·0295	0·750		
0·0180		77		0·0307	0·78		
0·0189	0·48			0·0310		68	
0·0197	0·500			0·0313			$\frac{1}{32}$
0·0200		76		0·0315	0·800		
0·0205	0·52			0·0320		67	
0·0210		75		0·0323	0·82		
0·0217	0·550			0·0330		66	
0·0225		74		0·0335	0·850		
0·0228	0·58			0·0346	0·88		
0·0236	0·600			0·0350		65	
0·0240		73		0·0354	0·900		

[*Continued on page* 306

Table L20 *continued.* **PROGRESSIVE DRILL SIZES**

in.	mm.	Wire No. or Letter	in.	in.	mm.	Wire No. or Letter	in.
0·0360		64		0·0785		47	
0·0362	0·92			0·0787	2·00		
0·0370		63		0·0807	2·05		
0·0374	0·950			0·0810		46	
0·0380		62		0·0820		45	
0·0386	0·98			0·0827	2·10		
0·0390		61		0·0846	2·15		
0·0394	1·000			0·0860		44	
0·0400		60		0·0866	2·20		
0·0410		59		0·0886	2·25		
0·0413	1·050			0·0890		43	
0·0420		58		0·0906	2·30		
0·0430		57		0·0925	2·35		
0·0433	1·100			0·0935		42	
0·0453	1·150			0·0938			$\frac{3}{32}$
0·0465		56		0·0945	2·40		
0·0469			$\frac{3}{64}$	0·0960		41	
0·0472	1·200			0·0965	2·45		
0·0492	1·250			0·0980		40	
0·0512	1·300			0·0984	2·50		
0·0520		55		0·0995		39	
0·0532	1·350			0·1004	2·55		
0·0550		54		0·1015		38	
0·0551	1·400			0·1024	2·60		
0·0571	1·450			0·1040		37	
0·0590	1·500			0·1043	2·65		
0·0595		53		0·1063	2·70		
0·0610	1·550			0·1065		36	
0·0625			$\frac{1}{16}$	0·1083	2·75		
0·0630	1·600			0·1094			$\frac{7}{64}$
0·0635		52		0·1100		35	
0·0650	1·650			0·1102	2·80		
0·0669	1·700			0·1110		34	
0·0670		51		0·1122	2·85		
0·0689	1·750			0·1130		33	
0·0700		50		0·1142	2·90		
0·0709	1·800			0·1160		32	
0·0728	1·850			0·1161	2·95		
0·0730		49		0·1181	3·0		
0·0748	1·900			0·1200		31	
0·0760		48		0·1220	3·10		
0·0768	1·950			0·1250			$\frac{1}{8}$
0·0781			$\frac{5}{64}$	0·1260	3·20		

Table L20 *continued.* **PROGRESSIVE DRILL SIZES**

in.	mm.	Wire No. or Letter	in.	in.	mm.	Wire No. or Letter	in.
0·1285		30		0·2008	5·10		
0·1299	3·30			0·2010		7	
0·1339	3·40			0·2031			$\frac{13}{64}$
0·1360		29		0·2040		6	
0·1378	3·50			0·2047	5·20		
0·1405		28		0·2055		5	
0·1406			$\frac{9}{64}$	0·2087	5·30		
0·1417	3·60			0·2090		4	
0·1440		27		0·2126	5·40		
0·1457	3·70			0·2130		3	
0·1470		26		0·2165	5·50		
0·1495		25		0·2188			$\frac{7}{32}$
0·1496	3·80			0·2205	5·60		
0·1520		24		0·2210		2	
0·1535	3·90			0·2244	5·7		
0·1540		23		0·2280		1	
0·1563			$\frac{5}{32}$	0·2283	5·80		
0·1570		22		0·2323	5·90		
0·1575	4·00			0·2340		A	
0·1590		21		0·2344			$\frac{15}{64}$
0·1610		20		0·2362	6·00		
0·1614	4·10			0·2380		B	
0·1654	4·20			0·2402	6·10		
0·1660		19		0·2420		C	
0·1693	4·30			0·2441	6·20		
0·1695		18		0·2460		D	
0·1719			$\frac{11}{64}$	0·2480	6·30		
0·1730		17		0·2500		E	$\frac{1}{4}$
0·1732	4·40			0·2520	6·40		
0·1770		16		0·2559	6·50		
0·1772	4·50			0·2570		F	
0·1800		15		0·2598	6·60		
0·1811	4·60			0·2610		G	
0·1820		14		0·2638	6·70		
0·1850	4·70	13		0·2656			$\frac{17}{64}$
0·1875			$\frac{3}{16}$	0·2660		H	
0·1890	4·80	12		0·2677	6·80		
0·1910		11		0·2717	6·90		
0·1929	4·90			0·2720		I	
0·1935		10		0·2756	7·00		
0·1960		9		0·2770		J	
0·1968	5·00			0·2795	7·10		
0·1990		8		0·2810		K	

[*Continued on page* 308

Table L20 *continued.* **PROGRESSIVE DRILL SIZES**

in.	mm.	Wire No. or Letter	in.	in.	mm.	Wire No. or Letter	in.
0·2813			$\frac{9}{32}$	0·3480		S	
0·2835	7·20			0·3504	8·90		
0·2874	7·30			0·3543	9·00		
0·2900		L		0·3580		T	
0·2913	7·40			0·3583	9·10		
0·2950		M		0·3594			$\frac{23}{64}$
0·2953	7·50			0·3622	9·20		
0·2969			$\frac{19}{64}$	0·3661	9·30		
0·2992	7·60			0·3680		U	
0·3020		N		0·3701	9·40		
0·3031	7·70			0·3740	9·50		
0·3071	7·80			0·3750			$\frac{3}{8}$
0·3110	7·90			0·3770		V	
0·3125			$\frac{5}{16}$	0·3780	9·60		
0·3150	8·00			0·3819	9·70		
0·3160		O		0·3858	9·80		
0·3189	8·10			0·3860		W	
0·3228	8·20			0·3898	9·90		
0·3230		P		0·3906			$\frac{25}{64}$
0·3268	8·30			0·3937	10·00		
0·3281			$\frac{21}{64}$	0·3970		X	
0·3307	8·40			0·3976	10·10		
0·3320		Q		0·4016	10·20		
0·3346	8·50			0·4040		Y	
0·3386	8·60			0·4055	10·30		
0·3390		R		0·4063			$\frac{13}{32}$
0·3425	8·70			0·4094	10·40		
0·3438			$\frac{11}{32}$	0·4130		Z	
0·3465	8·80						

Table L20 *continued.* **PROGRESSIVE DRILL SIZES**

in.	mm.	in.	in.	mm.	in.	in.	mm.	in.
0·4134	10·5		0·5394	13·7		0·7677	19·5	
0·4173	10·6		0·5433	13·8		0·7776	19·75	
0·4213	10·7		0·5469		35/64	0·7813		25/32
0·4219		27/64	0·5472	13·9		0·7874	20·0	
0·4252	10·8		0·5512	14·0		0·7969		51/64
0·4291	10·9		0·5610	14·25		0·7972	20·25	
0·4331	11·0		0·5625		9/16	0·8071	20·5	
0·4370	11·1		0·5709	14·5		0·8125		13/16
0·4375		7/16	0·5781		37/64	0·8169	20·75	
0·4409	11·2		0·5807	14·75		0·8268	21·0	
0·4449	11·3		0·5906	15·0		0·8281		53/64
0·4488	11·4		0·5938		19/32	0·8366	21·25	
0·4528	11·5		0·6004	15·25		0·8438		27/32
0·4531		29/64	0·6094		39/64	0·8465	21·5	
0·4567	11·6		0·6102	15·5		0·8563	21·75	
0·4606	11·7		0·6201	15·75		0·8594		55/64
0·4646	11·8		0·6250		5/8	0·8661	22·0	
0·4685	11·9		0·6299	16·0		0·8750		7/8
0·4688		15/32	0·6398	16·25		0·8760	22·25	
0·4724	12·0		0·6406		41/64	0·8858	22·5	
0·4764	12·1		0·6496	16·5		0·8906		57/64
0·4803	12·2		0·6563		21/32	0·8957	22·75	
0·4842	12·3		0·6594	16·75		0·9055	23·0	
0·4844		31/64	0·6693	17·0		0·9063		29/32
0·4882	12·4		0·6719		43/64	0·9154	23·25	
0·4921	12·5		0·6791	17·25		0·9219		59/64
0·4961	12·6		0·6875		11/16	0·9252	23·5	
0·5000	12·7	1/2	0·6890	17·5		0·9350	23·75	
0·5020	12·75		0·6988	17·75		0·9375		15/16
0·5039	12·8		0·7031		45/64	0·9449	24·0	
0·5079	12·9		0·7087	18·0		0·9531		61/64
0·5118	13·0		0·7185	18·25		0·9547	24·25	
0·5156		33/64	0·7188		23/32	0·9646	24·5	
0·5157	13·1		0·7283	18·5		0·9688		31/32
0·5197	13·2		0·7344		47/64	0·9744	24·75	
0·5236	13·3		0·7382	18·75		0·9842	25·0	
0·5276	13·4		0·7480	19·0		0·9844		63/64
0·5313		17/32	0·7500		3/4	1·000		1
0·5315	13·5		0·7579	19·25				
0·5354	13·6		0·7656		49/64			

GLOSSARY

	French	*German*
Allowance	Abaissement	Abmasse
Black or Unfinished	Brut	Roh
Bolt	Boulon	Bolzen
(,,) fully threaded	(,,) entièrement filetée	ohne Schaft
(,,) Hexagonal Headed	(,,) Tête hexagonale	Sechskantbolzen
(,,) partly threaded	(,,) partiellement filetée	mit Schaft
Bright or Finished	Usinée	Blank
Chamfered	chamfreinée	abgeschrägt
Clearance	jeu	Spiel; Spielraum
Cross section of Core	Section du noyau	Kernquerschnitt
Depth of engagement	Portage	Tragtiefe
,, of nut	Hauteur pour écrou	Höhe für Mutter
,, of Thread	Profondeur du filet	Gewindetiefe; Gangtiefe
Double Start	——	Zweigängig
Effective Diameter	diamètre effectif; diamètre sur flancs	Flankendurchmesser
Flat at Crest	aplatissement au sommet	Lückenbreite am Knopf
,, ,, Root	aplatissement au fond	,, ,, Fuss
Gas Thread	Filetage au pas du gas	Gas Rohrgewinde
Gauge Length (taper Threads)	Distance du diamètre de jauge au bout du tube	Abstand des Lehrdurchmesser vom Rohrende
Gauging Length	Longueur jaugé	Lehrenlängen
Major Diameter	Diamètre extèrieur	Aussendurchmesser
Maximum Dimension	Dimension maximum	Grösstmass
Minimum ,,	,, minimum	Kleinmass
Minor Diameter	Diamètre intèrieur	Kerndurchmesser
Multi-start (thread)	filets multiples	Mehrgängig
Nominal Diameter	diamètre nominal	Nenndurchmesser
Nut	écrou	Mutter
,, Hexagon	écrou six-pans	Sechskantmutter
,, Lock	écrou bas; contre écrou	Gegenmutter
,, Square	écrou carré	Vierkantmutter
,, Wing	écrou à ailettes	Flügelmutter
Pipe Thread	Filetage sur tubes	Rohrgewinde
Pitch	Pas	Steigung; Ganghöhe
Pitch Diameter (*Amer.*)	diamètre effectif: diamètre sur flancs	Flankendurchmesser
Radius at Crest or Root	Arrondi	Rundung; Abrundungsradius
Screw	Vis	Schrauben
,, Cheese Headed	Vis à tête cylindrique	Zylinderschrauben
,, Countersunk	Vis à tête fraisée plate	Senkschrauben
,, Raised Head	Vis à tête fraisée bombée	Linsensenkschraube
,, Roundhead	Vis à tête ronde	Halbrundschraube
Set Screw	——	Würfelschraube

Thread Angle	Angle de filet	Flankenwinkel
,, Form	Profile du filet	Gewindeform; Gewindeprofil
Threads per inch	Nombre de filets au pouce	Gangzahl auf 1 Zoll
Tolerances	Tolérances	Toleranzen
Triangular Height	Hauteur de triangle primitif	Dreieckshöhe
Truncation	Troncature	Abflachung
Washer	Rondelle	Unterlagscheibe
Width across flats	Largeur sur pans	Schlüsselweite
Wood screw	Vis à bois	Holzschraube
,, ,, Countersunk	Vis à bois à tête fraisé plate	Senkschraube

INDEX

INDEX—*continued*